Vehicle Body Refinishing

Vehicle Body Refinishing

NVQ Level 3

A. Robinson RTechEng, MInstBE, MIBCAM, AWeldI, MISME

Section Leader for Vehicle Bodywork and Vehicle Body Repair Course, Coordinator for CGLI 3440 Vehicle Body Repair, Coordinator for CGLI 398 Vehicle Body Competences Course, Coordinator for CGLI 385 Vehicle Bodywork, Gateshead College

Newnes
An imprint of Butterworth-Heinemann Ltd
Linacre House, Jordan Hill, Oxford OX2 8DP

A member of the Reed Elsevier plc group

OXFORD LONDON BOSTON
MUNICH NEW DELHI SINGAPORE SYDNEY
TOKYO TORONTO WELLINGTON

First published 1996

British Library Cataloguing in Publication Data
A catalogue record for this book is available from the British Library.

ISBN 0 7506 2270 9

Library of Congress Cataloguing in Publication Data
A catalogue record for this book is available from the Library of Congress.

Printed in Great Britain

Contents

Foreword

The Vehicle Builders and Repairers Association Ltd has always pioneered the need for well-trained staff and the highest standards of workmanship in the motor vehicle repair industry. It has become increasingly important, with the revolution in materials technology bringing major changes in vehicle repair equipment and techniques, that those who work in the body repair industry receive the proper skills training.

Over the last few years we have also seen a revolution in vocational training, with the advent of the NVQ establishing new standards for the industry.

Vehicle Body Refitting has been written specifically to provide a textbook for the level 3 NVQ. It contains all the information that students at this level will need, and provides step-by-step guidance aimed at a wide audience of part-time students working in firms of all sizes.

Vehicle Body Refitting is an important training manual for all those studying for, or teaching, National Vocational Qualifications in vehicle repair.

R.D. Nicholson FIMI
Director of Operations
Vehicle Builders and Repairers
Association Ltd

Preface

This book covers the underpinning knowledge to support the evidence of competency within NVQ qualifications and is intended for apprentices entering the trade, students already in the trade and mature tradesmen wishing to further their qualifications by studying for the City and Guilds of London Institute, National Vocational Qualifications (NVQ).

National Vocational Qualifications (NVQ) for the Motor Industry are part of a new structure of qualifications being put in place to support the efforts of employers and employees to raise the standards of performance and competence in the workplace.

The qualifications are based on a new approach to describing and assessing a person's performance while engaged at work. This involves identification of the competence and the definition of standards in the workplace to which a person is expected to perform.

The Motor Industry Training Standards Council (MITSC) was established in 1990 as a lead body for the United Kingdom retail motor industry and its associated sectors. The Council is committed to the development of occupational standards for the motor industry especially in the areas of vehicle body repair, vehicle body fitting, vehicle body refinishing and the coordination of the award of NVQ.

The NVQ awards are at five levels, so there are qualifications for those just starting their career (levels 1 and 2), those carrying out skilled work (level 3), and those at supervisory and management positions (levels 4 and 5).

In order for candidates to be awarded NVQ they must prove their ability in a number of units of occupational competency, each unit being certified upon its successful completion and building up a portfolio of evidence for accreditation in Vehicle Body Work to Level 2 and/or Level 3 within the NVQ framework.

Within the motor industry the following vehicle body units are necessary to achieve an NVQ qualification

CGLI 3440 Vehicle Body Fitting Level 2 (the following 6 units are required for this level) (see Figure 1)

A2-LH Replace vehicle body panels, panel sections and ancillary fittings.
A9-L2 Remove and replace vehicle components.
A10-L Augment vehicle bodywork to meet customer requirements.
A11-L Prepare new, used or repaired vehicles for customer use.
A12-G Maintain effective working relationships.
A13-G Maintain the health, safety and security of the working environment.

CGLI 3440 Vehicle Body Repair Level 3 (the following 6 units are required for this level) (see Figure 2)

A2-LH Replace vehicle body panels, panel sections and ancillary fittings.
A3-LH Repair body panels.
A10-L Augment vehicle bodywork to meet customer requirements.
A12-G Maintain effective working relationships.
A13-G Maintain the health, safety and security of the working environment.
A4-L Rectify body misalignment and repair/replace body sections.
OR
A4-LH Rectify chassis frame misalignment and repair/replace body sections.

(Note: Figures 1 and 2 include unit A6-G, 'Reinstate the cleanliness of the vehicle', as an additional unit.)

CGLI 3440 NVQ in Vehicle Refinishing Level 3 (the following 8 units are required for this level) (see Figure 3)

A15-LH Complete preparations for the refinishing of vehicle bodies/body sections.
A16-LH Refinishing vehicle bodies and body sections with finishing top coats.

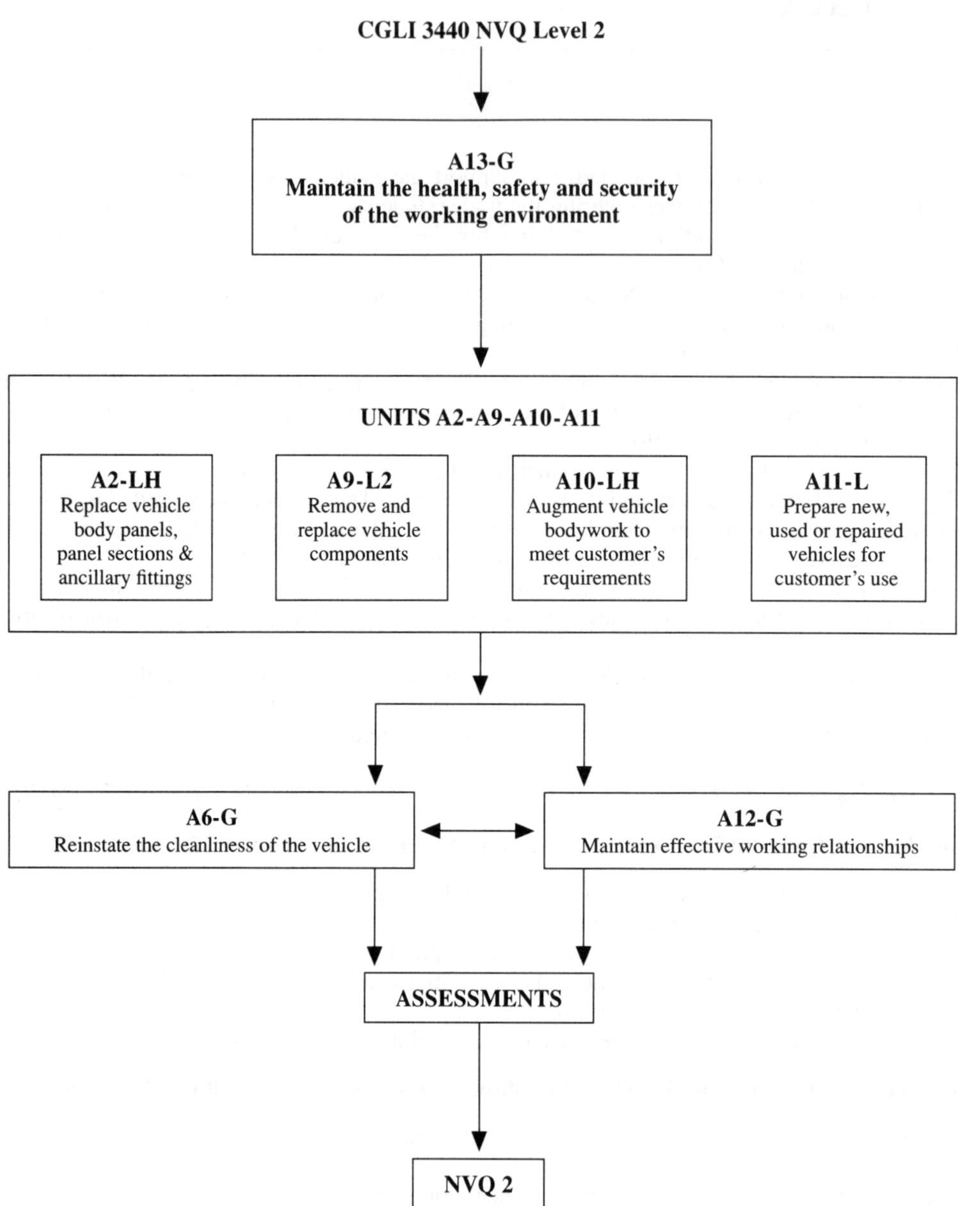

Figure 1 Route plan to NVQ Level 2

A17-LH Rectify vehicle paint finish defects.
A6-G Reinstate the cleanliness of the vehicle.
A10-L Augment vehicle bodywork to meet customer's requirements.
A11-L Prepare new, used or repaired vehicles for customer's use.
A12-G Maintain effective working relationships.
A13-G Maintain the health, safety and security of the working environment.

(Note: Further information about these units is available from MITSC or CGLI.)

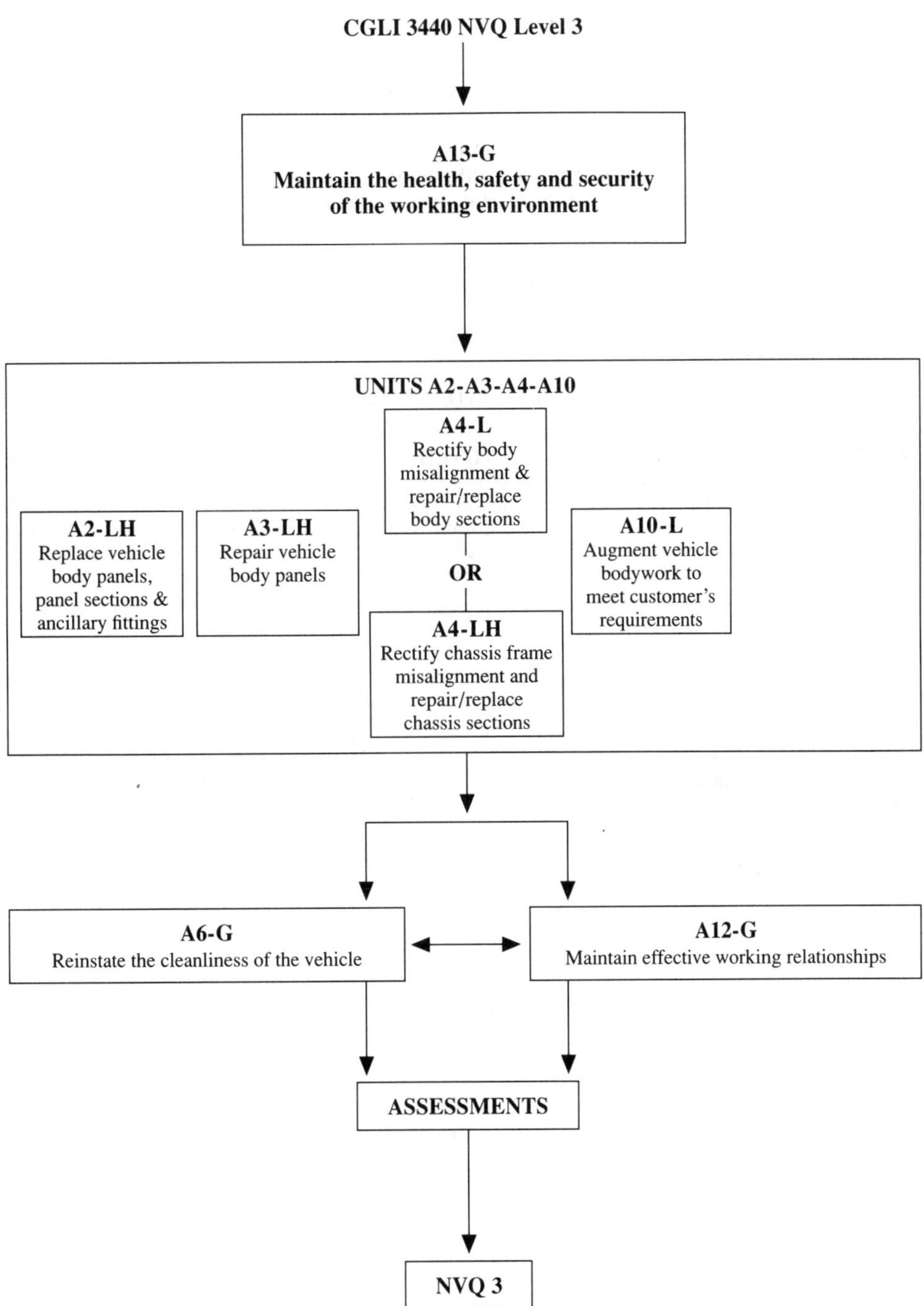

Figure 2 Route plan to NVQ Level 3

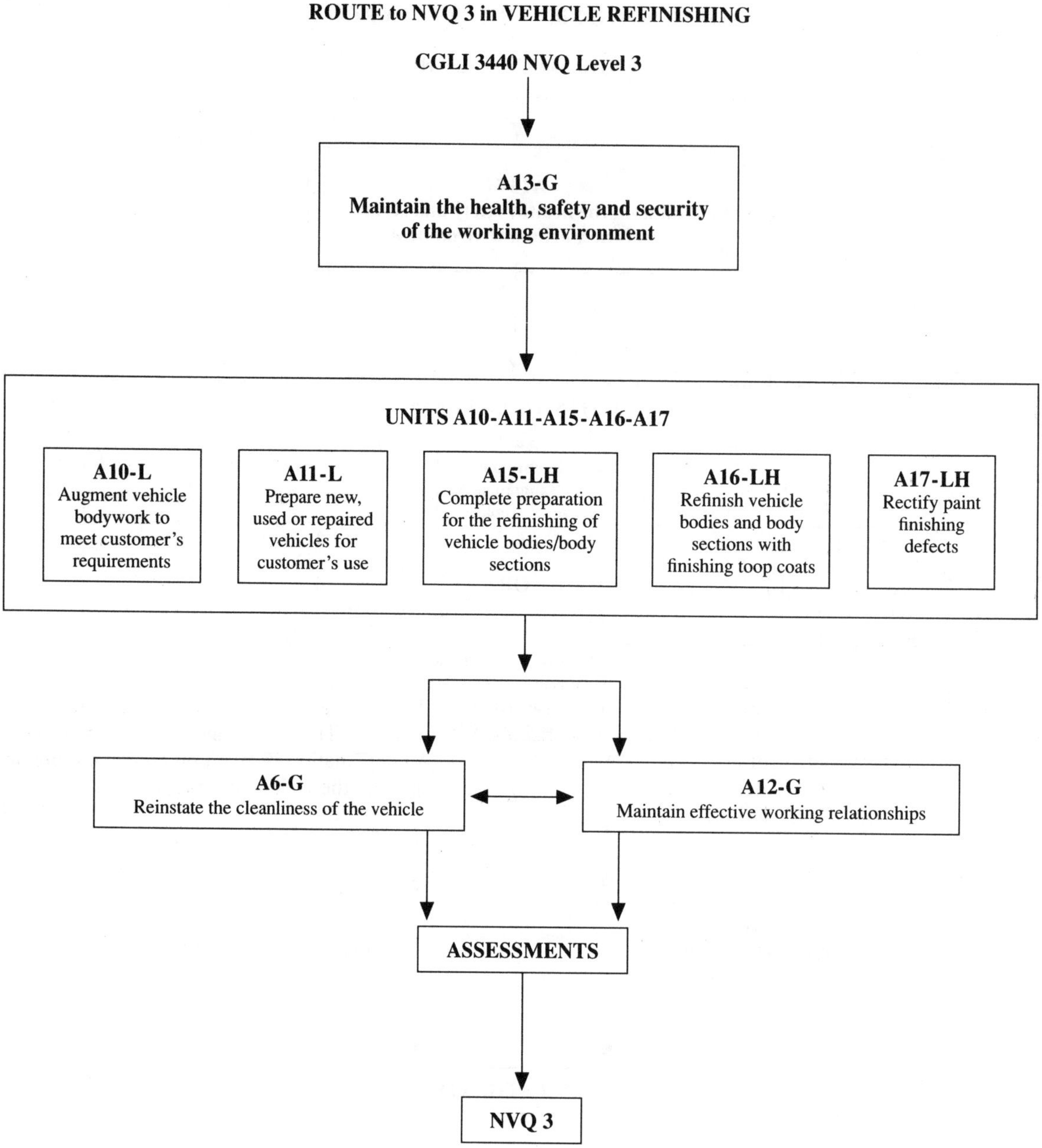

ROUTE to NVQ 3 in VEHICLE REFINISHING

CGLI 3440 NVQ Level 3

A13-G
Maintain the health, safety and security of the working environment

UNITS A10-A11-A15-A16-A17

A10-L Augment vehicle bodywork to meet customer's requirements

A11-L Prepare new, used or repaired vehicles for customer's use

A15-LH Complete preparation for the refinishing of vehicle bodies/body sections

A16-LH Refinish vehicle bodies and body sections with finishing toop coats

A17-LH Rectify paint finishing defects

A6-G Reinstate the cleanliness of the vehicle

A12-G Maintain effective working relationships

ASSESSMENTS

NVQ 3

Figure 3 Route plan to NVQ Level 3

NVQs for vehicle bodywork

This book is intended to supply underpinning knowledge for students working to achieve National Vocational Qualifications (NVQs) in Vehicle Body Refinishing/Repair through the Motor Industry Training Standards Council (MITSC).

CGLI has been accredited by NVQ as one of the awarding bodies for these qualifications under the following schemes:

Scheme No. 3440 Vehicle Body Fitting – Level 2 NVQ
Scheme No. 3440 Vehicle Body Repair – Level 3 NVQ
Scheme No. 3440 Vehicle Refinishing – Level 3 NVQ

NVQ Level 3 unit reference								
	A6	A10	A11	A12	A13	A15	A16	A17
Chapter 1 Health and safety in the working environment	√	√	√		√	√	√	√
Chapter 2 The design and construction of the car body		√	√					
Chapter 3 Materials used in vehicle bodies	√	√				√	√	√
Chapter 4 Automotive finishing and refinishing	√	√				√	√	√
Chapter 5 Company administration and personnel		√	√	√	√	√		
Note: Tick denotes element information may be found in this chapter								

Figure 4 Level 3 – Unit reference chart

The book begins by giving an insight into health, safety and personal protection which are essential parts of the employee's safe working environment, and covers personal hygiene, protective clothing, fire precautions, safety signs and general workshop safety.

It continues with a brief description of the history and development of the motor car and the evolution of design, and goes on to explain the early types of construction, explaining styling forms and vehicle classification leading up to the present day.

There follows a comprehensive study of the various materials which are used in the manufacture and repair of vehicles.

The finishing and refinishing of vehicles begins with the history of painting, and continues with the basic composition of paint, types of paint used for refinishing and the necessary spray painting equipment. The techniques of spray painting, burnishing and polishing are included. The importance of rust-proofing is explained and a description of common spray defects follows. An explanation is given on colour mixing and matching. Finally vehicle valeting is included as an important part of the refinishing process.

Company structure, staff duties and responsibilities are dealt with in addition to staff appointments and staff training. Different types of customer and customer relationships, together with the handling of customer complaints, are important parts of any bodyshop's activities.

Relevant Regulations and Acts which influence the administration of any company are included: Construction and Use Regulations, Vehicle Type Approval, Trade Description Act, Law of Contract, Supply of Goods and Services Act, and warranties given by the manufacturer, the VBRA and individual bodyshops.

A. Robinson

Acknowledgements

I wish to express my appreciation of the assistance given by my colleagues in the Body Work and Painting Section at Gateshead College, and especially that of Mr O. Carr (NCTEC Final, CGLI Final Motor Vehicle Painting and Industrial Finishing), who is a lecturer in Motor Vehicle Spray Painting and has helped to compile the information and illustrations in Chapter 4. I am also grateful for the efficient help given by the library staff of the college.

I wish to thank my wife Norma for the many hours spent working at the computer to assist me. Also I would like to take this opportunity to thank my son Andrew for the production of some of the photographs taken especially for this edition.

I would also like to take this opportunity of offering my sincere thanks to the following firms and/or their representatives who have so readily cooperated with me by permitting the reproduction of data, illustrations and photographs:

AGA Ltd
Al Welders Ltd
Akzo Coatings PLC
Alcan Design and Development
Alkor Plastics (UK) Ltd
Aluminium Federation Ltd
Ambi-Rad Ltd
ARO Welding Ltd
Aston Martin Lagonda Ltd
Auchard Development Co. Ltd
Autoglym
Autokraft Ltd
Auto-Quote
Avdel Ltd
Bayer UK Ltd
Berger Industrial Coatings
Black and Decker Ltd
Blackhawk Automotive Ltd
BOC Ltd
Bodyshop Magazine
Bodymaster UK
Bondaglass-Voss Ltd
Bostick Ltd
BP Chemicals (UK) Ltd
British Motor Industry Heritage Trust
British Standards Institution
British Steel Stainless
BSC Strip Product Group
Callow & Maddox Bros Ltd
Car Bench
Car-O-Liner UK Ltd
Celette UK Ltd
Cengar Universal Tool Company Ltd
Chicago Pneumatic Tool Co. Ltd
Chief Automotive Ltd
Chubb Fire Ltd

Citroen UK Ltd
City and Guilds of London Institute
Dana Distribution Ltd, Martyn Ferguson
Dataliner, Geotronics Ltd
Desoutter Automotive Ltd
DeVilbiss Automotive Refinishing Products
Dinol-Protectol Ltd
DRG Kwikseal Products
Du Pont (UK) Ltd
Duramix
Dunlop Adhesives
East Coast Traders
Edwards Pearson Ltd
ESAB Ltd
European Industrial Services
Facom Tools Ltd
Farécla Products Ltd
Fifth Generation Technology Ltd
Fire Extinguishing Trades Association
Ford Motor Company Ltd
Forest Fasteners
Frost Auto Restoration Techniques Ltd
Fry's Metals Ltd
George Marshall (Power Tools) Ltd
GE Plastics Ltd
GKN Screws and Fasteners Ltd
Glasurit Automotive Refinish
Glas-Weld Systems (UK) Ltd
Go-Jo Industries Europe Ltd
Gramos Chemicals International Ltd
Gray Campling Ltd
Harboran Ltd
Herberts
Hooper & Co. (Coachbuilders) Ltd
Huck UK Ltd
IBCAM Journal
ICI Autocolor
ICI Chemicals and Polymers Group
ICI Paints
Industrial and Trade Fairs Ltd
Jack Sealey Ltd
Jaguar Cars Ltd
John Cotton (Colne) Ltd
Kärcher
Kenmore Computing Services Ltd
Kroll (UK) Ltd
Land Rover Ltd
Lotus Engineering
Migatronic Welding Equipment Ltd
Minden Industrial Ltd
Morgan Motor Company Ltd
Motor Industry Training Standards Council (MITSC)
Murex Welding Products Ltd
3M Automotive Trades

National Adhesives & Resins Ltd
Nederman Ltd
Neill Tools Ltd
Nettlefolds Ltd
Olympus Welding & Cutting Technology Ltd
Owens-Corning Fiberglas (GB) Ltd
Parkway Consultancy (S. Hamill)
Partco Engineering
Permabond
Pickles Godfrey Design Partnership, Ernest W. Godfrey
Plastics and Rubber Institute
PPG Industries (UK) Ltd
Racal Safety Ltd
Reliant Motors Ltd
Renault UK Ltd
Rolls-Royce Motor Cars Ltd
Rover Group Ltd
Saab-Scania
Schlegel (UK) Ltd
Scott Bader Co. Ltd
Selson Machine Tool Co. Ltd
SIP (Industrial Products) Ltd
Spraybake Ltd
Standard Forms Ltd
Stanners Ltd
Sun Electric UK Ltd
Sykes-Pickavant Ltd
Teroson UK
The Institute of Materials
The Motor Insurance Repair Research Centre
The National Motor Museum
Thornley and Knight Ltd
Triplex Safety Glass
Tri-Sphere Ltd
TRW United-Car Ltd
Tucker Fasteners Ltd
UK Fire International Ltd
United Continental Steels Ltd
Vauxhall Motors Ltd
Vitamol Ltd
Volkswagen-Audi
Volvo Concessionaires Ltd
W. David and Sons Limited
Welding & Metal Fabrication
Welwyn Tool Co. Ltd
Wheelforce V. L. Churchill
Wholesale Welding Supplies Ltd

Abbreviations and symbols

ABS	acrylonitrile butadiene styrene
ABS	anti-lock braking system
AF	across flats (bolt size)
AC	alternating current
A	ampere
AFFF	aqueous film forming foam (fire fighting)
bar	10^6 dyn/cm^2; 10^5 N/m^2; 0.986,82 atm; 14.505 psi
BATNEEC	best available techniques not enabling excessive costs
BS	British Standard
BSI	British Standards Institution
BCF	bromochlorodifluoromethane (fire extinguishers)
C_d	aerodynamic drag coefficient
CO	carbon monoxide
CO_2	carbon dioxide
C	centigrade (Celsius)
cm	centimetre
cm^2	square centimetres
CAD	computer-aided design
CAE	computer-aided engineering
CAM	computer-aided manufacturing
CIM	computer-integrated manufacturing
COSHH	Control of Substances Hazardous to Health (Regulations)
cm^3	cubic centimetres
°	degree (angle or temperature)
dB(A)	noise level at the ear
DTI	dial test indicator
dia.	diameter
DC	direct current
dB	decibel
ECU	electronic control unit
EPA	Environmental Protection Act
EC	European Community
F	Fahrenheit
ft	foot
ft/min	feet per minute
gal	gallon (imperial)
GLS	general lighting service (lamp)
GRP	glass fibre reinforced plastic
g	gram (mass)
HASAWA	Health and Safety at Work Act
HSE	Health and Safety Executive

HT high tension (electrical)
HVLP high velocity low pressure (spray guns)

in inch
in^2 square inches
in^3 cubic inches
IFS independent front suspension
IR infrared
ID internal diameter
ISO International Organization for Standardization

kg kilogram (mass)
kW kilowatt

LH left hand
LHD left-hand drive
LHThd left-hand thread
l litre
LT low tension
lumen light energy radiated per second per unit solid angle by a uniform point source of 1 candela intensity
lux unit of illumination equal to 1 $lumen/m^2$

max. maximum
MAG metal active gas (welding)
MIG metal inert gas (welding)
m metre
mm millimetre
min. minimum
– minus (of tolerance)
′ minute (of angle)

(–) negative (electrical)
N m newton metre

Ibf pound force
Ib pound (mass)
Ibf ft pound force foot (torque)
Ihf/in^2 pound force per square inch

$L_{ep, d}$ noise exposure level, personal (daily)
$L_{cp, w}$ noise exposure level, personal (weekly)
no. Number

ozf ounce (force)
oz ounce (mass)
OD outside diameter

part no. part number
% percentage
PPE Personal Protection Equipment (Regulations)
pt pint (imperial)
\+ plus (tolerance)
± plus or minus

PVA	polyvinyl acetate
PVC	polyvinyl chloride
+	positive (electrical)
r	radius
ref.	reference
rev/min	revolutions per minute
RH	right hand
RHD	right-hand drive
″	second (angle)
SAE	Society of Automobile Engineers
std	standard
TIG	tungsten inert gas (welding)
VIN	vehicle identification number
VOCs	volatile organic compounds
V	volt

1

Health and safety in the working environment

The main responsibility for occupational health and safety lies with the employer. It is the employer who must provide a safe working environment, safe equipment and safety protection and must also ensure that all work methods are carried out safely.

The Health and Safety at Work Act (HASAWA) 1974 is a major piece of occupational legislation, which requires the employer to ensure, as far as is reasonably practicable, the health and safety of all staff and any other personnel who may be affected by the work carried out. The other two important regulations affecting bodyshops are the Control of Substances Hazardous to Health (COSHH) Regulations 1988 and the Environmental Protection Act (EPA) 1990.

1.1 PERSONAL SAFETY AND HEALTH PRACTICES

1.1.1 Skin care (personal hygiene) systems

All employees should be aware of the importance of personal hygiene and should follow correct procedures to clean and protect the skin in order to avoid irritants causing skin infections and dermatitis. All personnel should use a suitable barrier cream before starting work and again when recommencing work after a break. There are waterless hand cleaners available which will remove heavy dirt on skin prior to thorough washing. When the skin has been washed, after-work creams will help to restore its natural moisture.

Many paints, refinishing chemicals and bodyshop materials will cause irritation on contact with the skin and must be removed promptly with a suitable cleansing material. Paint solvents may cause dermatitis, particularly where skin has been in contact with peroxide hardeners or acid catalysts: these have a drying effect which removes the natural oils in the skin. There are specialist products available for the bodyshop which will remove these types of materials from the skin quickly, safely and effectively.

1.1.2 Hand protection

Body technicians and painters are constantly handling substances which are harmful to health. The harmful effect of liquids, chemicals and materials on the hands can be prevented, in many cases, by wearing the correct type of gloves. To comply with COSHH Regulations, vinyl disposable gloves must be used by painters to give skin protection against toxic substances. Other specialist gloves available are: rubber and PVC gloves for protection against solvents, oil and acids; leather gloves for hard wear and general repair work in the bodyshop; and welding gauntlets, which are made from specially treated leather and are longer than normal gloves to give adequate protection to the welder's forearms.

1.1.3 Protective clothing

Protective clothing is worn to protect the worker and his clothes from coming into contact with dirt, extremes of temperature, falling objects and chemical substances. The most common form of protective clothing for the body repairer is the overall, a one-piece boiler suit made from good quality cotton, preferably flame-proof. Worn and torn materials should be avoided as they can catch in moving machinery. Where it is necessary to protect the skin, closely fitted sleeves should be worn down to the wrist with the cuffs fastened. All overall buttons must be kept fastened, and any loose items such as ties and scarves should not be worn.

Protective clothing worn in the paint shop by the spray painters should be either good quality washable nylon garments, anti-static, and complete with hood, elasticated wrists and ankles; or low-linting disposable coveralls, which offer a liquid barrier protection from splashes, airborne dusts and paint overspray. The coveralls must withstand continuous exposure to a variety of chemicals and must be suitable for protection when using isocyanate-based two-pack paints; they also prevent the environment being contaminated by particles from the operator's clothing and hair.

(a)

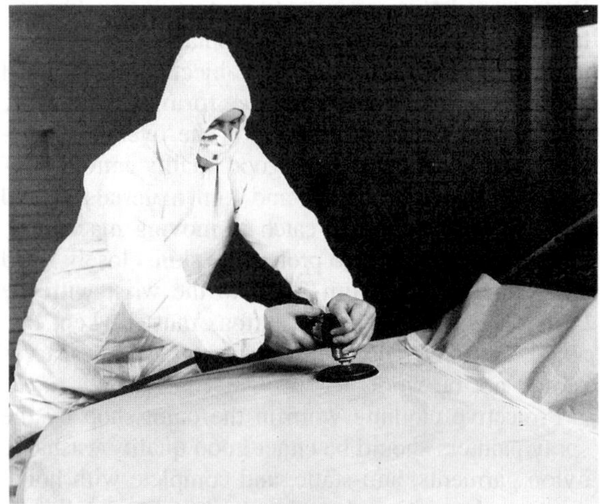

(b)

Figure 1.1 (a) Disposable protective coverall (b) protective coverall and face mask in use (*Gramos Chemicals International Ltd*)

They can be of the one-piece variety or can have separate disposable hoods (Figure 1.1a, b).

1.1.4 Head protection

Head protection is very important to the body worker when working underneath a vehicle or under its bonnet while it is being repaired. A light safety helmet, normally made from aluminium, fibreglass or plastic, should be worn if there is any danger from falling objects, and will protect the head from damage when working below vehicles. Hats and other forms of fabric headwear keep out dust, dirt and overspray and also prevent long hair (tied back) becoming entangled in moving equipment.

1.1.5 Eye and face protection

Eye protection is required where there is a possibility of eye injury from flying particles when using a grinder, disc sander, power drill or pneumatic chisel, or when removing glass windscreens or working underneath vehicles. Many employers are now requiring all employees to wear some form of safety glasses

Figure 1.2 Lightweight safety spectacles (*Racal Safety Ltd*)

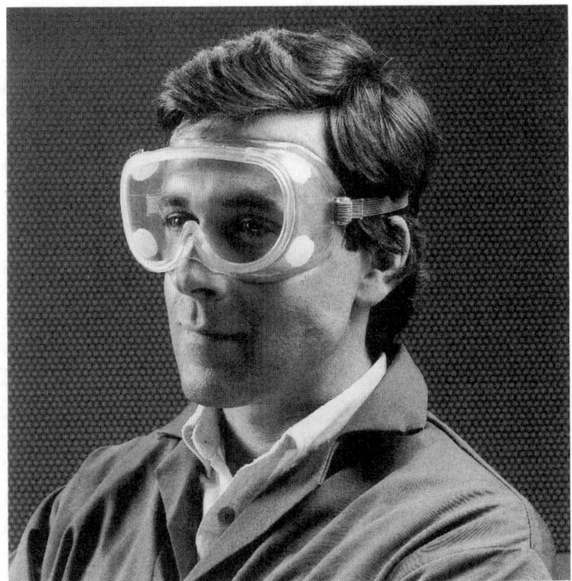

Figure 1.3 General-purpose safety goggles (*Racal Safety Ltd*)

when they are in either the repair or the paint areas of the bodyshop, because in any bodyshop location there is always the possibility of flying objects, dust particles, or splashing liquids entering the eyes. Not only is this painful but it can, in extreme cases, cause loss of sight. Eyes are irreplaceable: therefore it is advisable to wear safety goggles, glasses or face shields in all working areas.

The following types of eye protection are available:

Light-weight safety spectacles with adjustable arms and with side shields for extra protection. There is a choice of impact grades for the lenses (Figure 1.2).

General-purpose safety goggles with a moulded PVC frame which is resistant to oils, chemicals and water. These have either a clear acetate or a polycarbonate lens with BS impact grades 1 and 2 (Figure 1.3).

Face shields with an adjustable head harness and deep polycarbonate brow guard with replaceable swivel-up clear or anti-glare polycarbonate visor BS grade 1, which gives protection against sparks, molten metal and chemicals (Figure 1.4a, b).

Welding helmet or welding goggles with appropriate shaded lens to BS regulations. These must be worn at all times when welding. They will protect the eyes and face from flying molten particles of steel when gas welding and brazing, and from the harmful light rays generated by the arc when MIG/MAG, TIG or MMA welding (Figure 1.5a, b).

1.1.6 Foot protection

Safety footwear is essential in the bodyshop environment. Boots or shoes with steel toecaps will protect the toes from falling objects. Rubber boots will give protection from acids or wet conditions. Never wear defective footwear as this becomes a hazard in any workshop environment.

1.1.7 Respiratory protection (lungs)

One of the most important hazards faced by the bodyshop worker is that of potential damage to the lungs. Respirators are usually needed in body repair shops even though adequate ventilation is provided for the working areas. During welding, metal or paint preparation, or spraying, some form of protection is necessary. Under the COSHH Regulations, respiratory protection is essential and therefore must be used.

Respirators give protection against abrasive dusts, gases, vapours from caustic solutions and solvents, and spray mist from undercoats and finishing paint, by filtering the contaminated atmosphere before it is inhaled by the wearer. They may be either simple filtering devices, where the operator's lungs are used to draw air through the filter, or powered devices incorporating a battery-driven fan to draw contaminated air through the filters and deliver a flow of clean air to the wearer's face. There are four primary types of respirator available to protect bodyshop technicians:

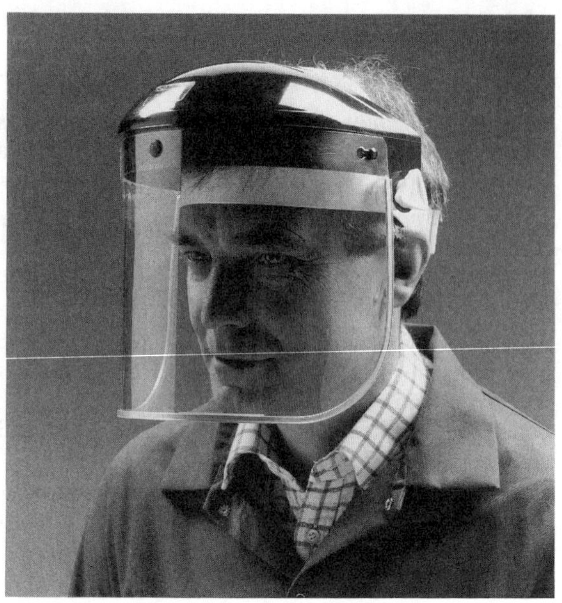

(a)

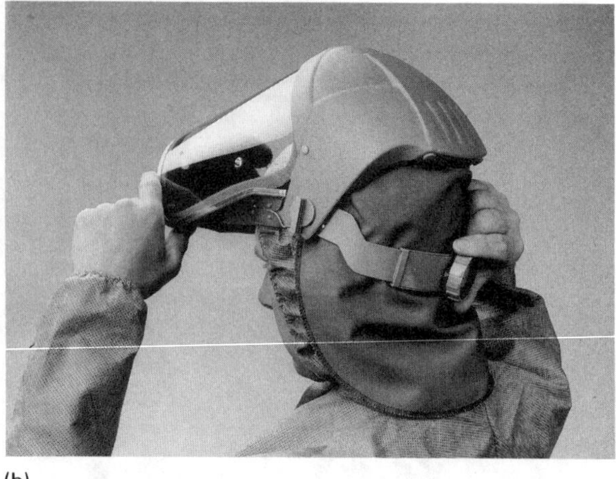

(b)

Figure 1.4 (a) Face shield (*Racal Safety Ltd*) (b) face shield with protective hood (*DeVilbiss Automotive Refinishing Products*)

(a)

(b)

Figure 1.5 (a) Standard visor-type welding helmet (b) standard welding goggles with hinged lenses (*Racal Safety Ltd*)

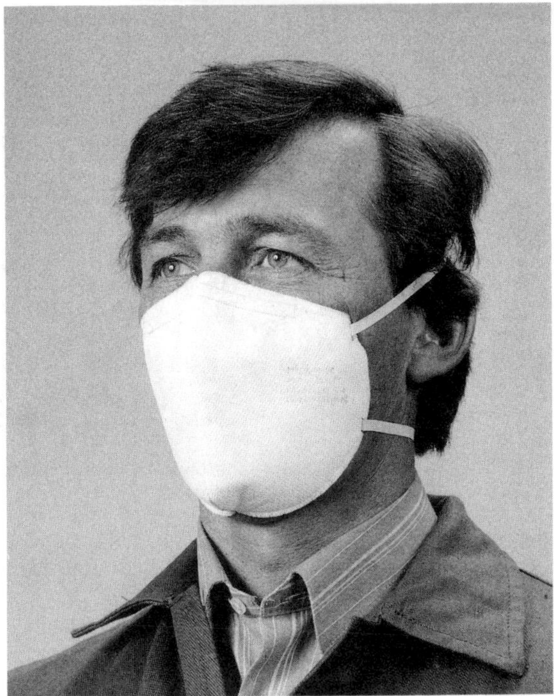

(a)

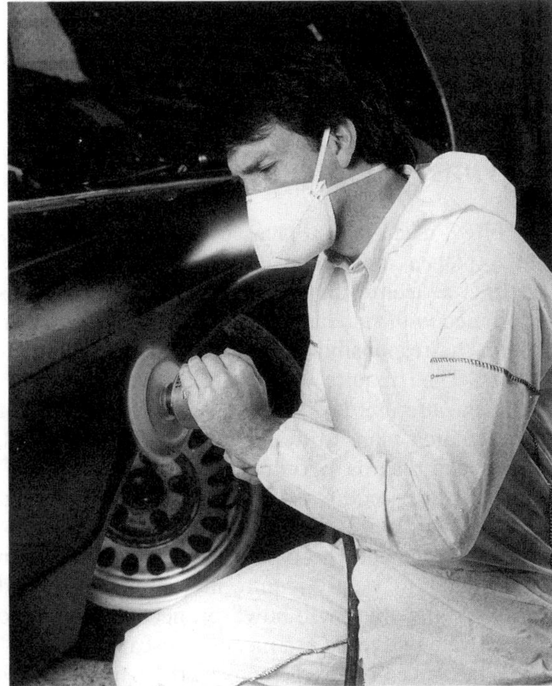

(b)

Figure 1.6 (a) Dust mask (b) dust mask in use (*Racal Safety Ltd*)

dust respirators, cartridge filter respirators, powered respirators, and constant flow air line breathing apparatus.

Dust respirators (masks)

The most basic form of respiratory protection is the disposable filtering half-mask, typically used when preparing or finishing bodywork such as by rubbing down or buffing, and where dust, mist and fumes are a problem. This face mask provides an excellent face seal while at the same time allowing the wearer to speak freely without breaking the seal. Breathing resistance is minimal, offering cool and comfortable use. Various types of mask are available for use in a variety of environments where contaminants vary from nuisance dust particles to fine dusts and toxic mists. These masks can only be used in atmospheres containing less than the occupational exposure limit of the contaminant (Figure 1.6a, b).

Cartridge filter mask

The cartridge filter or organic vapour type of respirator, which covers the nose and mouth, is equipped with a replacement cartridge that removes the organic vapours by chemical absorption. Some of these are also designed with a pre-filter to remove solid particles from the air before the air passes through the chemical cartridge. They are used in finishing operations with non-toxic paints, but not with isocyanate paints. For the vapour/particle respirator to function correctly it is essential that it fits properly against the face. Follow the manufacturer's instructions for changing the cartridges when spraying over a continuous period (Figure 1.7a, b).

Powered respirators

Powered respirators using canister filters offer protection against toxic dusts and gases. The respirator draws contaminated air through filters with a motor fan powered by a rechargeable battery and supplies clean air to the wearer's face. This avoids discomfort and fatigue caused by the effort of having to inhale air through filters, permitting longer working periods. These devices find great use both in the spray shop and in the repair shop when carrying out welding (Figure 1.8a, b).

Constant-flow air line breathing apparatus

The constant-flow compressed air line breathing apparatus is designed to operate from an industrial compressed air system in conjunction with the spray gun. Using a waist-belt-mounted miniature fixed-pressure regulator and a pre-filter, the equipment supplies breathing quality air through a small-bore hose

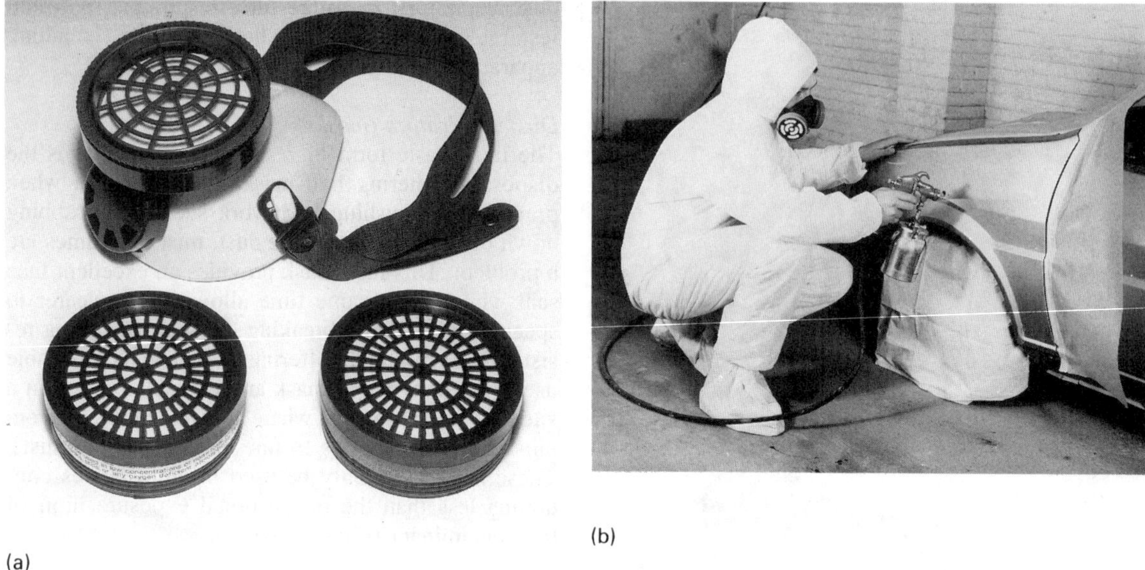

(a)

(b)

Figure 1.7 (a) Standard cartridge mask with filters (*Racal Safety Ltd*) (b) cartridge mask and protective coveralls being used while spraying (*Gramos Chemicals International Ltd*)

to a variety of face masks and visors to provide respiratory protection for paint spraying (such as with isocyanates), cleaning and grinding (Figure 1.9a, b). The COSHH Regulations have made it mandatory for all respiratory protection equipment to be both approved and suitable for the purpose, for the operatives to be correctly trained in the equipment's use and maintenance, and for proper records to be kept.

1.1.8 Ear protection

The Noise at Work Regulations 1989 define three action levels for exposure to noise at work:

1 A daily personal exposure of up to 85 dB(A). Where exposure exceeds this level, suitable hearing protection must be provided on request (Figure 1.10).
2 A daily personal exposure of up to 90 dB(A). Above this second level of provision, hearing protection is mandatory.
3 A peal; sound pressure of 200 pascals (140 dB).

Where the second or third levels are reached, employers must designate ear protection zones and require all who enter these zones to wear ear protection. Where the third level is exceeded, steps must be taken to reduce noise levels as far as is reasonably practicable. In every case where there is a risk of significant exposure to noise, assessment must be carried out and action taken to minimize hearing damage.

The first two noise action levels relate to exposure over a period (one day) and are intended to cater for the risks of prolonged work in noisy surroundings. The third level is related to sudden impact noises like those occurring in metal working procedures.

1.2 FIRE PRECAUTIONS

The Fire Precautions (Places of Work) Regulations 1992 replaced and extended the old Fire Precautions Act 1971 as from 1 January 1993. These Regulations are aligned with standard practice in EC Directives in placing the responsibility for compliance on the employers. They require employers not only to assess risks from fire, but now to include the preparation of an evacuation plan, to train staff in fire precautions, and to keep records. Workplaces with fewer than 20 employees may require emergency lighting points and fire warning systems. The self-employed who do not employ anyone but whose premises are regularly open to the public may only require fire extinguishers and warning signs; they will, however, need to be able to demonstrate that there is a means of escape in case of fire. Where five or more persons work on the premises as employees, all assessments need to be recorded in writing.

Most of these requirements were already covered by existing legislation. The prime differences are the

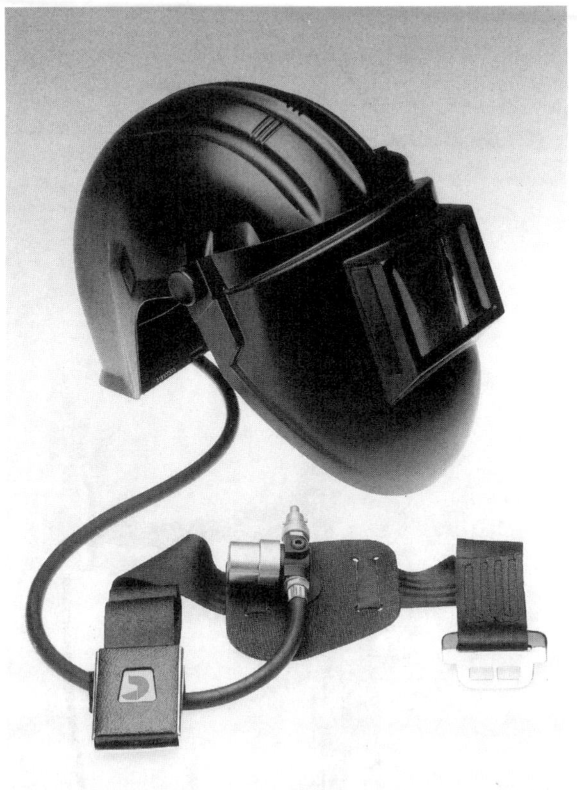

(a)

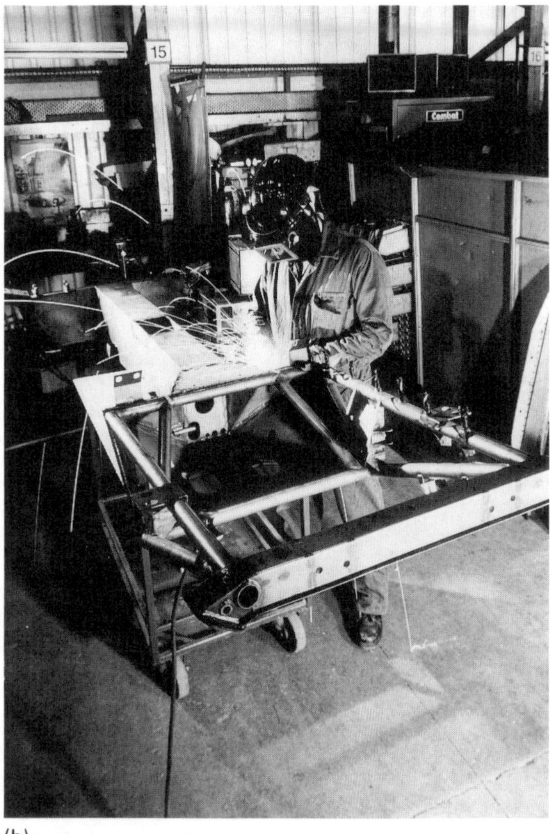

(b)

Figure 1.8 (a) Airstream welding helmet (powered respirator) (b) powered respirator in use in welding (*Recal Safety Ltd*)

recording of assessments, the provision of training, and the requirement that means of fighting fire, detecting fire and giving warning in case of fire be maintained in good working order.

1.2.1 What is fire?

Fire is a chemical reaction called combustion (usually oxidation resulting in the release of heat and light). To initiate and maintain this chemical reaction, or in other words for an outbreak of fire to occur and continue, the following elements are essential (Figure 1.11):

Fuel A combination substance, either solid, liquid or gas.

Oxygen Usually air, which contains 21 per cent oxygen.

Heat The attainment of a certain temperature (once a fire has started it normally maintains its own heat supply).

1.2.2 Methods of extinction

Because three ingredients are necessary for fire to occur, it follows logically that if one or more of these ingredients is removed, fire will be extinguished. Basically three methods are employed to extinguish a fire: removal of heat (cooling); removal of fuel (starving); and removal or limitation of oxygen (blanketing or smothering).

Removal of heat

If the rate of heat generation is less than the rate of dissipation, combustion cannot continue. For example, if cooling water can absorb heat to a point where more heat is being absorbed than generated, the fire will go out.

Removal of fuel

This is not a method that can be applied to fire extinguishers. The subdividing of risks can starve a fire, prevent large losses and enable portable extinguishers

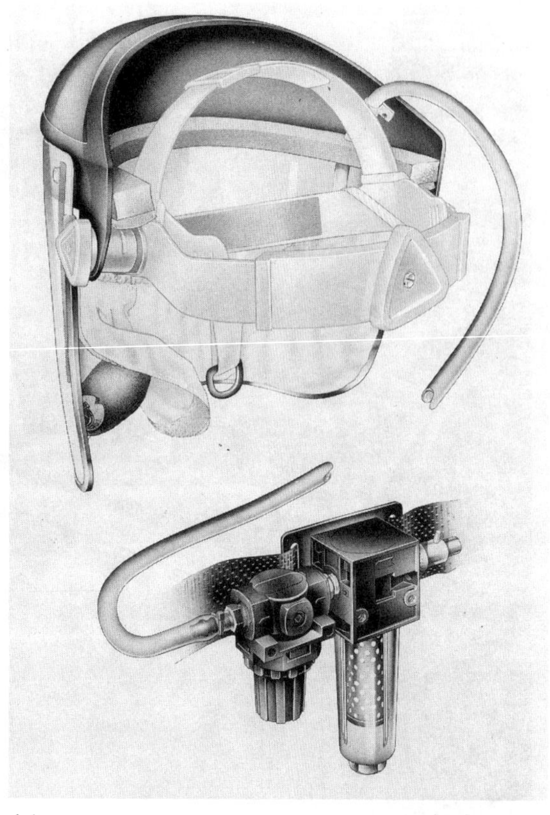

(a)

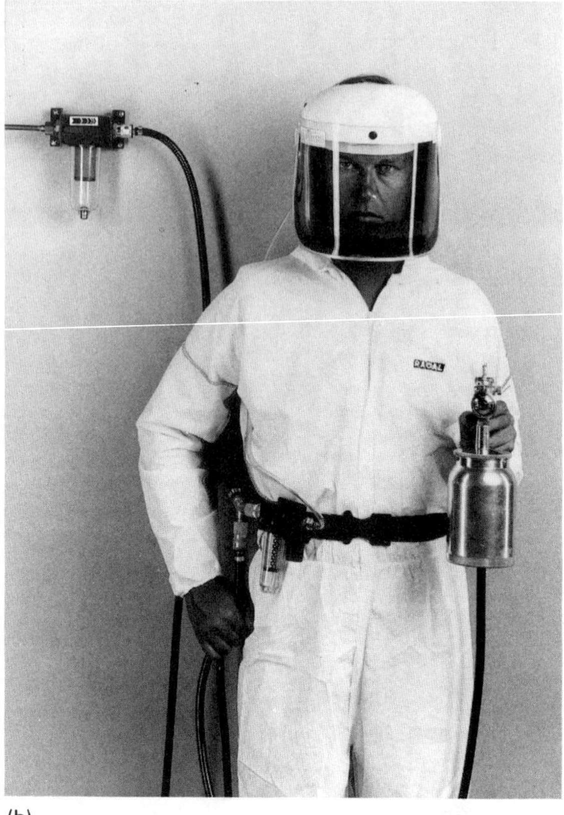

(b)

Figure 1.9 (a) Visionair constant-flow breathing apparatus (b) operator wearing complete constant-flow breathing apparatus (*Racal Safety Ltd*)

Figure 1.10 Ear protectors (*Racal Safety Ltd*)

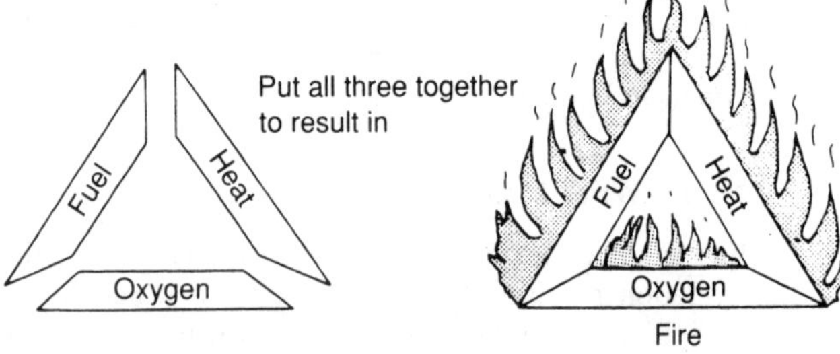

Figure 1.11 The fire triangle (*Chubb Fire Ltd*)

to retain control; for example, part of a building may be demolished to provide a fire stop.

The following advice can contribute to a company's fire protection programme:

1 What can cause fire in this location, and how can it be prevented?
2 If fire starts, regardless of cause, can it spread?

3 If so, where to?
4 Can anything be divided or moved to prevent such spread?

Removal or limitation of oxygen
It is not necessary to prevent the contact of oxygen with the heated fuel to achieve extinguishment. It will be found that where most flammable liquids are concerned, reducing the oxygen in the air from 21 to 15 per cent or less will extinguish the fire. Combustion becomes impossible even though a considerable proportion of oxygen remains in the atmosphere. This rule applies to most solid fuels although the degree to which oxygen content must be reduced may vary. Where solid materials are involved they may continue to burn or smoulder until the oxygen in the air is reduced to 6 per cent. There are also substances which carry within their own structures sufficient oxygen to sustain combustion.

1.2.3 Fire risks in the workshop

Fire risks in the vehicle body repair shop cover all classes of fire: class A, i.e. paper, wood and cloth; class B, i.e. flammable liquids such as oils, spirits, alcohols, solvents and grease; class C, i.e. flammable gases such as acetylene, propane, butane; and also electrical risks. It is essential that fire is detected and extinguished in the early stages. Workshop staff must know the risks involved and should be aware of the procedures necessary to combat fire. Bodyshop personnel should be aware of the various classes of fire and how they relate to common workshop practice.

Class A fires: wood, paper and cloth
Today wood is not used in cars, although there are exceptions. Cloth materials are used for some main trim items and are therefore a potential fire hazard. The paper used for masking purposes is a prime area of concern. Once it has done its job and is covered in overspray it is important that it is correctly disposed of, ideally in a metal container with a lid, and not scrunched up and thrown on the floor to form the potential start of a deep-seated fire.

Class B fires: flammable liquids
Flammable liquids are the stock materials used in the trade for all body refinishing processes: gun cleaner to clean finish coats, cellulose to the more modern finishes, can all burn and produce acrid smoke.

Class C fires: gases
Not many cars run on liquid propane gas (LPG), but welding gases or propane space heaters not only burn but can be the source of ignition for A or B fires.

Electrical hazards
Electricity is not of itself a class of fire. It is, however, a potential source of ignition for all of the fire classes mentioned above.

The Electricity at Work Regulations cover the care of cables, plugs and wiring. In addition, in the bodyshop the use of welding and cutting equipment produces sparks which can, in the absence of good housekeeping, start a big fire. Training in how to use fire fighting equipment can stop a fire in its early stages. Another hazard is the electrical energy present in all car batteries. A short-circuit across the terminals of a battery can produce sufficient energy to form a weld and in turn heating, a prime source of ignition. When tackling a car fire a fireman will always try to disconnect the battery, as otherwise any attempt to extinguish a fire can result in the re-ignition of flammable vapours.

Body filler
A further possible source of ignition to be aware of in general use in the body repair business is the mixing of two materials to use as a body filler. The result of mixing in the wrong proportions can give rise to an exothermic (heat releasing) reaction; in extreme cases the mix can ignite.

1.2.4 General precautions to reduce fire risk

(a) Good housekeeping means putting rubbish away rather than letting it accumulate.
(b) Read the manufacturer's material safety data sheets so that the dangers of flammable liquids are known.
(c) Only take from the stores sufficient flammable material for the job in hand.
(d) Materials left over from a specific job should be put back into a labelled container so that not only you but anyone (and this may be a fireman) can tell what the potential risk may be.
(e) Take care when welding that sparks or burning underseal do not cause a problem, especially when working in confined areas of vehicles.
(f) Be extremely careful when working close to plastic fuel lines.
(g) Petrol tanks are a potential hazard: supposedly empty tanks may be full of vapour. To give some idea of the potential problem, consider one gallon of petrol: it will evaporate into 33 ft^3 of neat vapour, which will mix with air to form 2140 ft^3 of flammable vapour. Thus the average petrol tank needs only a small amount of petrol to give a tank full of vapour waiting to ignite and explode.

The key to fire safety is:

1 Take care.
2 Think.
3 Train staff in the correct procedures before things go wrong.
4 Ensure that these procedures are written down, understood and followed by all personnel within the workshop.

1.2.5 Portable extinguishers: types and uses

The colour codes for each type of appliance are as follows:

Red for water.
Cream for foam.
Black for CO_2.
Blue for powder.
Green for halon (BCF).

Figures 1.12a–c show various types of extinguisher.

Water
Water is the most widely used extinguisher agent.

With portable extinguishers, a limited quantity of water can be expelled under pressure and its direction controlled by a nozzle.

There are basically two types of water extinguishers. The gas (CO_2) cartridge-operated extinguisher, when pierced by a plunger, pressurizes the body of the extinguisher, thus expelling the water and producing a powerful jet capable of rapidly extinguishing class A fires. In stored pressure extinguishers the main body is constantly under pressure from dry air or nitrogen, and the extinguisher is operated by opening the squeeze grip discharge valve. These extinguishers are available with 6 litre or 9 litre capacity bodies and thus provide alternatives of weight and accessibility.

Foam
Foam is an agent most suitable for dealing with flammable liquid fires. Foam is produced when a solution of foam liquid and water is expelled under pressure through a foam-making branch pipe at which point air is entrained, converting the solution into a foam.

Foam extinguishers can be pressurized either by a CO_2 gas cartridge or by stored pressure. The standard capacities are 6 and 9 litres.

Spray foam
Unlike conventional foams, aqueous film forming foam (AFFF) does not require to be fully aspirated in order to extinguish fires. Spray foam extinguishers expel an AFFF solution in an atomized form which is suitable for use on class A and class B fires. AFFF is a fast and powerful means of tackling a fire and seals the surfaces of the material, preventing re-ignition. The capacity can be 6 or 9 litres, and operation can be by CO_2 cartridge or stored pressure.

Carbon dioxide
Designed specifically to deal with class B, class C and electrical fire risks, these extinguishers deliver a powerful concentration of carbon dioxide gas under great pressure. This not only smothers the fire very rapidly, but is also non-toxic and is harmless to most delicate mechanisms and materials.

Dry powder
This type of extinguisher is highly effective against flammable gases, open or running fires involving flammable liquids such as oils, spirits, alcohols, solvents and waxes, and electrical risks. The powder is contained in the metal body of the extinguisher from which it is supplied either by a sealed gas cartridge, or by dry air or nitrogen stored under pressure in the body of the extinguisher in contact with the powder.

Dry powder extinguishers are usually made in sizes containing 1 to 9 kg of either standard powder or (preferably and more generally) all-purpose powder, which is suitable for mixed risk areas.

Vaporizing liquid (halon 1211, BCF)
Portable extinguishers of this type are manufactured in sizes ranging from 1 to 15 kg. They are particularly effective for dealing with class B fires and with fires started by an electrical source.

Halon 1211 (bromochlorodifluoromethane, BCF) has a low toxicity level, is considered to be non-corrosive and has a long storage life. It is clean to use and leaves no residue, thus rendering it harmless to delicate fabrics and machinery. However, owing to the contribution of halons to atmospheric ozone depletion most companies have decided to cease production of halon 1211.

Choosing and siting portable extinguishers
Because there is such a variety of fire risks in bodyshops, it is important to analyse these risks separately and (with the help of experts such as fire officers) to choose the correct fire fighting medium to deal with each possible fire situation. It should be noted that portable fire extinguishers are classified as first-aid fire fighting and are designed for ease of operation in

Types of modern fire extinguishers
B.S.I.DD48:1976

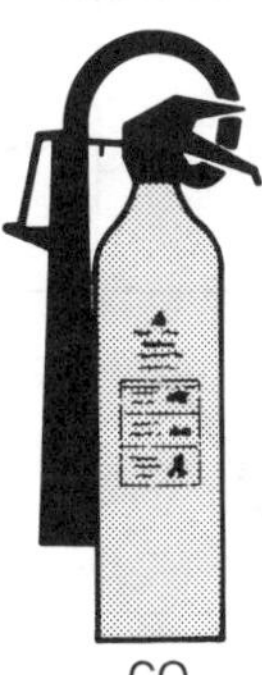

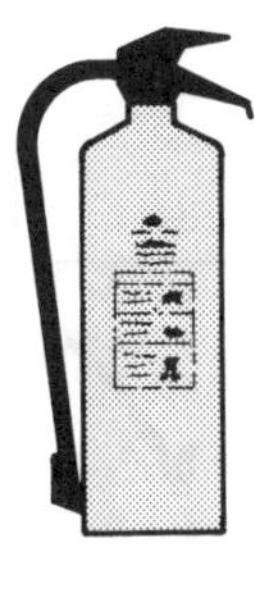

Water	Foam	Fire blanket	Powder	CO_2	BCF
Use for wood, paper, fabrics etc	Use for flammable liquids, oils, fats, spirits etc	Use for smothering	Use for all fires flammable liquids and gases	Use for electrical and flammable liquid fires	Use for electrical and flammable liquid fires
Do not use on electrical or flammable liquid fires	Do not use on electrical fires				

(a)

(b)

(c)

Figure 1.12 (a) Types of portable fire extinguisher (b) types of fire fighting equipment (*UK Fire International Ltd*) (c) portable fire extinguishers suitable for a bodyshop (*Chubb Fire Ltd*)

Class of fire	Water	Foam (AFFF)	CO_2 gas	Powder
A Paper Wood Textile Fabric	✓	✓		✓
B Flammable liquids		✓	✓	✓
C Flammable gases			✓	✓
Electrical hazards			✓	✓
Vehicle protection		✓		✓

Figure 1.13 Which extinguisher to use (*Chubb Fire Ltd*)

an emergency. It is important to realize that because they are portable they have only a limited discharge. Therefore their siting, together with an appreciation of their individual characteristics, is fundamental to their success in fighting fire (Figure 1.13).

1.3 SAFETY SIGNS IN THE WORKSHOP

It is a legal requirement that all safety signs used in a bodyshop comply with BS 5378:Part 1.

Each of these signs is a combination of colour and design, within which the symbol is inserted. If additional information is required, supplementary text may be used in conjunction with the relevant symbol, provided that it does not interfere with the symbol. The text can be in an oblong or square box of the same colour as the sign, with the text in the relevant contrasting colour, or the box can be white and the text black.

BS 5378 divides signs into four categories (Figure 1.14):

Prohibition Prohibition signs have a red circular outline and crossbar running from top left to bottom right on a white background (Figure 1.15a). The symbol displayed on the sign must be black and placed centrally on the background, without obliterating the crossbar. The colour red is associated with 'stop' or 'do not'.

Warning Warning signs have a yellow triangle with a black outline (Figure 1.15b). The symbol or text used on the sign must be black and placed centrally on the background. This combination of black and yellow identifies caution.

Mandatory Mandatory signs have a blue circular background (Figure 1.15c). The symbol or text used must be white and placed centrally on the background. Mandatory signs indicate that a specific course of action is to be taken.

Safe condition The safe condition signs provide information for a particular facility (Figure 1.15d) and have a green square or rectangular background to accommodate the symbol or text, which must be in white. The safety colour green indicates 'access' or 'permission'.

Fire safety signs are specified by BS 5499, which gives the characteristics of signs for fire equipment, precautions and means of escape in case of fire (Figure

Key to British and European Standard safety signs

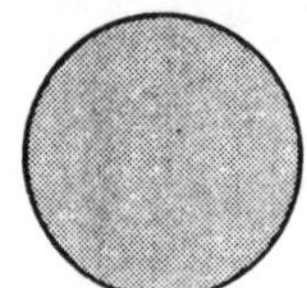

Figure 1.14 Standard safety signs

Figure 1.15 (a) Prohibition signs (b) warning signs (c) mandatory signs (d) safe condition signs

1.16). It uses the basic framework concerning safety colours and design adopted by BS 5378.

1.4 GENERAL SAFETY PRECAUTIONS IN THE WORKSHOP

The Health and Safety at Work Act imposes on employers a statutory duty to ensure safe working conditions and an absence of risk in the use of equipment and the handling of materials, and to comply with Regulations regarding safe working practices in order to reduce to a minimum the hazards to health and safety associated with vehicle body repair work. To skilled and experienced operators this does not mean that any additional restrictions are imposed on their activities, but merely that they should carry out their

Mandatory signs

Eye protection must be worn

Head protection must be worn

Hearing protection must be worn

Respiratory protection must be worn

Foot protection must be worn

Hand protection must be worn

(c)

Safe condition signs

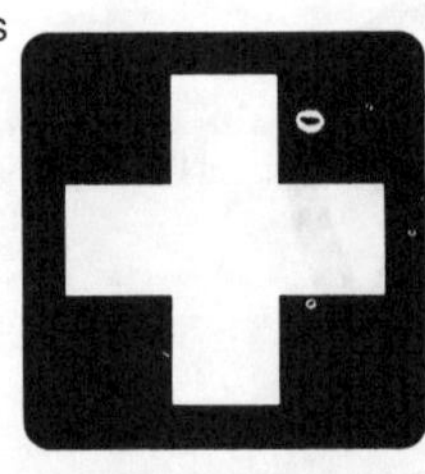

First aid

Indication of direction (may be used in conjunction with A.4.1)

Emergency eye wash

Emergency telephone

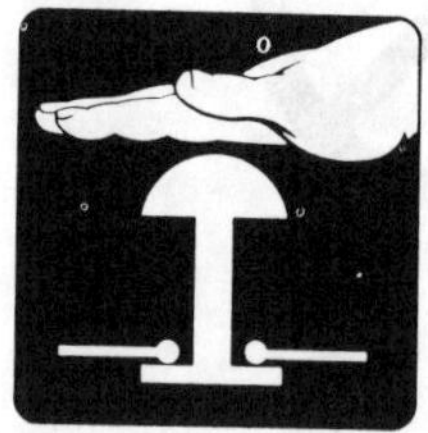

Emergency stop push-button

(d)

Figure 1.15 (continued)

Figure 1.16 Fire signs

tasks with constant regard for the health and safety of themselves and their fellow workers.

Particular hazards may be encountered in the bodyshop, and safety precautions associated with them are as follows:

1 Do wash before eating, drinking or using toilet facilities to avoid transferring the residues of sealers, pigments, solvents, filings of steel, lead and other metals from the hands to the inner parts and other sensitive areas of the body.
2 Do not use kerosene, thinners or solvents to wash the skin. They remove the skin's natural protective oils and can cause dryness and irritation or have serious toxic effects.
3 Do not overuse waterless hand cleaners, soaps or detergents, as they can remove the skin's protective barrier oils.
4 Always use barrier cream to protect the hands, especially against fuels, oils, greases, hydrocarbon solvents and solvent-based sealers.
5 Do follow work practices that minimize the contact of exposed skin and the length of time liquids or substances stay on the skin.
6 Do thoroughly wash contaminants such as used engine oil from the skin as soon as possible with soap and water. A waterless hand cleaner can be used when soap and water are not available. Always apply skin cream after using waterless hand cleaner.
7 Do not put contaminated or oily rags in pockets or tuck them under a belt, as this can cause continuous skin contact.
8 Do not dispose of dangerous fluids by pouring them on the ground, or down drains or sewers.
9 Do not continue to wear overalls which have become badly soiled or which have acid, oil, grease, fuel or toxic solvents spilt over them. The effect of prolonged contact from heavily soiled overalls with the skin can be cumulative and life threatening. If the soilants are or become flammable from the effect of body temperature, a spark from welding or grinding could envelop the wearer in flames with disastrous consequences.
10 Do not clean dusty overalls with an air line: it is more likely to blow the dust into the skin, with possible serious or even fatal results.
11 Do wash contaminated or oily clothing before wearing it again.
12 Do disguard contaminated shoes.
13 Wear only shoes which afford adequate protection to the feet from the effect of dropping tools and sharp and/or heavy objects on them, and also from red hot and burning materials. Sharp or hot objects could easily penetrate unsuitable footwear such as canvas plimsolls or trainers. The soles of the shoes should also be maintained in good condition to guard against upward penetration by sharp or hot pieces of metal.
14 Ensure gloves are free from holes and are clean on the inside. Always wear them when handling materials of a hazardous or toxic nature.
15 Keep goggles clean and in good condition. The front of the glasses or eyepieces can become obscured by welding spatter adhering to them. Renew the glass or goggles as necessary. Never use goggles with cracked glasses.
16 Always wear goggles when using a bench grindstone or portable grinders, disc sanders, power saws and chisels.
17 When welding, always wear adequate eye protection for the process being used. MIG/MAG welding is particularly high in ultraviolet radiation which can seriously affect the eyes.
18 Glasses, when worn, should have 'safety' or 'splinter-proof' glass or plastic lenses.
19 Always keep a suitable mask for use when dry flatting or working in dusty environments and when spraying adhesive, sealers, solvent carried waxes, and paints.
20 In particularly hostile environments such as when

using volatile solvents or isocyanate materials, respirators or fresh air fed masks must be worn.

21 Electric shock can result from the use of faulty and poorly maintained electrical equipment or misuse of equipment. All electrical equipment must be frequently checked and maintained in good condition. Flexes, cables and plugs must not be frayed, cracked, cut or damaged in any way. Equipment must be protected by the correctly rated fuse.

22 Use low-voltage equipment wherever possible (110 volts).

23 In case of electric shock:
   (a) Avoid physical contact with the victim.
   (b) Switch off the electricity.
   (c) If this is not possible, drag or push the victim away from the source of the electricity using non-conductive material.
   (d) Commence resuscitation if trained to do so.
   (e) Summon medical assistance as soon as possible.

1.5 DUST AND FUME EXTRACTION (EXTRACTION AND ARRESTMENT SYSTEMS)

Polluted air is often invisible to the naked eye. However, the effect it can have on the health of a workforce and the overall efficiency of an organization can be dramatic. The most effective way to purify air is to capture airborne pollutants at source and, depending on individual applications, either to recirculate fresh and preheated air or vent the pollutants away from the working environment to a safe collection point. The systems should actually be extraction and arrestment systems and should extract the pollutant materials and collect them in a safe and manageable form (Figure 1.17).

Figure 1.17 Portable dust extraction system used in the workshop (*Nederman Ltd*)

In the context of bodyshops, the main problems are fillers and paint dust from the rubbing down and flatting processes, paint and solvent fumes from the wiping down and painting processes, and welding fumes.

Filler and paint dust generated in the preparation area is best collected as soon as it is produced by using off-the-tool extraction (Figure 1.18). This can either be by portable units serving one or two operators, or by fixed systems with a central extraction unit serving a number of fixed extraction points located in the workshop (Figure 1.19).

Specially designed extraction equipment can be tailor-made to an individual bodyshop for the removal of dust and fumes. Gases, powders and chemical vapours are all types of hazardous elements to which fume extraction can be applied. The range of self-supporting arms, combined with the versatility to mount the system on ceilings, floors, benches and walls, make the access to applications unlimited. Furthermore, the easily manoeuvred suction hoods create extraction right at the source of the problem. The fans are designed to draw the polluted air through the extraction arms, dispersing the fumes via the assembly ducting. Alternatively, to recirculate the purified air, an electrostatic unit can be employed to eliminate harmful particles and utilize existing pre-heated air.

A range of vehicle exhaust extraction systems is available: a choice can be made from a simple drop system through to the rail system (Figure 1.20), which allows vehicles to be driven while maintaining at-source extraction with both advanced infrared remote controlled and electrically motor driven reels.

At-source extraction of welding fumes is far more energy efficient than using central ventilation systems.

Harmful fumes can be captured and disposed of in a safe and simple way regardless of the welding environment. Irritation, fever, poisoning and fibrosis are a few of the effects that can be minimized with the extraction of fumes from the welding operations. Many

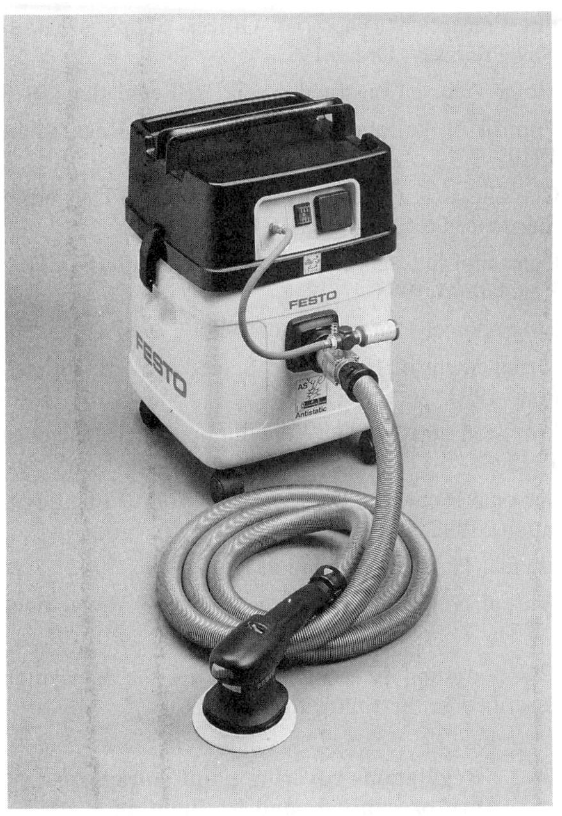

Figure 1.18 Tool extraction system (*Minden Industrial Ltd*)

Figure 1.19 Central extraction unit (*Minden Industrial Ltd*)

welding processes create noxious and harmful fumes which can be eliminated with portable welding smoke eliminators (Figure 1.21). These provide complete extraction where confined areas pose a problem, especially in body repair workshops. There is a wide range of lightweight smoke eliminators which take up very little floor space and can be carried from job to job. The filters in these portable extraction systems can be changed very quickly, and some have an alarm light fitted to warn the user that the filter needs replacing.

It is a specific requirement of the COSHH Regulations that equipment is maintained in efficient working order. The physical system must be regarded as part of a broader health and safety housekeeping policy aimed at keeping the whole area dust and fume free.

Figure 1.20 Vehicle exhaust extraction using a rail system (*Nederman Ltd*)

1.6 BODYSHOPS AND LEGAL REQUIREMENTS

Health and safety legislation has made the vehicle body repair industry increasingly aware of the need to

Figure 1.21 Portable fume and smoke extraction system (*Nederman Ltd*)

provide adequate facilities for employees, both as a legal duty and to improve the working environment. Within this framework of a safe working environment the employer must also promote efficient work methods, which together should result in improved productivity.

1.6.1 Statutory legislation

Petroleum (Consolidation) Act 1928

Petroleum (Mixtures) Orders 1929 and 1947

Factory Act 1961

Weights and Measures Act 1963

Fire Precautions Act 1971

Highly Flammable Liquids and Liquefied Petroleum Gases Regulations 1972

Road Traffic Act 1972 (MOT)

Abrasive Wheel Regulations 1974

Control of Pollution Act 1974

Protection of Eyes Regulations 1974

Health and Safety at Work Act 1974

Fire Precautions (Factories, Offices, Shops and Railway Premises) Order 1976

Motor Vehicle Construction and Use Regulations 1978

Control of Pollution (Special Wastes) Regulations 1980

Classification, Packaging and Labelling of Dangerous Substances Regulations 1984

Control of Substances Hazardous to Health (COSHH) Regulations 1988

Environment Protection Act 1990

Management of Health and Safety at Work Regulations 1992

Provision and Use of Work Equipment Regulations 1992

Personal Protection Equipment at Work (PPE) Regulations 1992

Manual Handling Operation Regulations 1992

Workplace (Health, Safety and Welfare) Regulations 1992

The bodyshop is most affected by the legislation described in the following sections.

1.6.2 Regulations covering paint storage, fire precautions and fire fighting

The storage of paints with a flashpoint below 22°C is governed by the Petroleum (Consolidation) Act and the statutory rules and orders 1929. Those with a flash point of 22–32°C are governed by the Highly Flammable Liquids and Liquefied Petroleum Gases Regulations 1972.

Requirements of the Petroleum (Consolidation) Act 1928

Paints, thinners, and other products governed by this Act will either be labelled 'Petroleum mixture giving off an inflammable heavy vapour' or have a flame symbol and the words 'Highly flammable', indicating a flashpoint below 22°C.

The Act requires that the storeroom should be licensed and constructed to approved standards.

Requirements of the Highly Flammable Liquids Regulations 1972

Paints governed by these Regulations will be labelled 'Flammable'. More precise flashpoint details can be obtained from the relevant product data sheet, which indicates products with a flashpoint range between 22 and 32°C.

Products with a flashpoint greater than 32°C do not fall within the Regulations, but other Regulations require that products with a flashpoint up to 55°C have a 'Flammable' warning.

1.6.3 Health and safety legislation

The Health and Safety at Work Act 1974 places a duty upon employers to ensure, so far as is reasonably practicable, safe working conditions and the absence of risks to health in connection with the use, handling, storage and transport of articles and substances.

The Act places a statutory duty on employers to have a declared safety policy for a business in which more than five persons are employed. The Act does not specify what you should do; it merely provides the framework in which you should operate, together with establishing the Health and Safety Executive and the Health and Safety Commission. The Act regulates all working methods. The importance of this Act cannot be overemphasized; no working methods may be employed that can be seen to be a health or safety hazard to employees.

First, a company safety policy must be established and a safety committee formed. The committee should consist of members with specialized knowledge of the risks of a particular area, i.e. bodyshop, paint shop, parts department, offices. The chairman of the committee should be a senior member of the company.

Safety committee's main objectives

1 To ensure that the company's premises outside and inside are safe and healthy.
2 To set up an administrative system to maintain a safe working environment.
3 To set up some kind of information system on all matters relating to health and safety.
4 To ensure that employees receive adequate training in health and safety matters.

Regular inspections of all equipment as outlined in the safety practice code are now a legal requirement. Failure to meet this requirement can result in imprisonment or unlimited fines or both.

Employer's duties
To provide and maintain:

1 Safe equipment.
2 Safe place of work with safe access and exit.
3 Safe systems of work.
4 Healthy working environment.
5 Welfare arrangements (washing facilities, eating facilities, first aid).
6 Proper training with time off for this purpose.

Employee's duties

1 To take reasonable care for his/her own safety and the safety of others who may be affected by his/her acts.
2 To cooperate with his/her employer or any person on whom a duty or requirement is imposed.

The company should establish, in conjunction with the safety committee, its own code of practice on safety procedures within its establishment, and all personnel should receive a copy.

Objectives of the Act

1 To repeal, replace or modify existing regulations.
2 To maintain or improve standards of health and welfare of people at work.
3 To protect people or other workers against any risk to health or safety arising out of the activities of people at work.
4 To control the keeping and use of dangerous substances including explosive and highly flammable materials (paints, solvents, resins).
5 To control the emissions of toxic substances into the atmosphere (dust, fumes, smoke, gases).

1.6.4 Control of Substances Hazardous to Health (COSHH) Regulations 1988

Since 1974 a number of Regulations have been introduced which describe in detail the requirements for specific safe working practices. One of the most important is the Control of Substances Hazardous to Health (COSHH) Regulations 1988.

COSHH is a major piece of health and safety legislation. It tightens the general obligations of an employer under Section 2 of the Health and Safety at Work Act by specifying comprehensive rules on how substances should be controlled. COSHH applies to all substances in all forms, including gases, vapours, solids, dusts, liquids, and even micro-organisms. It also covers mixtures and preparations.

A substance may be a 'hazard to health' if the substance itself is harmful or if:

1 There are hazards from impurities.
2 Dust or fumes are generated during use.
3 It is dangerous when used in combination with other materials.

COSHH attempts to follow the principle of good occupational hygiene practice:

1 Assess the hazard (Regulation 6).
2 Control it (Regulations 7, 8, 12).
3 Maintain the control (Regulations 9, 10, 11).

HEALTH SURVEILLANCE RECORD

EMPLOYEE'S NAME	TYPE OF SURVEILLANCE	RESULT	DATE CARRIED OUT
MR JOHN SMITH	RESPIRATORY FUNCTION SCREENING.	WITHIN NORMAL LIMITS.	30-1-93
MR DEREK BROWN	CHEST X-RAY	WITHIN NORMAL LIMITS.	6-3-93
MR WILLIAM BLACK	AUDIOMETRIC HEARING TEST.	CATEGORY 5 (NORMAL)	20-4-93

Figure 1.22 Health surveillance record (*Akzo Coatings PLC*)

TRAINING RECORD

EMPLOYEE'S NAME	TYPE OF TRAINING	DATE CARRIED OUT
MR JOHN SMITH	SAFETY TRAINING ON THE USE OF AIR-FED MASKS	1-3-93
MR DEREK BROWN	TRAINING IN THE USE OF PORTABLE FIRE-FIGHTING EQUIPMENT FOR THE WORKSHOP	10-4-93
MR WILLIAM BLACK	TRAINING IN THE SAFE USAGE OF MIG WELDING EQUIPMENT.	5-5-93

Figure 1.23 Training record (*Akzo Coatings PLC*)

Assessment: Regulation 6

An employer shall not carry on any work which is liable to expose any employees to any substance hazardous to health unless he has made a suitable and sufficient assessment of the risk created, and the steps needed to achieve and maintain control.

Control of exposure: Regulation 7

Exposure must be prevented or adequately controlled. 'Adequately controlled' means that repeated exposure must not cause damage to health. Maximum exposure limits must not be exceeded or employers may be prosecuted.

Use of controls: Regulation 8

Employers should take all reasonable steps to ensure that control measures are properly used, and employees shall make full and proper use of the controls provided.

Maintenance of controls: Regulation 9

Controls must be maintained in efficient working order and good repair. Thorough examinations and tests must be made at suitable intervals. Records of tests and repairs must be kept for five years.

Exposed monitoring: Regulation 10

Air sampling is required to monitor exposure every 12 months if:

1 It is needed to maintain control.
2 It is needed to protect health.
3 Listed carcinogens are in use.

Health surveillance: Regulation 11

Health surveillance is required at least every 12 months if:

1 Valid health surveillance techniques exist.
2 Specified substances are in use.

Surveillance may include biological monitoring, clinical examinations and review of health records. Records must be kept for 30 years (Figure 1.22).

Information, instruction and training: Regulation 12

The employees must be told:

1 The risk to health.
2 The results of the tests.
3 The control measures to be used.

They must be given appropriate instruction and training. Any person carrying out work under the Regulations must be competent to do so (Figure 1.23).

Carrying out a COSHH assessment

Gather information

1 What is used, handled, stored.
2 Intermediates, by-products, wastes.

Identify hazards (a hazard is the potential to cause harm)

1 Nature of hazard.
2 Route of exposure.
3 Possible interactions.

Evaluate the risk (risk is the likelihood that a substance will cause harm in the actual circumstances of use)

1 How is the substance used?
2 How is it controlled?
3 What is the level of exposure?

Decide on controls

1 Type of controls (substitution, local exhaust ventilation (LEV)).
2 Maintenance and testing procedures.
3 Air monitoring.
4 Health surveillance.
5 Record keeping.

Record assessment A record must be kept of all assessments other than in the most simple cases that can be easily repeated (Table 1.1, Figure 1.24).

Review Assessments should be reviewed for example when:

1 There are any significant changes to the process, e.g. plant changes, volume of production.
2 New substances are introduced.
3 Ill-health occurs.
4 New information becomes available on hazards or risks.

1.6.5 Environment Protection Act 1990

The Environment Protection Act 1990, like the Health and Safety at Work Act, allows the Secretary of State to introduce Regulations to control, among other things, the release of harmful substances to air, water and land. The legislation is wide in scope, with the long-term aim of minimizing environment pollution. The paint application industry is among the third largest group of environment polluters, and it is because of this that the EPA introduces many refinishing industry specific controls. The object of the Act is to reduce and eventually eliminate the use of volatile organic compounds (VOCs) in surface coatings, using the best available techniques not entailing excessive costs (BATNEECs). VOCs are substances which react with nitrous oxides in sunlight to create low-level ozone. This photochemical air pollution causes damage to vegetation and can also cause serious breathing difficulties in humans and animals. All users will have to use BATNEECs to prevent, minimize and render harmless any release of VOCs. If there is alleged contravention, the user will have to prove that no better techniques were available.

Bodyshops purchasing more than 2 tonnes of solvent per year will be required to register under the classification of Part B 'The Respraying of Motor Vehicles' of the Act. Any bodyshop currently operating below these levels is not required to register.

Existing bodyshops which registered by the deadline set for their area of the country (in England and Wales by 30 September 1992, and in Scotland by 31 March 1993) have until October 1998 to fulfil all their legal obligations under the Act.

Two separate pollution control regimes are established in Part 1 of the Act to control industrial processes falling into the categories A and B. Category A is integrated pollution control (IPC) operated by HMIP in England and Wales and HMIPI in Scotland. Category B is local authority air pollution control.

Environmental Protection (Prescribed Processes and Substances) Regulations 1991

These identify the respraying of road vehicles as a category B process (local authority air pollution control) if the process may result in the release into the air of any particular matter or of any volatile organic compound (solvents) where the process is likely to involve the use of 2 tonnes or more of organic solvents in any 12-month period. Cleaning agents, fillers, stoppers, primers, gunwash and many other products in the refinishing of motor vehicles all contribute to this figure. The figure should represent the amount consumed in the process, i.e. lost through evaporation, spillage and transfer.

If the process involves the use of 2 tonnes of organic solvents in any 12-month period then the bodyshop operator must seek authorization from his or her local authority to carry out the process. If the operator of the bodyshop calculates the solvent usage to be less than 2 tonnes, then the best advice would be to monitor solvent usage in order to demonstrate the fact to the local authority.

First check with the local authority, which will require information about your operation such as:

1 Who and where you are.
2 What you do and how you do it.
3 What is released to air, how much, and where from.
4 What you do to prevent, control and monitor such releases.
5 Evidence or proposals that the objectives of BATNEEC will be met.

Upgrading of existing bodyshops to comply with the Act

The engineering controls that the EPA imposes upon equipment are very significant, particularly regarding spray booths if emission exceeds the 2 tonnes VOCs limit. For spray booths already working there is a requirement to submit an upgrading programme to the local authority within 12 months of the initial registration. That programme must be implemented by 1 April 1998.

The requirements for upgrading are as follows:

Table. 1.1 Guide to safe working practices: to be filled in by the bodyshop. The example gives a typical process carried out in most paint shops every day. The initial assessment has identified there is a serious risk to people's health from the way the hazardous substance is being used. Immediate steps must be introduced to control and reduce the risks. Failure to comply to the approved code of practice by an employer may lead to prosecution

			Precautions				
Activity	**Hazard**	**Risk**	**Personal**	**General**	**Products used**	**Products contain**	**Remarks**
Wet sanding of fillers and stoppers	Wet sludges of unknown composition	Skin contact with unknown materials	Gloves	Carry out in a properly ventilated area			
		Eye contact with splashes of unknown materials	Goggles or visor				
		Remarks:	Conclusions:	Conclusions:			
Priming Mixing (hardener addition and or thinning)	Solvents and, depending on the particular product, isocyanates, acids or polyamides may be present	Inhalation of solvent vapours	Respirator suitable for solvent vapours where extraction inadequate	Carrying out in a well ventilated area or with local exhaust ventilation to minimize vapour build-up. NB air-fed respirators are not required when mixing isocyanate containing products at ambient temperatures under conditions of good ventilation	Washfiller 580	Zinc chromate <5% MEL Butyl acetate 10–25% OES Plus blends of additional solvents	Product substitution not possible. Issue staff with respirator and eye protection Arrange for extraction to be installed in mixing room
		Skin contact with solvents	Gloves				
		Eye contact with vapours, solvents or paint splashes	Goggles or visor				
		Remarks: High risk Carried out three times daily	Conclusions: Strong smell of solvent from acid hardener No respirator or eye protection used whilst mixing	Conclusions: No LEV in mixing area	Washfiller phosphoric hardener	Ethanol 15–30% OES Methyl isobutyl ketone 30–45% OES + Plus blend of additional solvents including phosphoric acid	Reduce mixing from three times a day to once daily

Table 1.1 Continued

Activity	Hazard	Risk	Precautions: Personal	Precautions: General	Products used	Products contain	Remarks
Application	Spray mist containing solvent and, depending on the particular product, isocyanates, zinc or strontium chromates, polyamides or acids	Inhalation of solvent vapours or spray mists	Respirator suitable for solvent vapours and particulates Air-fed respirators *must* be worn (i) whenever isocyanates are present or (ii) when zinc or strontium chromate containing primers are used and the ventilation in the spray enclosure is inadequate for the particular job	All spraying *must* be carried out in a properly ventilated spray enclosure	As above	As above	Issue replacement charcoal respirator, set up inspection and maintenance records. No spraying to be done outside spray booth
		Skin contact with solvent vapour or spray mists	Gloves				
		Eye contact with solvent vapour or spray mists	Goggles or visor				
		Remarks: High risk Process carried out daily	Conclusions: Cartridge mask used but in poor condition	Conclusions: Spraying in open workshop Poor extraction			

COSHH ASSESSMENT RECORD

RECORD No.

PROCESS IDENTIFICATION ....

....

PROCESS LOCATIONS ....

FREQUENCY OF USE ....

NUMBER OF OPERATORS ....

SUBSTANCES/PRODUCTS USED	SUPPLIER	HAZARDS	OES	MEL

CONTROLS

PRECAUTIONS CURRENTLY USED ....

HOW EFFECTIVE ARE THEY ....

LIKELY HAZARD ....

– INHALATION ....

– SKIN CONTAMINATION ....

– EYE CONTACT ....

ACTION REQUIRED ....

....

NEW CONTROLS INTRODUCED ....

....

....

IS A MAINTENANCE SCHEDULE REQUIRED	YES	NO

....

HEALTH SURVEILLANCE REQUIRED	YES	NO

....

CARRIED OUT BY ....

SIGNATURE .... DATE ....

Figure 1.24 COSHH assessment record (*Akzo Coatings PLC*)

1 Filtration systems have to ensure that the concentration of paint particulate matter in the final discharge to atmosphere from the spray booth does not exceed 10 mg/m^3.
2 Manufacturers or companies upgrading booths will be required to provide a guarantee confirming that the equipment conforms to the emissions limit.
3 The vent velocity of the extract duct must achieve a minimum of 15 m/s for dry filter systems.
4 The vent velocity of the extract duct must achieve a maximum of 9 m/s where a wet method is used.
5 Pressure control systems must be provided which will shut down the spray booth if it is overpressurized and activate an audio alarm.
6 Exhaust duct openings must not be fitted with plates, caps or cowls that could act as restrictors or deflectors.
7 Exhaust ducting (the chimney) must be a minimum of 8 metres above ground level, but it also has to be minimum of 3 metres higher than the roof height of any nearby building which is within a distance of five times the uncorrected chimney height.
8 Preventive measures must be taken for fugitive emissions of odour, fumes and particulate matter from mechanical operations like welding, grinding or sanding. All these activities should take place in a building.
9 Shot blasting emissions must not exceed 50 mg/m^3.
10 Paint mixing and equipment cleaning are to take place in an adequately ventilated area.
11 An automatic totally enclosed machine for cleaning spray guns and equipment must be provided.
12 Spray gun testing must be into an extracted area.

Waste management: EPA Part 11
Section 34 of the EPA makes it a criminal offence to 'treat, keep or dispose of controlled waste in a manner likely to cause pollution of the environment or to harm human health'. In order to meet this stated aim, the Part 11 provisions impose a 'duty of care' on everybody involved in the chain of waste management. This means that if you are involved in the creation of waste, then you are responsible for its safe and proper disposal. Typical relevant bodyshop waste includes dirty solvent, paint residues, empty cans and dirty rags. Those subject to the duty of care must try to achieve the following four things:

1 To prevent any other person committing the offence of disposing of 'controlled waste', or treating it, or storing it without a waste management licence; or breaking the conditions of a licence; or in a manner likely to cause pollution or harm to health.
2 To prevent the escape of waste which is, or at any time has been, under their control (this has implications for waste storage facilities and waste containment).
3 To ensure that if waste is transferred it goes only to an 'authorized person' or to a person for 'authorized transport purposes'.
4 When waste is transferred, to make sure that there is also transferred a written description of the waste, which is good enough to enable each person receiving it to avoid committing any offence under 1, and to comply with the duty at 2 to prevent the escape of waste.

Under the code of practice, the waste must be:

1 Identified (paint, solvent, paper and tape, dust, loaded extract filter media, scrap metal, tyres and batteries).
2 Categorized.
3 Kept in appropriate containers (external skip, internal container, paper baler, can crusher, solvent/paint 'closed loop' system, parts for recycling).
4 Collected and disposed of by a registered operator (transfer note signed by the disposer and collector, and a written description of the waste).
5 Documented at all stages.

1.6.6 Management of Health and Safety at Work Regulations 1992

These Regulations set out broad general duties which apply to almost all work activities in Great Britain. They are aimed mainly at improving health and safety management, and can be seen as a way of making more explicit what is required of employers under the HSW Act. The main provisions are designed to encourage a more systematic and better organized approach to deal with health and safety.

The Regulations require that employers should:

1 Assess the risk to health and safety of employees and of anyone else who may be affected by the company's work activity. This is so that the necessary preventive and protective measures can be identified. Employers with five or more employees will have to record the significant findings of the assessment.
2 Make arrangements for putting into practice the health and safety measures that follow from the assessment.
3 Provide appropriate health surveillance for employees where the risk assessment shows it to be necessary.
4 Appoint competent people either from inside the organization or from outside to help devise and

apply measures needed to comply with duties under health and safety law.
5 Set up emergency procedures.
6 Provide employees with information they can understand about health and safety matters.
7 Cooperate with other employers sharing the work site.
8 Make sure that employees have adequate health and safety training and are capable enough at their jobs to avoid risk.
9 Provide temporary workers with some particular health and safety information to meet special needs.

The Regulations will also:

1 Place duties on the employees to follow health and safety instructions and report danger.
2 Extend the current law which requires the employer to consult employees' safety representatives and provide facilities for them.

1.6.7 Provision and Use of Work Equipment Regulations 1992

These regulations will place general duties on employers, and list minimum requirements for work equipment to deal with selected hazards whatever the industry.

In general the Regulations will make explicit what is already somewhere in the law or is good practice. If equipment has been well chosen and well maintained there should be little need to do more than follow the guidance on the Regulations. Some older equipment may need to be upgraded to meet the minimum requirements, but the company will have until 1997 to do any necessary work.

Work equipment will be broadly defined to include everything from a hand tool, through machines of all kinds, to a complete plant. Use of the equipment will include starting, stopping, repairing, modifying, installing, dismantling, programming, setting, transporting, maintaining, servicing and cleaning.

The general duties will require the employer to:

1 Make sure that equipment is suitable for the use that will be made of it.
2 Take into account the working conditions and hazards in the workplace when selecting equipment.
3 Ensure that equipment is used only for operations for which and under conditions for which it is suitable.
4 Ensure the equipment is maintained in an efficient state, in efficient working order and in good repair.
5 Give adequate information, instruction and training.
6 Provide equipment that conforms with EC product safety Directives (see below).

Specific requirements will cover:

1 Guarding of dangerous parts of machinery (replacing the current law on this).
2 Protection against specific hazards, i.e. falling/ejected articles and substances, rupture/disintegration of work equipment parts, equipment catching fire or overheating, unintended or premature discharge of articles and substances, explosion.
3 Working equipment parts and substances at high or very low temperatures.
4 Control systems and control devices.
5 Isolation of equipment from the source of energy.
6 Stability of equipment.
7 Lighting.
8 Maintaining operations.
9 Warnings and markings.

The Regulations implement an EC Directive aimed at the protection of workers. There are other Directives setting out conditions which much new equipment (especially machinery) will have to satisfy before it can be sold in the EC member states. They will be implemented in the UK by Regulations made by the Department of Trade and Industry.

1.6.8 Personal Protective Equipment (PPE) at Work Regulations 1992

These Regulations set out in legislation sound principles for selecting, providing, maintaining and using PPE. They do not replace recently introduced law dealing with PPE, for example COSHH or noise at work regulations.

PPE is defined as all equipment designed to be worn or held to protect against a risk to health or safety. This includes most types of protective clothing and equipment such as eye, foot and head protection, safety harnesses, life jackets and high-visibility clothing. PPE should be relied upon only as a last resort, but where risks are not adequately controlled by other means the employer will have a duty to ensure that suitable PPE is provided free of charge for employees exposed to these risks. The Regulations say what is meant by suitable PPE, a key point in making sure that it effectively protects the wearer. PPE will only be suitable if it is appropriate for the risks and the working conditions, takes account of workers' needs, and fits properly, gives adequate protection, and is compatible with any other item of PPE worn.

Employers also have duties to:

1 Assess the risks and PPE intended for issue, to ensure that it is suitable.
2 Maintain, clean and replace PPE.

3 Provide storage for PPE when it is not being used.
4 Ensure that PPE is properly used.
5 Give training, information and instruction to employees on its use and how to look after it.

PPE is also subject to a separate EC Directive on design, certification and testing, and will be marked by the manufacturer with a CE mark.

1.6.9 Manual Handling Operations Regulations 1992

These will apply to any manual handling operations which may cause injury at work. These operations will be identified by the risk assessment carried out under the Management of Health and Safety at Work Regulations 1992. They will include not only lifting of loads, but also lowering, pushing, pulling, carrying, or moving them, whether by hand or other bodily force.

Employers will have to take three key steps:

1 Avoid hazardous manual handling operations wherever reasonably practicable. Consider whether the load must be moved at all and, if it must, whether it can be moved mechanically.
2 Assess adequately any hazardous operations that cannot be avoided. An ergonomic assessment should look at more than just the weight of the load. Employers should consider the shape and size of the load, the way the task is carried out, the handler's posture, the working environment (cramped or hot), the individual's capacity, and the strength required. Unless the assessment is very simple, a written record will be needed.
3 Reduce the risk of injury as far as reasonably practicable. A good assessment will not only show whether there is a problem but will also point to where the problem lies.

1.6.10 Workplace (Health, Safety and Welfare) Regulations 1992

The Regulations will cover many aspects of health, safety and welfare in the workplace. Some of them are not explicitly mentioned in the current law, though they are implied in the general duties of the Health and Safety at Work Act.

The Regulations will set out general requirements in four broad areas.

Working environment

1 Temperature in indoor workplaces.
2 Ventilation.
3 Lighting including emergency lighting.
4 Room dimensions and space.
5 Suitability of workstations and seating.

Safety

1 Safe passage of pedestrians and vehicles (traffic routes must be wide enough and marked where necessary, and there must be enough of them).
2 Windows and skylights (safe opening, closing and cleaning).
3 Transparent and translucent doors and partitions (use of safety material and marking).
4 Doors, gates, escalators (safety devices).
5 Floors (construction and maintenance, obstructions and slipping and tripping hazards).
6 Falling a distance into a dangerous substance.
7 Falling objects.

Facilities

1 Toilets.
2 Washing, eating and changing facilities.
3 Clothing storage.
4 Drinking water.
5 Rest area (and arrangements to protect people from the discomfort of tobacco smoke).
6 Rest facilities for pregnant women and nursing mothers.

Housekeeping

1 Maintenance of workplace, equipment and facilities.
2 Cleanliness.
3 Removal of waste materials.

Employers will have to make sure that any workplace within their control complies with the Regulations. Existing workplaces will have until 1996 to comply.

QUESTIONS

1 State five basic rules concerning dress and behaviour which demonstrate personal safety in the workshop environment.
2 List five necessary precautions for safety in the workshop and describe each one briefly.
3 What is meant by a skin care system as used in the workshop?
4 Explain the importance of eye and face protection in the workshop environment.
5 Explain the importance of protective clothing for a body repairer and a paint sprayer.
6 Explain the significance of headwear and footwear while working in the workshop.

7 Name the four types of respirator used in a bodyshop.
8 Why have the COSHH Regulations made the use of respiratory equipment mandatory?
9 State the minimum noise level at which ear protection must be used.
10 With the aid of a diagram, explain the fire triangle.
11 Name the three methods of fire extinction.
12 Explain the three classifications of fire.
13 Identify the correct colour code for the following fire extinguishers: water, foam, CO_2, powder, halon.
14 Name the four categories of safety signs used in the workshop.
15 Sketch and identify a safety sign used in a bodyshop.
16 Name the items of personal safety equipment that should be used when operating the following power tools: power saw, power chisel, disc sander.
17 State the essential personal safety precautions to be taken before working under a vehicle which is on a hoist.
18 Give practical reasons for wearing safety gloves in the workshop.
19 Identify the type of fire extinguisher that must be used when dealing with a solvent fire.
20 Explain the precautions which must be taken when handling toxic substances in a workshop environment.
21 When dealing with a petrol fire, which would be the correct type of fire extinguisher to use?
22 Explain the importance of the use of a barrier cream.
23 Explain the following abbreviations: COSHH, EPA, HASAWA.
24 State the health hazards associated with the use of GRP for repairs.
25 Explain the importance of dust and fume extraction systems.
26 Name five of the main health and safety legislation Acts which affect the workshop.
27 Discuss volatile organic compounds (VOCs) and their effect on the environment.
28 Which Act applies to waste management?
29 Explain the importance of the Personal Protective Equipment (PPE) at Work Regulations 1992.
30 Explain what is meant by the Manual Handling Operation Regulations 1992.

2

The design and construction of the car body

2.1 DEVELOPMENT OF THE MOTOR CAR BODY

2.1.1 Brief history

The first motor car bodies and chassis frames, made between 1896 and 1910, were similar in design to horse-drawn carriages and, like the carriages, were made almost entirely of wood.

The frames were generally made from heavy ash, and the joints were reinforced by wrought iron brackets which were individually fitted. The panels were either cedar or Honduras mahogany about 9.5 mm thick, glued, pinned or screwed to the framework. The tops, on cars which had them, were of rubberized canvas or other fabrics. Some bodies were built with closed cabs, and the tops were held in place by strips of wood bent to form a solid frame. About 1921 the Weymann construction was introduced, in which the floor structure carried all the weight of the seating, and the body shell, which was of very light construction, was attached to the floor unit. Each joint in the shell and between the shell and the floor was made by a pair of steel plates, one on each side of the joint and bolted through both pieces of timber, leaving a slight gap between the two pieces. The panelling was of fabric; first canvas, then a layer of wadding calico and finally a covering of leather cloth. This form of construction allowed flexibility in the framing and made a very light and quiet body frame, but the outer covering had a very short life.

As the demand for vehicles increased it became necessary to find a quicker method of production. Up to that time steel had been shaped by hand, but it was known that metal in large sheets could be shaped using simple die tools in presses, and machine presses were introduced to the steel industry to form steel sheets into body panels. Initially the sheets were not formed into complex shapes or contours, and the first bodies were very square and angular with few curves. The frame and inner construction was still for the most part made of wood, as shown in Figure 2.1. About 1923 the first attempts were made to build all-steel bodies, but these were not satisfactory as the design principles used were similar to those which had been adopted for the timber-framed body. The real beginning of the all-steel body shell came in 1927, when presses became capable of producing a greater number of panels and in more complex shapes; this was the dawn of the mass production era. During the 1930s most of the large companies which manufactured motor vehicles adopted the use of metal for the complete construction of the body shell, and motor cars began to be produced in even greater quantities.

Owing to the ever-increasing demand for private transport, competition increased between rival firms, and in consequence their body engineers began to incorporate features which added to the comfort of the driver and passengers. This brought about the development of the closed cars or saloons as we know them today. The gradual development of the shape of the motor car body can be clearly seen in Figure 2.2, which shows a selection of Austin vehicles from 1909 to 1992.

The inner construction of the head roof of these saloons was concealed by a headlining. Up to and including the immediate post-war years, this headlining was made from a woollen fabric stitched together and tacked into position on wooden frames. However, the more recently developed plastic and vinyl materials were found to be more suitable than fabric, being cheaper and easier to clean and fit. They are fitted by stretching over self-tensioning frames which are clipped into position for easy removal, or alternatively the headlining is fastened into position with adhesives.

Comfort improved tremendously with the use of latex foam rubber together with coil springs in the seating, instead of the original plain springing. The general interior finish has also been improved by the introduction of door trim pads, fully trimmed dash panels and a floor covering of either removable rubber or carpeting.

Then came the general use of celluloid for windows instead of side curtains, and next a raising and

Figure 2.1 Timber constructed bodies: (a) De Dion Works bodyshop, Finchley, *c*.1923 (b) Gordon England Ltd, 1922 (*National Motor Museum, Beaulieu*)

lowering mechanism for the windows. Nowadays the windscreen and door glasses are made of laminated and/or toughened safety glass. The window mechanism in use today did not begin to develop until well into the 1920s.

Mudguards, which began as wooden or leather protections against splattered mud, grew into wide splayed deflectors in the early part of the twentieth century and then gradually receded into the bodywork, becoming gracefully moulded into the streamlining of the modern motor car and taking the name of wings. Carriage steps retained on earlier models gave place to running boards which in their turn disappeared altogether.

Between 1890 and 1906 steering was operated by a tiller (Figure 2.3). This was followed by the steering wheel which is in current use. The position of the gear lever made an early change from the floor to the steering column, only to return to some convenient place on the floor.

Some of the first vehicles, or horseless carriages as they were known, carried no lights at all; then carriage candle lamps made their appearance. Later came oil lamps, acetylene lamps and finally the electric lighting system, first fitted as a luxury extra and ultimately becoming standard and finally obligatory equipment which must conform with legislation of the day.

When windscreens were first introduced such accessories as windscreen wipers and washers were unknown. Then came the single hand-operated wiper, followed by the suction wiper and finally electrically driven wipers.

The design of the wheels was at first dictated by fashion. It was considered necessary for the rear wheels to be larger than the front, a legacy from the elegant horse-drawn carriages. Wooden spokes and iron tyres were the first wheels to appear, and with both rear and front wheels of the same dimensions. Then came the wooden-spoked artillery wheel with pneumatic tyre (Figure 2.4). The artillery wheel gave way to wire-spoked wheel, and this in turn to the modern disc wheel with tubeless tyres.

Great strides have been made in the evolution of the motor car since 1770, when Cugnot's steam wagon travelled at 3 mile/h (4.8 km/h), to the modern vehicle which can carry driver and passengers in silence, comfort and safety at speeds which at one time were thought to be beyond human endurance: indeed, special vehicles on prepared tracks are now approaching the speed of sound.

It must be borne in mind that the speed of the vehicle is governed by (a) the type of power unit, (b) its stability and manoeuvring capabilities and (c) its shape, which is perhaps at present one of the most important features in high-speed travel. Whatever the mechanical future of the car, we may rest assured that the shape of the motor car body will continue to change as technical progress is made (Figure 2.5).

2.1.2 Highlights of motor vehicle history

The idea of a self-propelled vehicle occurs in Homer's *Iliad*. Vulcan, the blacksmith of the gods, in one day made 20 tricycles which 'self-moved obedient to the beck of the gods'. The landmarks in more modern motor vehicle history are as follows:

1688 Ferdinand Verbiest, missionary in China, made a model steam carriage using the steam turbine principle.

Figure 2.2 Development of the Austin car body 1909 to 1992

1970 Austin Maxi 1800

1973 Austin Allegro

1975 Austin Princess

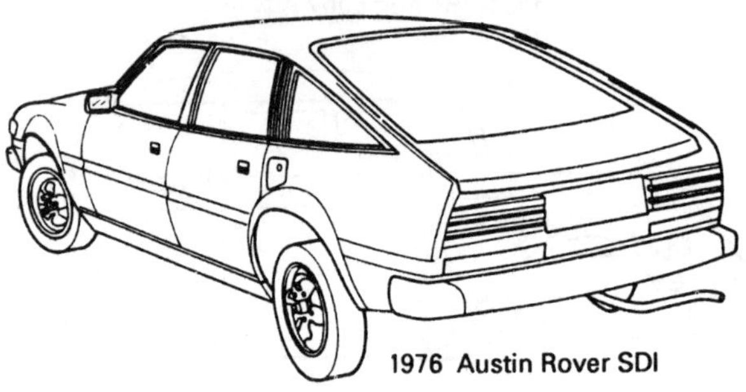

1976 Austin Rover SDI

1980 Austin Metro

1983 Austin Maestro

1984 Austin Montego

1986 Austin Rover 200

1987 Austin Rover Sterling

Figure 2.2 (continued)

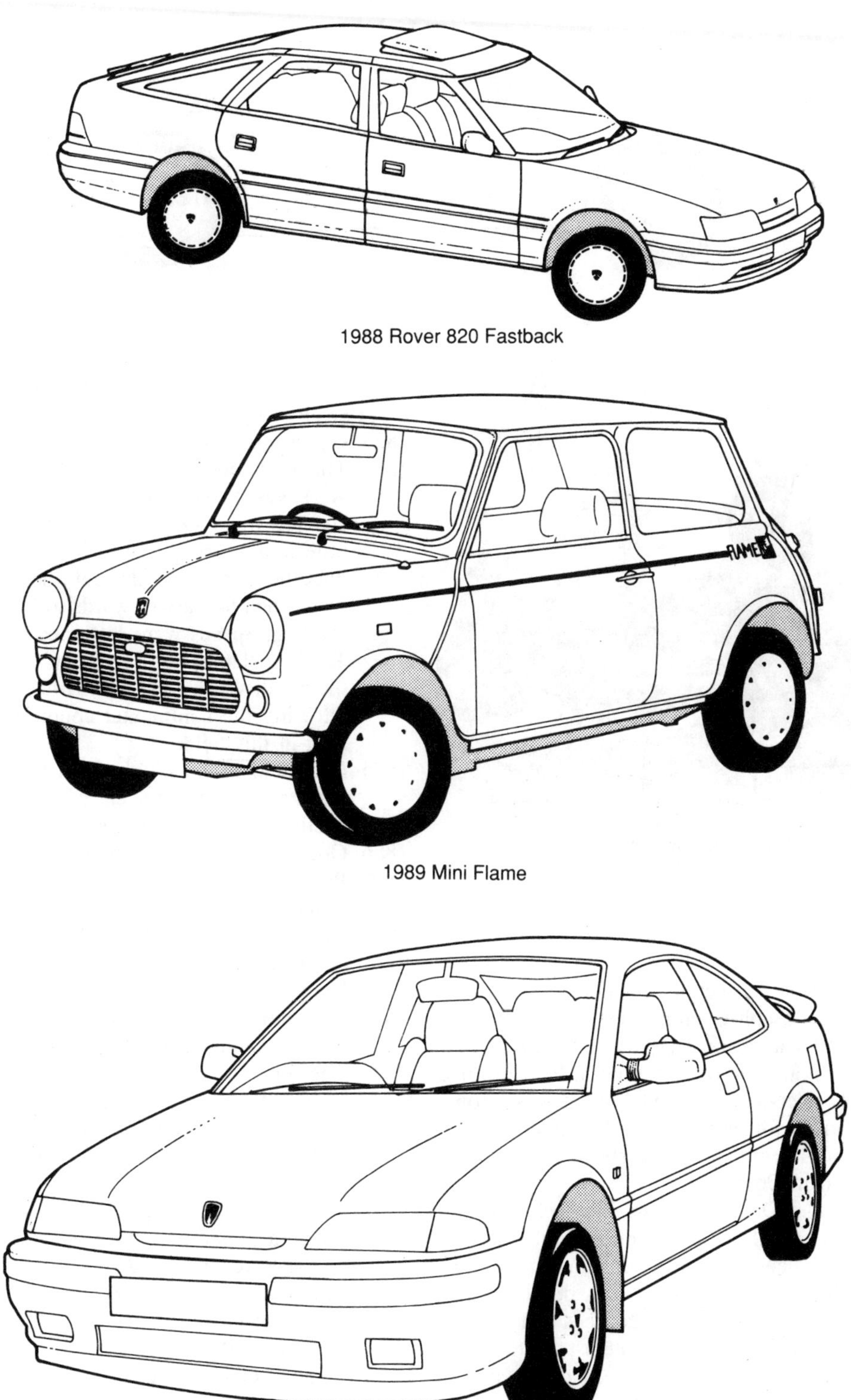

1988 Rover 820 Fastback

1989 Mini Flame

1992 Rover 220 Coupé

Figure 2.2 (continued)

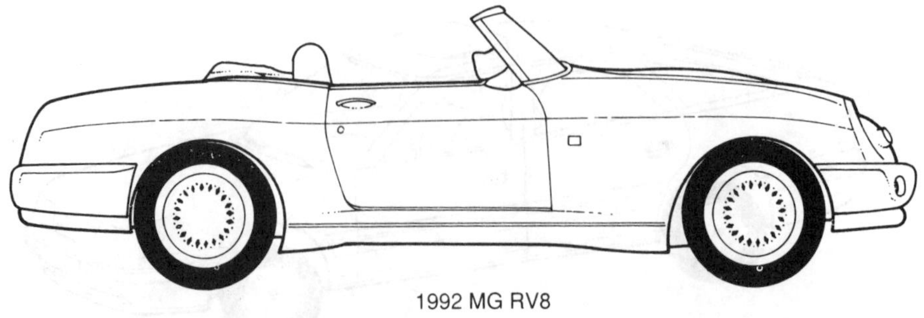

Figure 2.2 (continued)

Figure 2.3 First Vauxhall, tiller operated, 1903 (*Vauxhall Motors Ltd*)

1740 Jacques de Vaucansen showed a clockwork carriage in Paris.

1765 Watt developed the steam engine.

1765 Nicholas Joseph Cugnot, a French artillery officer, built a steam wagon which carried four people at a speed of 2.25 mile/h. It overturned in the streets of Paris and Cugnot was thrown into prison for endangering the populace.

1803 Richard Trevithick built a steam carriage and drove it in Cornwall.

1831 Sir Charles Dance ran a steam coach (built by Sir Goldsworthy Gurney) on a regular service from Gloucester to Cheltenham. Sometimes they did four round trips a day, doing 9 miles in 45 minutes. The steam coaches were driven off the road by the vested interests of the stage coach companies, which increased toll charges and piled heaps of stones in the roads along which the steam coaches passed. This, combined with the problems of boilers bursting and mechanical breakdowns and the advent of the railways, contributed to the withdrawal of the steam coaches.

1859 Oil was discovered in USA.

1865 The Locomotive Act of 1865 (the Red Flag Act) was pushed through by the railway and coach owners. One of the stipulations was that at least three people must be employed to conduct the locomotive through the streets, one of whom had to walk 60 yards in front carrying a red flag. Speeds were restricted to 2–4 mile/h. This legislation held back the development of the motor vehicle in Great Britain for 31 years, allowing the continental countries to take the lead in this field.

1885 Karl Benz produced his first car. This is recognized as being the first car with an internal combustion engine as we know it.

1886 Gottlieb Daimler also produced a car.

1890 Panhard and Levasser began making cars in France.

1892 Charles and Frank Duryea built the first American petrol-driven car, although steam cars had been in use long before this.

1895 First motor race in Paris.
First Automobile Club formed in Paris.

1896 The repeal of the Red Flag Act. This is commemorated by the London to Brighton veteran car run. The speed limit was raised to 12 mile/h and remained at that until 1903, when the 20 mile/h limit in built-up areas was introduced. There was much persecution of motorists by police at this time, which led to the formation of the RAC and the AA.

1897 The RAC was formed, largely through the efforts of F. R. Simms, who also founded the SMMT in 1902.

1899 Jenatzy set the world speed record of 66 mile/h.

1900 Steering wheel replaces tiller.
Frederick Lanchester produced his first car, a 10 hp model. He had built an experimental phaeton in 1895.

Figure 2.4 1905 and 1909 Vauxhalls with wooden-spoked artillery wheels with pneumatic tyres (*Vauxhall Motors Ltd*)

Figure 2.5 Twin concept car with interchangeable engines (petrol/electric motor drive module) (*Vauxhall Motors Ltd*)

1901 Front-mounted engine.
Mercedes car produced.

1902 Running board.
Serpollet did a speed of 74 km/h in a steam car.

1903 Pressed steel frames.
First windshield.
The Motor Car Act resulted in considerable persecution of the motorist for speeding, number plates and lights, so much so that the motoring organizations paid cyclists to find police speed traps.

1904 Folding windshield.
Closed saloon-type body.
A petrol car reached 100 mile/h and, in the same year, a Stanley steam car achieved a speed of 127 mile/h. Stanley steam cars used paraffin in a multitube boiler and had a chassis made from hickory.

Rolls-Royce exhibited their first car in Paris. The motoring press were impressed with its reliability.
Veteran cars are cars up to and including this year.

2.1.3 Terms used to describe early vehicle body styles

In the history of the motor car there has been some ambiguity in the names used to describe various types of body styles, built by coachbuilders from different countries. The following terms relate to the vehicles produced during the period 1895–1915, and show the derivation of the terminology used to describe the modern vehicle.

Berlina Rarely used before the First World War. A closed luxury car with small windows which allowed the occupants to see without being seen.

Cab A term taken directly from the days of the horse-drawn carriages. Used to describe an enclosed vehicle which carried two passengers, while the driver was situated in front of this compartment and unprotected.

Cabriolet Used towards the end of the period. Describes a car with a collapsible hood and seating two or four people.

Coupé A vehicle divided by a fixed or movable glass partition behind the front seat. The driver's position was only partially protected by the roof while the rear compartment was totally enclosed and very luxurious.

Coupé cabriolet or double cabriolet A long vehicle having the front part designed as a coupé and the rear part designed as a cabriolet. There were often two supplementary seats.

Coupé chauffeur A coupé with the driving position completely covered by an extension of the rear roof.

Coupé de ville A coupé having the driving position completely open.

Coupé limousine A vehicle having a totally enclosed rear compartment and the front driving position closed on the sides only.

Double Berlina A longer version of the Berlina but having the driving position separated from the rear part of the vehicle.

Double landaulet A longer version of the landaulet. It had two permanent seats plus two occasional seats in the rear and a driving position in front.

Double phaeton A phaeton which had two double seats including the driver's seat.

Double tonneau A longer version of the tonneau in which the front seats were completely separated from the rear seats.

Glass saloon A large closed vehicle similar to a double Berlina but with enlarged windows.

Landau A cabriolet limousine having only the roof behind the rear windows collapsible.

Landaulet or landaulette A small landau having only two seats in the closed collapsible roof portion.

Limousine A longer version of the coupé with double side windows in the rear compartment.

Limousine chauffeur A limousine with an extended rear roof to cover the driving position.

Phaeton A term from the days of the horse-drawn carriage. In early motoring it was used to describe a lightweight car with large spoked wheels, one double seat and usually a hood.

Runabout An open sporting type of vehicle with simple bodywork and two seats only.

Tonneau An open vehicle having a front bench seat and a semicircular rear seat which was built into the rear doors.

Saloon A vehicle having the driving seat inside the enclosed car but not separated from the rear seat by a partition.

Torpedo A long sports vehicle having its hood attached to the windscreen.

Victoria Another term derived from the era of horses. The Victoria was a long, luxurious vehicle with a separate driving position and a large rear seat. It was equipped with hoods and side screens.

Wagon saloon A particularly luxurious saloon used for official purposes.

2.1.4 Vehicle classification

There are many ways in which motor vehicles may be classified into convenient groups for recognition. Much depends on such factors as the manufacturer, the make of the car, the series and the body type or style. Distinctive groups of passenger vehicle bodies include the following:

1 Small-bodied mass-produced vehicles.
2 Medium-bodied mass-produced vehicles.
3 Large-bodied mass-produced vehicles.
4 Modified mass-produced bodywork to give a standard production model a more distinctive appearance.
5 Specially built vehicles using the major components of mass-produced models.
6 High-quality coach-built limousines (hand made).

7 Sports and GT bodywork (mass produced).
8 Specially coach-built sports cars (hand made).

Styling forms include the following:

Saloon The most popular style for passenger vehicles is the two-door or four-door saloon. It has a fully enclosed, fixed-roof body for four or more people. This body style also has a separate luggage or boot compartment (Figure 2.6a).

Hatchback This body style is identified by its characteristic sloping rear tailgate, which is classed as one of the three or five doors. With the rear seats down there is no division between the passenger and luggage compartments and this increases the luggage carrying capacity of the vehicle (Figure 2.6b).

Estate This type of vehicle is styled so that the roof extends to the rear to give more luggage space, especially when the rear seats are lowered (Figure 2.6c).

Sports coupé and coupé A sports coupe is a two-seater sports car with a fixed roof and a high performance engine. A coupé is a two-door, fixed-roof, high-performance vehicle with similar styling but with two extra seats at the rear, and is sometimes referred to as a '2-plus-2' (Figure 2.6d).

Convertible or cabriolet This can have either two or four doors. It has a soft-top folding roof (hood) and wind-up windows, together with fully enclosed or open bodywork (Figure 2.6e).

Sports This is a two-seater vehicle with a high performance engine and a folding or removable roof (hood) (Figure 2.6f).

Limousine This vehicle is characterized by its extended length, a high roofline to allow better headroom for seating five passengers comfortably behind the driver, a high-quality finish and luxurious interiors (Figure 2.6g).

2.1.5 The evolution of design

When the first motor cars appeared, little attention was paid to their appearance; it was enough that they ran. Consequently the cars initially sold to the public mostly resembled horse-drawn carriages with engines added. Henry Ford launched his Model T in 1908, and it sold on its low price and utility rather than its looks. However, the body design of this car had to be changed over its 19-year production span to reflect changes in customer taste.

The 1930s saw greater emphasis on streamlining design. Manufacturers began to use wind tunnels to eliminate unnecessary drag-inducing projections from their cars. One of the dominant styling features of the 1950s and 1960s was the tail fin, inspired by the twin tail fins of the wartime Lockhead Lightning fighter aircraft. Eventually a reaction set in against such excesses and the trend returned to more streamlined styling.

In creating cars for today's highly competitive car market, designers have to do far more than just achieve a pleasing shape. National legal requirements determine the positions of lamps, direction indicators and other safety-related items, while the buying market has become much more sophisticated than before. Fuel economy, comfort, function and versatility are now extremely important.

2.2 CREATION OF A NEW DESIGN FROM CONCEPT TO REALIZATION

The planning, design, engineering and development of a new motor car is an extremely complex process. With approximately 15 000 separate parts, the car is the most complicated piece of equipment built using mass production methods.

Every major design project has its own design team led by a design manager, and they stay with the project throughout. The size of the team varies according to the progress and status of the project. The skill and judgement of the trained and experienced automotive designer is vital to the creation of any design concept.

To assist in the speed and accuracy of the ensuing stages of the design process (the implementation), some of the most advanced computer-assisted design equipment is used by the large vehicle manufacturers. For example, computer-controlled measuring bridges that can automatically scan model surfaces, or machines that can mill surfaces, are linked to a computer centre through a highly sophisticated satellite communication network. The key terms in computer equipment are as follows:

Computer-aided design (CAD) Computer-assisted design work, basically using graphics.

Computer-aided engineering (CAE) All computer-aided activities with respect to technical data processing, from idea to preparation for production, integrated in an optimum way.

Computer-aided manufacturing (CAM) Preparation of production and analysis of production processes.

Computer-integrated manufacturing (CIM) All computer-aided activities from idea to serial production.

The use of CAE is growing in the automotive industry and will probably result in further widespread

Figure 2.6 Vehicle styling forms: (a) saloon (b) hatchback (c) estate (d) coupé (e) convertible (f) sports (g) limousine

changes. Historically, the aerospace industry was the leader in CAE development. The three major motor companies of GM, Ford and Chrysler started their CAE activities as soon as computers became readily available in the early 1960s. The larger automotive companies in Europe started CAE activities in the early 1970s – about the same time as the Japanese companies.

Each new project starts with a series of detailed paper studies, aimed at identifying the most competitive and innovative product in whichever part of the market is under review. Original research into systems and concepts is then balanced against careful analysis of operating characteristics, features performance and economy targets, the projected cost of ownership and essential dimensional requirements. Research into competitors' vehicles, market research to judge tastes in future years, and possible changes in legislation are all factors that have to be taken into account by the product planners when determining the specification of a new vehicle.

The various stages of the design process are as follows:

1 Vehicle styling, ergonomics and safety.
2 Production of scale and full-size models.
3 Engine performance and testing.
4 Wind tunnel testing.
5 Prototype production.
6 Prototype testing.
7 Body engineering for production.

2.2.1 Vehicle styling

Styling

Styling has existed from early times. However, the terms 'stylist' and 'styling' originally came into common usage in the automotive industry during the first part of the twentieth century.

The automotive stylist needs to be a combination of artist, inventor, craftsman and engineer, with the ability to conceive new and imaginative ideas and to bring these ideas to economic reality by using up-to-date techniques and facilities. He must have a complete understanding of the vehicle and its functions, and a thorough knowledge of the materials available, the costs involved, the capabilities of the production machinery, the sources of supply and the directions of worldwide changes. His responsibilities include the conception, detail, design and development of all new products, both visual and mechanical. This includes the exterior form, all applied facias, the complete interior, controls, instrumentation, seating, and the colours and textures of everything visible outside and inside the vehicle.

Figure 2.7 Style artist at work (*Ford Motor Company Ltd*)

Styling departments vary enormously in size and facilities, ranging from the individual consultant stylist to the comprehensive resources of major American motor corporations like General Motors, which has more than 2000 staff in their styling department at Detroit. The individual consultant designer usually provides designs for organizations which are too small to employ full-time stylists. Some act as an additional brain for organizations that want to inject new ideas into their own production. Among the famous designers are the Italians Pininfarina (Lancia, Ferrari, Alfa), Bertone (Lamborghini), Ghia (Ford) and Issigonis (Mini).

The work of the modern car stylist is governed by the compromise between his creativity and the world of production engineering. Every specification, vehicle type, payload, overall dimensions, engine power and vehicle image inspire the stylist and the design proposals he will make. Initially he makes freehand sketches of all the fundamental components placed in their correct positions. If the drawing does not reduce the potential of the original ideas, he then produces more comprehensive sketches of this design, using colours to indicate more clearly to the senior executives the initial thinking of the design (Figure 2.7). Usually the highly successful classic designs are the work of one outstanding individual stylist rather than of a team.

The main aim of the designer is to improve passenger comfort and protection, vision, heating and ventilation. The styling team may consider the transverse engine as a means of reducing the space occupied by the mechanical elements of the car. Front-wheel drive eliminates the driveshaft and tunnel and the occupants can sit more comfortably. Certain minimum standards are laid down with regard to seat widths, knee room and headroom. The interior dimensions of the car are part of the initial specifications and not subject to much modification. Every centimetre of space is considered in the attempt to provide the maximum interior capacity for the design. The final dimensions of the interior and luggage space are shown in a drawing, together with provision for the engine and remaining mechanical assemblies.

Ergonomics

Ergonomics is a fundamental component of the process of vehicle design. It is the consideration of human factors in the efficient layout of controls in the driver's environment. In the design of instrument panels, factors such as the driver's reach zones and his field of vision, together with international standards, all have to be considered. Legal standards include material performance in relation to energy absorption and deformation under impact. The vision and reach zones are geometrically defined, and allow for the elimination of instrument reflections in the windshield.

Basic elements affecting the driver's relationship to the instrument panel controls, instruments, steering wheel, pedals, seats and other vital elements in the car are positioned for initial evaluation using the 'manikin', which is a two- and three-dimensional measuring tool developed as a result of numerous anthropometric surveys and representing the human figure. Changes are recorded until the designer is satisfied that an optimum layout has been achieved.

2.2.2 Safety

With regard to bodywork, the vehicle designer must take into account the safety of the driver, passengers and other road users. Although the vehicle cannot be expected to withstand collision with obstacles or other vehicles, much can be done to reduce the effects of collision by the use of careful design of the overall shape, the selection of suitable materials and the design of the components. The chances of injury can be reduced both outside and inside the vehicle by avoiding sharp-edged, projecting elements.

Every car should be designed with the following crash safety principles in mind:

1 The impact from a collision is absorbed gradually by controlled deformation of the outer parts of the car body.
2 The passenger area is kept intact as long as possible.
3 The interior is designed to reduce the risk of injury.

Safety-related vehicle laws cover design, performance levels and the associated testing procedures; requirements for tests, inspections, documentation and records for the process of approval; checks that standards are being maintained during production; the issue of safety-related documentation; and many other requirements throughout the vehicle's service life.

Primary or active safety

This refers to the features designed into the vehicle which reduce the possibility of an accident. These include primary design elements such as dual circuit braking systems, anti-lock braking systems, high aerodynamic stability and efficient bad weather equipment, together with features that make the driver's environment safer, such as efficient through ventilation, orthopaedic seating, improved all-round vision, easy-to-read instruments and ergonomic controls.

An anti-lock braking system (ABS) enhances a driver's ability to steer the vehicle during hard braking. Sensors monitor how fast the wheels are rotating and feed data continuously to a microprocessor in the vehicle to signal that a wheel is approaching lock-up. The computer responds by sending a signal to apply and release brake pressure as required. This pumping action continues as long as the driver maintains adequate force on the brake pedal and impending wheel lock condition is sensed.

The stability and handling of the vehicle are affected by the width of the track and the position of the centre of gravity. Therefore the lower the centre of gravity and the wider the track, the more stable is the vehicle.

Secondary or passive safety

If a crash does happen, secondary safety design should protect the passengers by:

1 Making sure that, in the event of an accident, the occupants stay inside the car.
2 Minimizing the magnitude and duration of the deceleration to which they are subjected.
3 Restraining the occupants so that they are not injured by secondary impacts within the car, and, if they do strike parts of the inside of the vehicle, making sure that there is sufficient padding to prevent serious injury.
4 Designing the outside of the vehicle so that the least possible injury is caused to pedestrians and others who may come into contact with the outside of the vehicle.

The primary concern is to develop efficient restraint systems which are comfortable to wear and easy to use. Manufacturers are now fitting automatic seatbelt

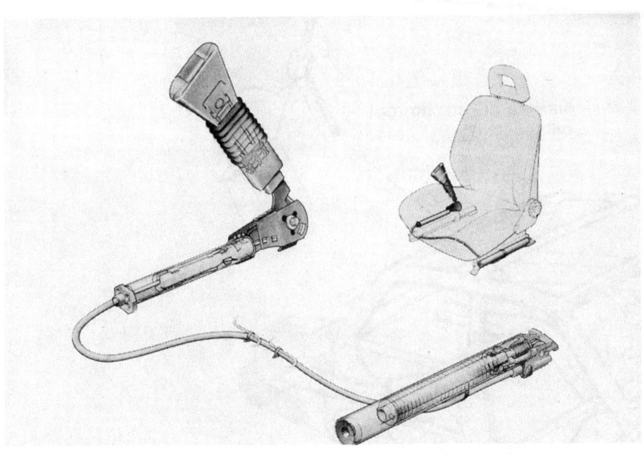

Figure 2.8 Automatic seatbelt tensioner (*Vauxhall Motors Ltd*)

tensioners. These automatic 'body lock' front seatbelt tensioners reduce the severity of head injuries by 20 per cent with similar gains in chest protection. In impacts over 12 mile/h (20 km/h) the extra tension in the seatbelt buckle triggers a sensor which tightens the lap and diagonal belts in 22 milliseconds, that is before the occupant even starts to move. In addition, because it operates at low speeds, it covers a broad spectrum of accident situations. Anti-submarining ramps built into the front seats further aid safety by reducing the possibility of occupants sliding under the belt (Figure 2.8). There are also engineering features such as impact energy-absorbing steering columns, head restraints, bumpers, anti-burst door locks, and self-aligning steering wheels. Anti-burst door locks are to prevent unrestrained occupants from falling out of the vehicle, especially during rollover. The chances of survival are much reduced if the occupant is thrown out. Broad padded steering wheels are used to prevent head or chest damage. Collapsible steering columns also prevent damage to the chest and abdomen and are designed to prevent the steering column being pushed back into the passenger compartment while the front end is crumpling. The self-aligning steering wheel is designed to distribute force more evenly if the driver comes into contact with the steering wheel during a crash. This steering wheel has an energy absorbing hub which incorporates six deformable metal legs. In a crash, the wheel deforms at the hub and the metal legs align the wheel parallel to the chest of the driver to help spread the impact and reduce chest, abdomen and facial injuries.

Body shells are now designed to withstand major collision and rollover impacts while absorbing shock by controlled deformation of structure in the front and rear of the vehicle. Vehicle design and accident prevention is based on the kinetic energy relationship of damage to a vehicle during a collision. Energy is absorbed by work done on the vehicle's materials by elastic deformation. This indicates that, to be effective, bumpers and other collision-absorbing parts of a vehicle should be made of materials such as foam-filled plastics and heavy rubber sections. Data indicates that long energy-absorbing distances should be provided in vehicle design, and the panel assemblies used for this purpose should have a lower stiffness than the central section or passenger compartment of the vehicle. The crumple zones are designed to help decelerate the car by absorbing the force of collision at a controlled rate, thereby cushioning the passengers and reducing the risk of injury (Figure 2.9). The safety cage (or safety cell) is the central section of the car body which acts as the passenger compartment. To ensure passenger safety, all body apertures around the passenger area should be reinforced by box-type profiles; seats should be secured rigidly to the floor; and heavy interior padding should be used around the dashboard areas. A strengthened roof construction, together with an anti-roll bar, afford additional protection in case of overturning (Figure 2.10).

Figure 2.9 Crumple zones (*Volvo Concessionaires Ltd*)

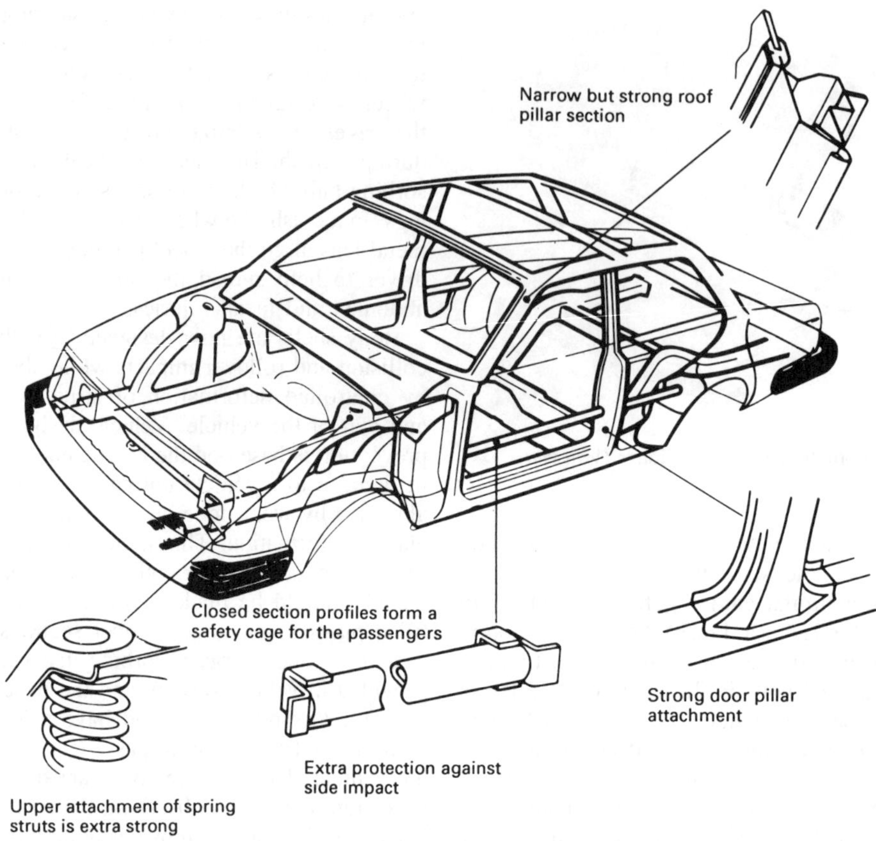

Figure 2.10 Safety cage (*Volvo Concessionaires Ltd*)

To counteract side impact manufacturers are now fitting, in both front and rear doors, lateral side supports in the form of twin high-strength steel tubular beams, which are set 90 mm apart to reduce the risk of the vehicle riding over the beams during side collision. These beams absorb the kinetic energy produced when the vehicle is struck from the side. To improve further the body structure the BC-pillars are being reinforced at the points of attachment to the sill and roof, again giving more strength to the safety cage and making it stronger and safer when the vehicle is involved in collision (Figure 2.11a, b).

Visibility in design is the ability to see and be seen. In poor visibility and after dark, light sources must be relied upon. The lights on vehicles now are much more efficient than on earlier models. The old tungsten filament lamp has given way to quartz-halogen lamps which provide much better illumination. The

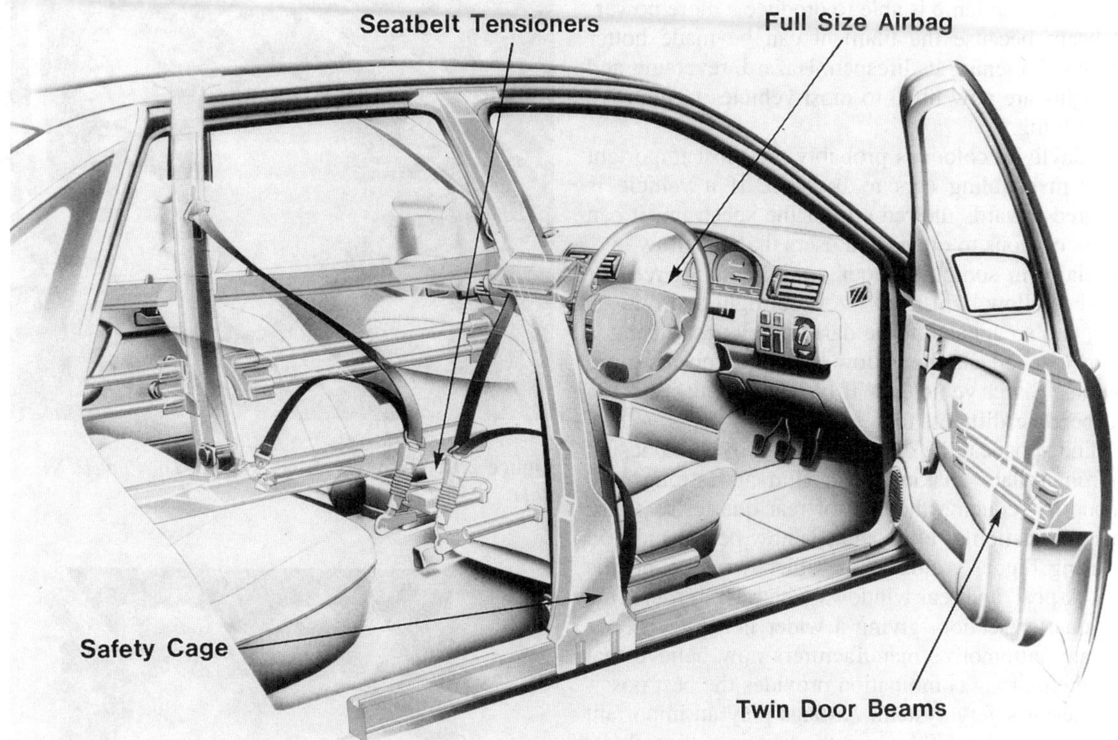

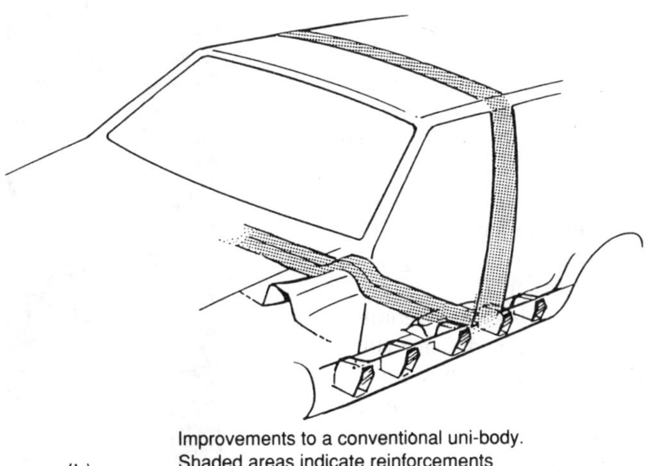

Figure 2.11 (a) Safety features included in the safety cage (*Vauxhall Motors Ltd*) (b) reinforced BC-pillar and anti-roll bar (*Volvo Concessionaires Ltd*)

quartz-halogen lamp is able to produce a more powerful beam because the filament can be made hotter without shortening its lifespan. Hazard, reversing and fog lights are now fitted to most vehicles to improve safe driving.

In daylight, colour is probably the most important factor in enabling cars to be seen. If a vehicle is coloured towards the red end of the spectrum, it can be less obvious to other road users than a yellow one, especially in sodium vapour street lights: a red car absorbs yellow light from the street light and reflects little, and so appears to be dark in colour, whereas a yellow car reflects the yellow light and appears more obvious. Silver vehicles will blend into mist and fog and become difficult to see.

Blind spots can be diminished first by good design of front pillars, making them slim and strong, and second by reducing the area of rear quarter sections. This elimination of blind spots is now being achieved by using bigger windscreens which wrap round the front A-post, and rear windows which wrap round the rear quarter section, giving a wider field of vision.

Many automotive manufacturers now believe that a seatbelt/airbag combination provides the best possible interior safety system. Airbags play an important safety role in the USA since the wearing of seatbelts is not compulsory in many of the states. As competition to manufacture Europe's safest car increases, more manufacturers including those in the UK are starting to fit airbags. These Eurobags, or facebags as they are now called, since their main function in Europe and the UK is to protect the face rather than the entire body in the event of collision, are less complex than their USA counterparts.

The first automotive airbags were made more than 20 years ago using nylon-based woven fabrics, and these remain the preferred materials among manufacturers. Nylon fabrics for airbags are supplied in two basic designs depending on whether the airbag is to protect the driver or the front passenger. The driver's airbag is housed in the steering wheel and requires special attention because of the confined space (Figure 2.12). The passenger's airbag system has a compartment door, located in front of the passenger in the dash area, which must open within 10 milliseconds and deploy the airbag within 30 milliseconds. The vehicle has a crash sensor which signals the airbags to deploy on impact (Figure 2.13).

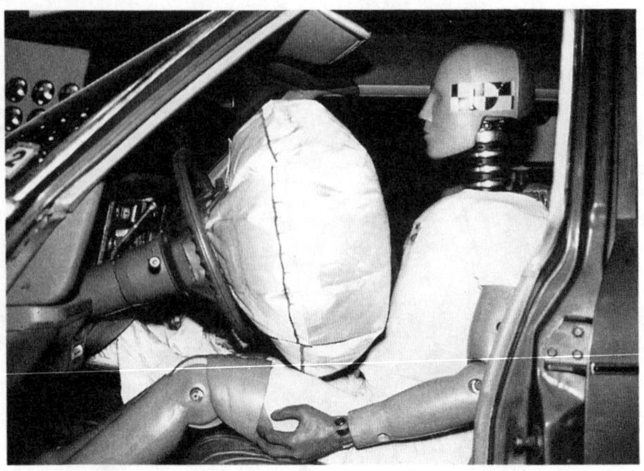

Figure 2.12 Driver's airbag system (*Du Pont (UK) Ltd*)

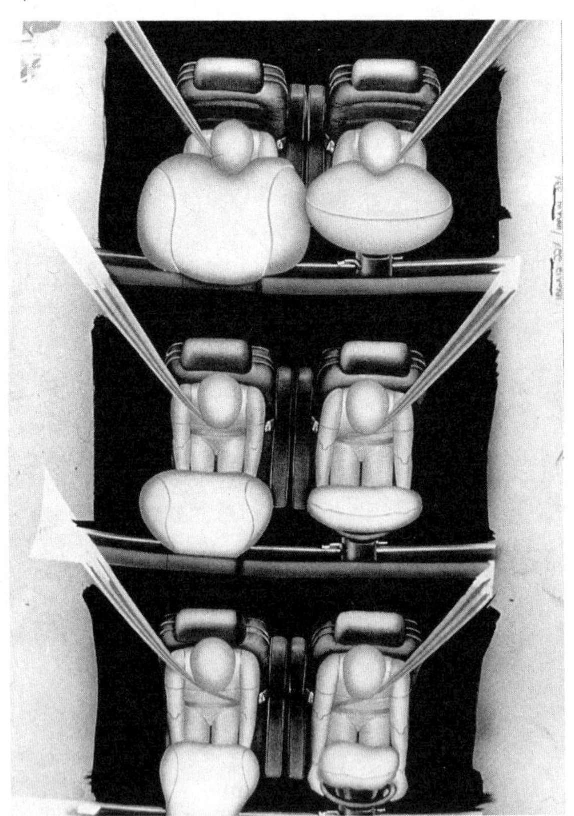

Figure 2.13 Driver and front passenger airbag systems in use (*Du Pont (UK) Ltd*)

2.2.3 Production of models

Scale models

Once the initial designs have been accepted, scale models are produced for wind tunnel testing to

Figure 2.14 Scale model maker at work (*Ford Motor Company Ltd*)

determine the aerodynamic values of such a design. These models are usually constructed of wood and clay to allow for modifications to be made easily. At the same time, design engineering personnel construct models of alternative interiors so that locations of instruments can be determined.

A $\frac{1}{4}$ or $\frac{3}{8}$ scale model is produced from the stylist's drawings to enable the stylist designer to evaluate the three-dimensional aspect of the vehicle. These scale models can look convincingly real (Figure 2.14). The clay surfaces are covered with thin coloured plastic sheet which closely resembles genuine painted metal. Bumpers, door handles and trim strips are all cleverly made-up dummies, and the windows are made of Plexiglass. The scale models are examined critically and tested. Changes to the design can be made at this stage.

Full-size models

A full-size clay model is begun when the scale model has been satisfactorily modified. It is constructed in a similar way to the scale model but uses a metal, wood and plastic frame called a buck. The clay is placed on to the framework by professional model makers, who create the final outside shape of the body to an accuracy of 0.375 mm. The high standard of finish and detail results in an exact replica of the future full-size vehicle (Figure 2.15). This replica is then evaluated by the styling management and submitted to top management for their approval. The accurate life-size model is used for further wind tunnel testing and also to provide measurements for the engineering and production departments. A scanner, linked to a computer, passes over the entire body and records each and every dimension (Figure 2.16). These are stored and can be produced on an automatic drafting machine. The same dimensions can also be projected on the screen of a graphics station; this is a sophisticated computer-controlled video system showing three-dimensional illustrations, allowing design engineers either to smooth the lines or to make detail alterations. The use of computers or CAD allows more flexibility and saves a lot of time compared with the more conventional drafting systems.

Figure 2.15 Full-size clay model (*Ford Motor Company Ltd*)

Figure 2.16 Checking dimensional accuracy of the full-size model (*Ford Motor Company Ltd*)

At the same time as the exterior model is being made, the interior model is also being produced accurately in every detail (Figure 2.17). It shows the seating arrangement, instrumentation, steering wheel, control unit location and pedal arrangements. Colours and fabrics are tried out on this mock-up until the interior styling is complete and ready for approval.

2.2.4 Engine performance and testing

Development engineers prepare to test an engine in a computer-linked test cell to establish the optimum settings for best performance, economy and emission levels. With the increasing emphasis on performance

Figure 2.17 Interior styling model (*Ford Motor Company Ltd*)

with economy, computers are used to obtain the best possible compromise. They are also used to monitor and control prolonged engine testing to establish reliability characteristics. If current engines and transmissions are to be used for a new model, a programme of refining and adapting for the new installation has to be initiated. However, if a completely new engine, transmission or driveline configuration is to be adopted, development work must be well in hand by this time.

2.2.5 Aerodynamics and wind tunnel testing

Aerodynamics is an experimental science whose aim is the study of the relative motions of a solid body and the surrounding air. Its application to the design of a car body constitutes one of the chief lines of the search for energy economy in motor vehicles.

In order to move over flat ground, a car must overcome two forces:

1 Resistance to tyre tread motion, which varies with the coefficient of tyre friction over the ground and with the vehicle's mass.
2 Aerodynamic resistance, which depends on the shape of the car, on its frontal area, on the density of the air and on the square of the speed.

One of the objects of aerodynamic research is to reduce the latter: in other words to design a shape that will, for identical performance, require lower energy production. An aerodynamic or streamlined body allows faster running for the same consumption of energy, or lower consumption for the same speed. Research for the ideal shape is done on reduced-scale models of the vehicle. The models are placed in a wind tunnel, an experimental installation producing wind of a certain quality and fitted with the means for

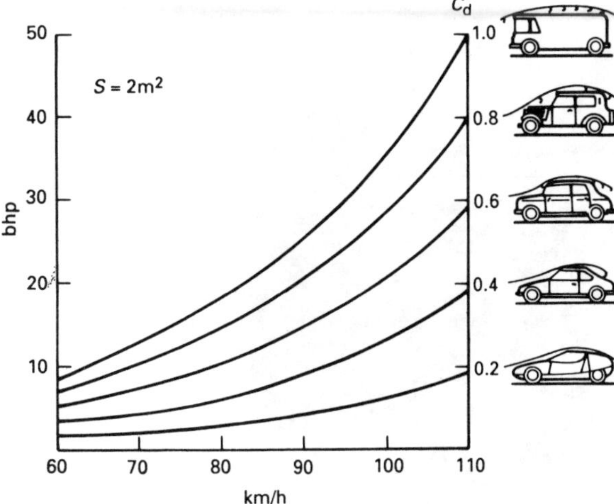

Figure 2.18 Theoretical drag curves for four types of vehicle, all reduced for comparison purposes to a front section of 2 m^2. Since air resistance increases in proportion to the square of the speed, a truck with C_d 1.0 requires 35 bhp at 100 km/h, whereas a coupé with C_d 0.2 requires only 7 bhp

measuring the various forces due to the action of the wind on the model or the vehicle. Moreover, at a given cruising speed, the more streamlined vehicle has more power left available for acceleration: this is a safety factor.

The design of a motor car body must, however, remain compatible with imperatives of production, of overall measurements and of inside spaciousness. It is also a matter of style, for the coachwork must be attractive to the public. This makes it impossible to apply the laws of aerodynamics literally. The evolution of the motor car nevertheless tends towards a gradual reduction in aerodynamic resistance.

Aerodynamic drag

The force which opposes the forward movement of an automobile is aerodynamic drag, in which air rubs against the exterior vehicle surfaces and forms disturbances about the body, thereby retarding forward movement. Aerodynamic drag increases with speed; thus if the speed of a vehicle is doubled, the corresponding engine power must be increased by eight times. Engineers express the magnitude of aerodynamic drag using the drag coefficient C_d. The coefficient expresses the aerodynamic efficiency of the vehicle: the smaller the value of the coefficient, the smaller the aerodynamic drag.

Figure 2.18 illustrates the improvements in aerodynamic drag coefficient achieved by alterations to the shape of vehicles. Over the years the value of C_d has been reduced roughly as follows:

1910	0.95	1960	0.40
1920	0.82	1970	0.36
1930	0.56	1980	0.30
1940	0.45	1990	0.22
1950	0.42	1995	0.20 (probable)

During the wind tunnel test all four wheels of the car rest on floating scales connected to a floor balance, which has a concrete foundation below the main floor area. The vehicle is then subjected to an air stream of up to 112 mile/h; the sensitive balances register the effect of the headwind on the vehicle as it is either pressed down or lifted up from the floor, pushed to the left or right, or rotated about its longitudinal axis. The manner in which the forces affect the vehicle body and the location at which the forces are exerted depends upon the body shape, underbody contours and projecting parts. The fewer disturbances which occur as air moves past the vehicle, the lower its drag. Threads on the vehicle exterior as well as smoke streams indicate the air flow, and enable test engineers to see where disturbance exists and where air flows are interrupted or redirected, and therefore where reshaping of the body is necessary in order to produce better aerodynamics (Figures 2.19 and 2.20).

2.2.6 Prototype production

The new model now enters the prototype phase. The mock-ups give way to the first genuine road-going vehicle, produced with the aid of accurate drawings and without complex tooling and machinery. The prototype must accurately reproduce the exact shape, construction and assembly conditions of the final production body it represents if it is to be of any value in illustrating possible manufacturing problems and accurate test data. The process begins with the issue of drawing office instructions to the experimental prototype workshop. Details of skin panels and other large pressings are provided in the form of tracings or as photographic reproductions of the master body drafts. As the various detailed parts are made, by either simple press tools or traditional hand methods, they are spot welded into minor assemblies or subassemblies; these later become part of a major assembly to form the completed vehicle body.

2.2.7 Prototype testing

While still in the prototype stage, the new car has to face a number of arduous tests. For these tests a mobile laboratory is connected to the vehicle by a cable, which

Figure 2.19 Wind tunnel testing of a prototype: front view (*Ford Motor Company Ltd*)

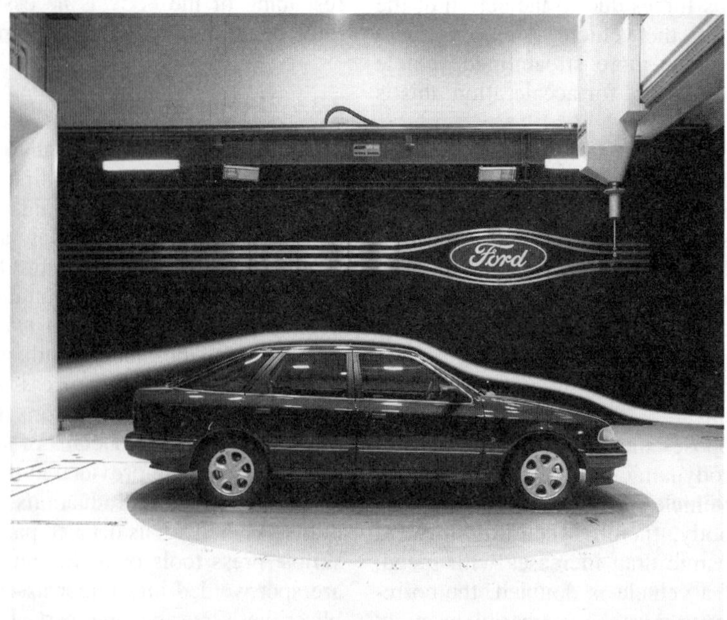

Figure 2.20 Wind tunnel testing of a prototype: side view (*Ford Motor Company Ltd*)

(a) (b) (c) (d)

Figure 2.21 Basic frontal crash and side impact (angled sideswipe) tests (*Vauxhall Motors Ltd*)

transmits signals from various sensors on the vehicle back to the onboard computer for collation and analysis. The prototype will also be placed on a computer-linked simulated rig to monitor, through controlled vibrations, the stresses and strains experienced by the driveline, suspension and body.

Crash testing (Figure 2.21) is undertaken to establish that the vehicle will suffer the minimum of damage or distortion in the event of an impact and that the occupants are safely installed within the strong passenger compartment or safety cell. The basic crash test is a frontal crash at 30 mile/h (48 km/h) into a fixed barrier set perpendicularly to the car's longitudinal axis. The collision is termed 100 per cent overlap, as the complete front of the car strikes the barrier and there is no offset (Figure 2.22). The main requirement is that the steering wheel must not be moved back by more than 120 mm, but there is no requirement to measure the force to which the occupants will be subject in a collision. The manufacturers use anthropometric dummies suitably instrumented with decelerometers and strain gauges which collect relevant data on the effect of the collision on the dummies. A passenger car side impact test aimed at

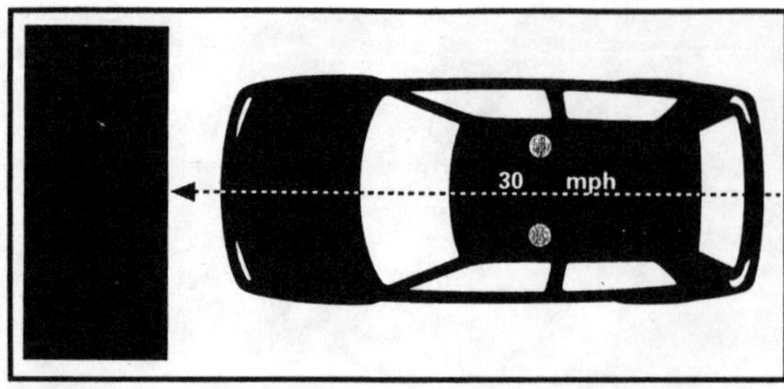

Figure 2.22 Standard frontal impact test

Figure 2.23 Volvo crash test (*Volvo Concessionaires Ltd*)

Figure 2.24 Road testing a prototype (*Ford Motor Company Ltd*)

reducing chest and pelvic injuries has been legal in the USA from 1993. This stricter standard requires that a new vehicle must pass a full-scale crash test designed to simulate a collision at an intersection in which a car travelling at 15 mile/h is hit in the side by another car travelling at 30 mile/h. This test is called an angled sideswipe: the displacement is 27 degrees forward from the perpendicular of the test vehicle's main axis. The test is conducted by propelling a movable deformable barrier at 33.5 mile/h into the side of a test car occupied by dummies in the front and rear seats. The dummies are wired with instruments to predict the risk potential of human injury. Volvo do a very unusual promotional crash test which involves propelling a car from the top of a tall building (Figure 2.23).

Extensive durability tests are undertaken on a variety of road surfaces in all conditions (Figure 2.24). Vehicles are also run through water tests (Figure 2.25) and subjected to extreme climatic temperature changes to confirm their durability.

The final stages are now being reached: mechanical specifications, trim levels, engine options, body styles and the feature lists are confirmed.

2.2.8 Body engineering for production

The body engineering responsibilities are to simulate the styling model and overall requirements laid down by the management in terms of drawings and specification. The engineering structures are designed for production, at a given date, at the lowest possible tooling cost and to a high standard of quality and reliability.

As competition between the major car manufacturers increases, so does the need for lighter and more effective body structures. Until recently the choice of section, size and metal gauges was based on previous experience. However, methods have now been evolved which allow engineers to solve problems with complicated geometry on a graphical display computer which can be constructed to resemble a body shape (Figure 2.26). Stiffness and stress can then be computed from its geometry, and calculations made of the load bearing of the structures using finite-element methods (Figure 2.27).

With the final specifications approved, the new car is ready for production. At this stage an initial batch of cars is built (a pilot run) to ensure that the plant facilities and the workforce are ready for the start of full production. When the production line begins to turn out the brand new model, every stage of production is carefully scrutinized to ensure quality in all the vehicles to be built.

2.3 METHODS OF CONSTRUCTION

The steel body can be divided into two main types: those which are mounted on a separate chassis frame, and those in which the underframe or floor forms an

Figure 2.25 Water testing a prototype (*Ford Motor Company Ltd*)

integral part of the body. The construction of today's mass-produced motor car has changed almost completely from the composite, that is conventional separate chassis and body, to the integral or mono unit. This change is the result of the need to reduce body weight and cost per unit of the total vehicle.

2.3.1 Composite construction (conventional separate chassis)

The chassis and body are built as two separate units (Figure 2.28). The body is then assembled on to the chassis with mounting brackets, which have rubber-bushed bolts to hold the body to the rigid chassis. These flexible mountings allow the body to move slightly when the car is in motion. This means that the car can be dismantled into the two units of the body and chassis. The chassis assembly is built up of engine, wheels, springs and transmission. On to this assembly is added the body, which has been pre-assembled in units to form a complete body shell (Figure 2.29).

2.3.2 Integral (mono or unity) construction

Integral body construction employs the same principles of design that have been used for years in the aircraft industry. The main aim is to strengthen without unnecessary weight, and the construction does not employ a conventional separate chassis frame for attachment of suspension, engine and other chassis and transmission components (Figure 2.30). The major

Figure 2.26 Three-dimensional graphics display of a scale model (*Ford Motor Company Ltd*)

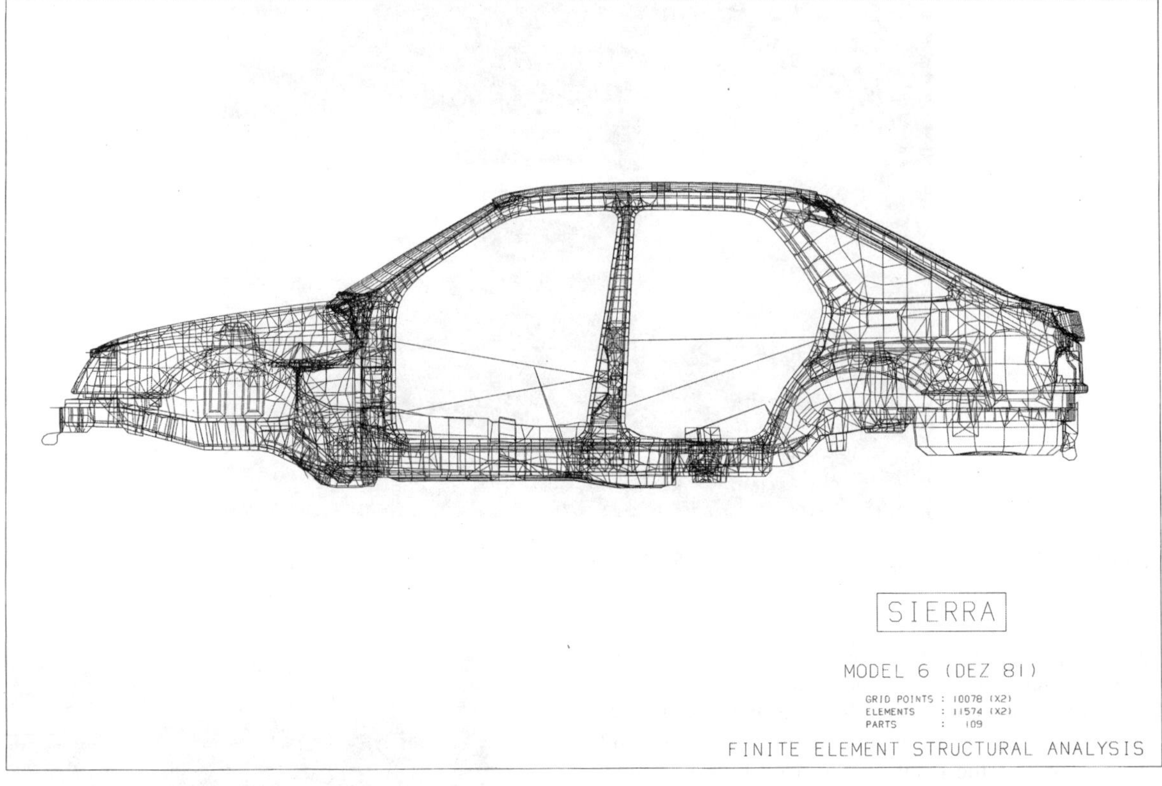

Figure 2.27 Finite-element structural analysis (*Ford Motor Company Ltd*)

difference between composite and integral construction is hence the design and construction of the floor (Figure 2.31). In integral bodies the floor pan area is generally called the underbody. The underbody is made up of formed floor sections, channels, boxed sections, formed rails and numerous reinforcements. In most integral underbodies a suspension member is incorporated in both the front and rear of the body. The suspension members have very much the same appearance as the conventional chassis frame from the underside, but the front suspension members end at the cowl or bulkhead and the rear suspension members end just forward of the rear boot floor. With the floor pan, side rails and reinforcements welded to them, the suspension members become an integral part of the underbody, and they form the supports for engine, front and rear suspension units and other chassis components. In the integral body the floor pan area is usually of heavier gauge metal than in the composite body, and has one or more box sections and several channel sections which may run across the floor either from side to side or from front to rear; this variety of underbody construction is due largely to the difference in wheelbase, length and weight of the car involved. A typical upper body for an integral constructed car is very much the same as the conventional composite body shell; the major differences lie in the rear seat area and the construction which joins the front wings to the front bulkhead or cowl assembly. The construction in the area to the rear of the back seat is much heavier in an integral body than in a composite body. The same is true of the attaching members for the front wings, front bulkhead and floor assembly, as these constructions give great strength and stability to the overall body structure.

2.3.3 Semi-integral methods of construction

In some forms of integral or mono assemblies, the entire front end or subframe forward of the bulkhead is joined to the cowl assembly with bolts. With this construction, the bolts can be easily removed and the entire front (or in some cases rear) subframe can be replaced as one assembly in the event of extensive damage.

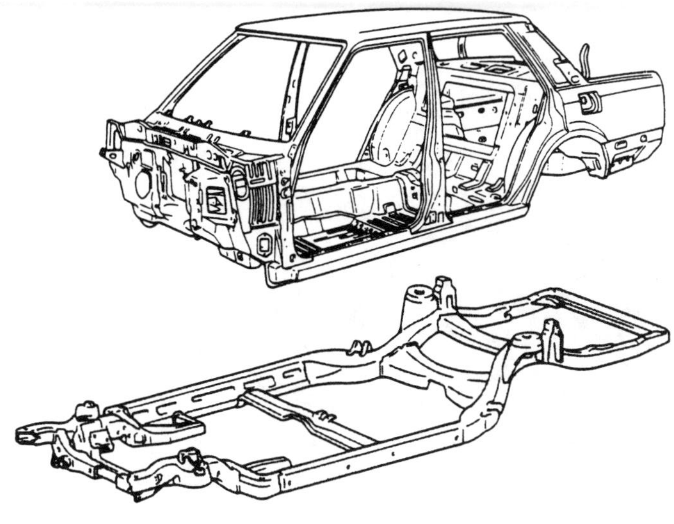

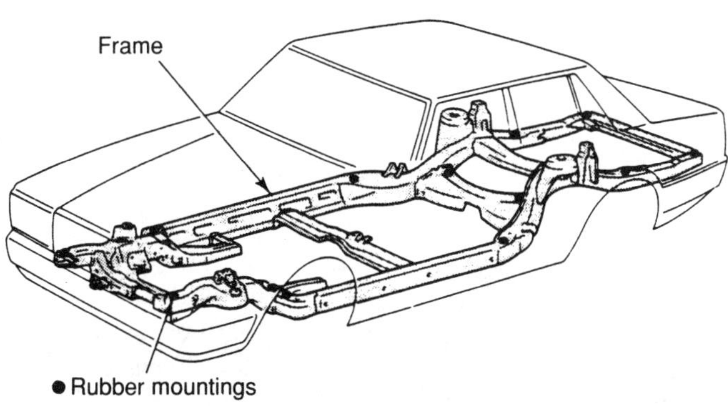

Figure 2.28 Composite construction (conventional separate chassis)

2.3.4 Glass fibre composite construction

This method of producing complex shapes involves applying layers of glass fibre and resin in a prepared mould. After hardening, a strong moulding is produced with a smooth outer surface requiring little maintenance. Among the many shapes available in this composite material are lorry cabs, bus front canopies, container vehicles, and the bodies of cars such as the Lotus. The designers, styled the GRP body so that separately moulded body panels could be used and overlapped to hide the attachment points. This allows the panels to be bolted directly to the supporting square-section steel tube armatures located on the main chassis frame. The inner body, which rests directly on the chassis frame and which forms the base for all internal trim equipment, is a complex GRP moulding. The windscreen aperture is moulded as a part of the inner body, and incorporates steel reinforcing hoops which are braced directly to the chassis. The boot compartment is also a separate hand-laid GRP moulding, as are the doors and some of the other panels. Most of the body panels are secured by self-tapping bolts which offer very positive location and a useful saving in assembly time (see Figures 2.32 and 2.33).

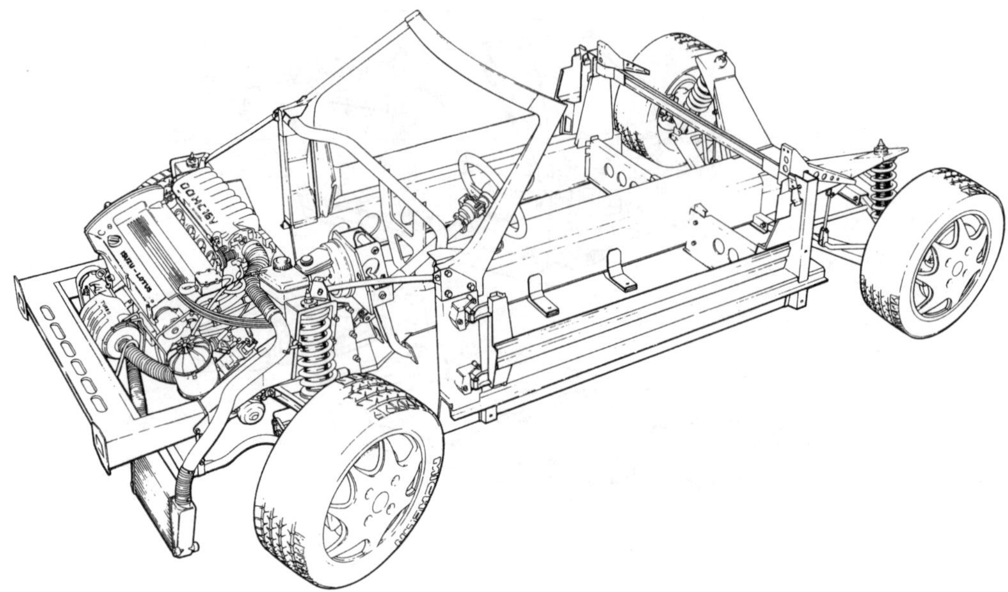

Figure 2.29 Composite construction showing a Lotus Elan chassis before fitting the body (*Lotus Engineering*)

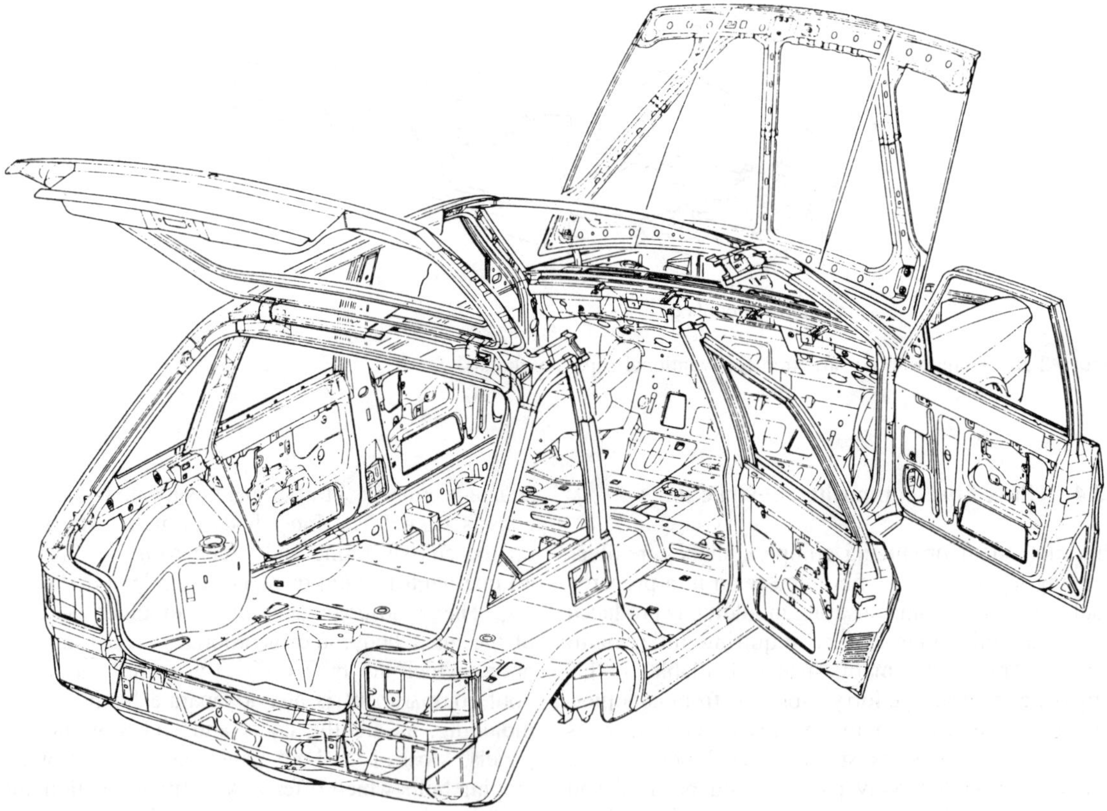

Figure 2.30 Complete body assembly of Austin Maestro (*Austin Rover Group Ltd*)

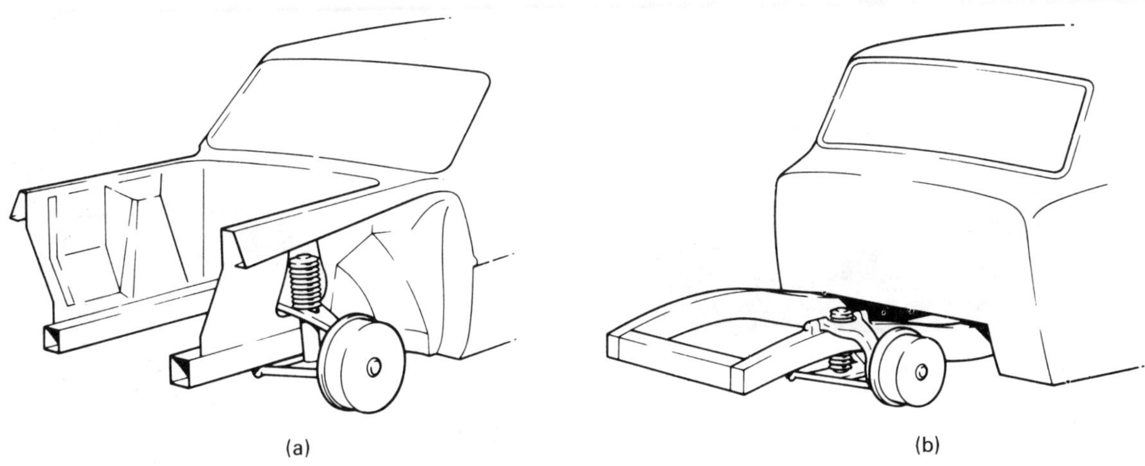

Figure 2.31 Front end construction: (a) integral or mono and (b) composite

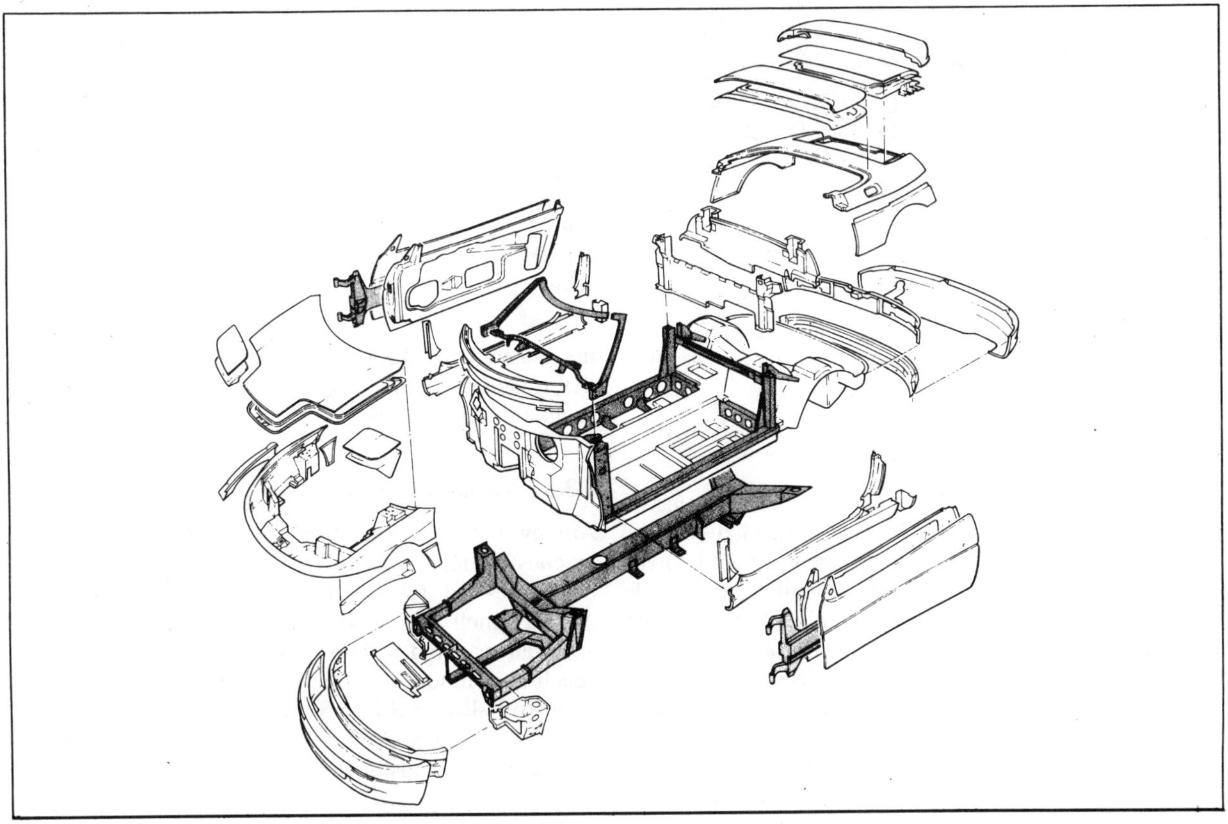

Figure 2.32 Motor body panel assembly using GRP: Lotus Elan (*Lotus Engineering*)

Figure 2.33 Complete Lotus Elan SE body shell (*Lotus Engineering*)

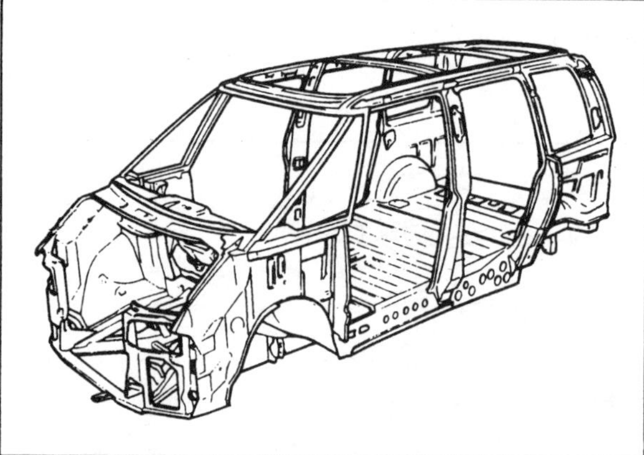

Figure 2.34 Espace high-rise car with galvanized skeletal steel body shell (*Renault UK Ltd*)

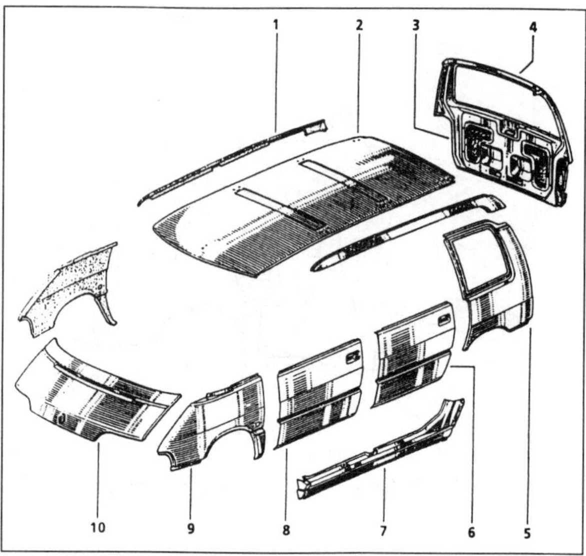

Figure 2.35 Espace high-rise car showing composite panel cladding (*Renault UK Ltd*). Plastic parts are made from a composite material based on polyester resin: pre-impregnated type (SMC) for parts 1, 4, 5, 6, 7, 8, 9, 10; injected resin type for parts 2, 3

Parts bonded to chassis:	Detachable parts:
1 Body top	3 Tailgate lining
2 Roof	4 Tailgate outer panel
5 Rear wing	6 Rear door panel
9 Front wing	8 Front door panel
7 Sill	10 Bonnet

2.3.5 Galvanized body shell clad entirely with composite skin panels

Renault have designed a high-rise car which has a skeletal steel body shell (Figure 2.34), clad entirely with composite panels. After assembly the complete body shell is immersed in a bath of molten zinc, which applies an all-over 6.5 micron (millionth of a metre) coating. The process gives anti-rust protection, while the chemical reaction causes a molecular change in the steel which strengthens it. Lighter-gauge steel can therefore be used without sacrificing strength, resulting in a substantial weight saving even with the zinc added.

Skin panels are formed in reinforced polyester sheet, made of equal parts of resin, fibreglass and mineral filler. The panels are joined to the galvanized frame and doors by rivets or bonding as appropriate. The one-piece high-rise tailgate is fabricated entirely from polyester with internal steel reinforcements (Figure 2.35). Damage to panels through impact shocks is contained locally and absorbed through destruction of the material, unlike the steel sheet which transmits deformation. Accident damage and consequent repair costs are thus reduced.

2.3.6 Variations in body shape

Among motor car manufacturers there are variations in constructional methods which result in different body types and styles. Figure 2.36 illustrates four types of body shell – a saloon with a boot, a hatchback, an estate car and a light van. Figure 2.37 shows a coach-built limousine of extremely high quality, built on a Rolls-Royce Silver Spirit chassis by the coach-builders Hooper & Co. This vehicle has been designed for the use of heads of state and world-ranking VIPs.

2.4 BASIC BODY CONSTRUCTION

A typical four-door saloon body can be likened to a hollow tube with holes cut in the sides. The bulkhead towards the front and rear completes the box-like form and assists in providing torsional stability. The roof,

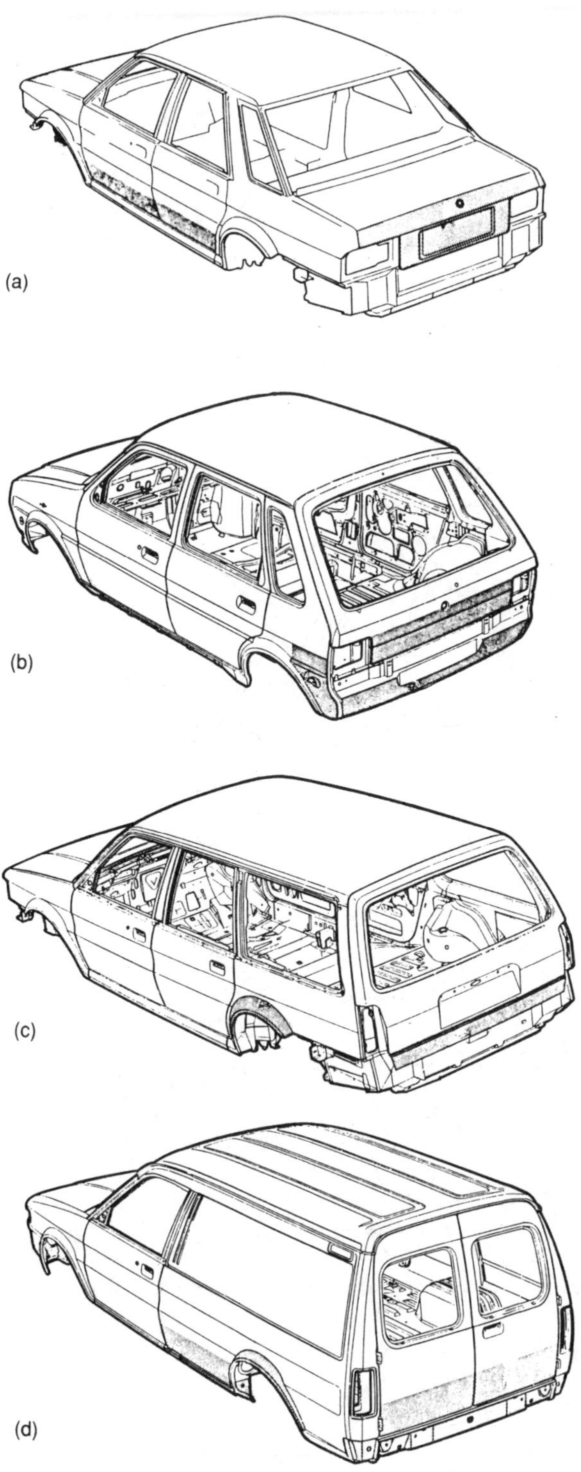

Figure 2.36 Body shell variations: (a) saloon with boot (b) hatchback (c) estate car (d) light van (*Rover Group Ltd*)

Figure 2.37 Coach-built limousine: Emperor State Landaulette (*Hooper & Co (Coach-builders) Ltd*)

even if it has to accommodate a sunshine roof, is usually a quite straightforward and stable structure; the curved shape of the roof panel prevents lozenging (going out of alignment in a diamond shape). The floor is a complete panel from front to rear when assembled, and is usually fitted with integral straightening ribs to prevent lozenging. With its bottom sides or sill panels, wheel arches, cross members and heelboard, it is the strongest part of the whole body. The rear bulkhead, mainly in the form of a rear squab panel, is again a very stable structure. However, the scuttle or forward bulkhead is a complex structure in a private motor car. Owing to the awkward shape of the scuttle and the accommodation required for much of the vehicle's equipment, it requires careful designing to obtain sufficient strength. Body sides with thin pillars, large windows and door openings are inherently weak, requiring reinforcing with radiusing corners to the apertures to give them sufficient constructional strength.

A designer in a small coach-building firm will consider methods necessary to build the body complete with trim and other finishing processes. The same job in a mass production factory may be done by a team of designers and engineers all expert in their own particular branch of the project. The small manufacturer produces bodies with skilled labour and a minimum number of jigs, while the mass producer uses many jigs and automatic processes to achieve the necessary output. However, the problems are basically the same: to maintain strength and stability, a good standard of finish and ease of production.

Figure 2.38 shows the build-up details of a four door saloon, from the main floor assembly to the complete shell assembly. In the figure the main floor unit (1), commencing at the front, comprises a toeboard or pedal panel, although in some cases this may become

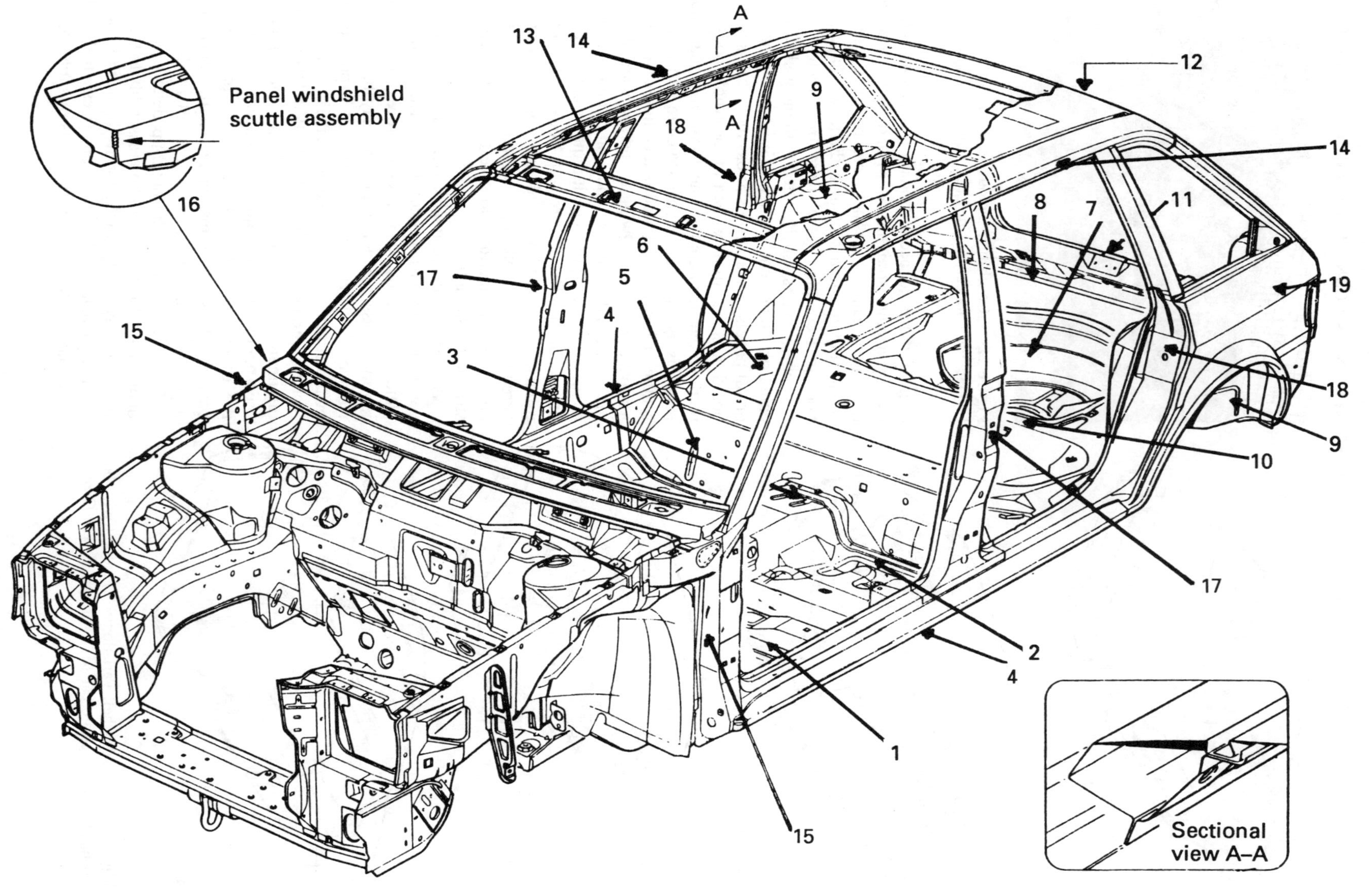

Panel windshield scuttle assembly
Sectional view A–A
A
A
1
2
3
4
4
5
6
7
8
9
9
10
11
12
13
14
14
15
15
16
17
17
18
18
19

a part of the scuttle or bulkhead. Apart from providing a rest for the front passengers' feet, it seals off the engine and gearbox from the body and connects the scuttle to the main floor. The main centre floor panel (2) should be sufficiently reinforced to carry the weight of the front seats and passengers. It may be necessary to have a tunnel running the length of the floor in the centre to clear the transmission system from the engine to the rear axle, and holes may have to be cut into the floor to allow access to the gearbox, oil filler and dipstick, in which case removable panels or large grommets would be fitted in these access holes (3).

The front end of the main floor is fixed to the toeboard panel and the sides of the main centre floor are strengthened by the bottom sills (4) and/or some form of side members which provide the necessary longitudinal strength. The transverse strength is provided by the cross members. The floor panel itself prevents lozenging, and the joints between side members and cross members are designed to resist torsional stresses.

The rear end of the floor is stiffened transversely by the rear seat heelboard (5). This heelboard also stiffens the front edges of the rear seat panel. In addition it often provides the retaining lip for the rear seat cushion, which is usually made detachable from the body. The heelboard together with the rear panel and rear squab panel forms the platform for the rear seat.

The rear seat panel (6) is reinforced or swaged if necessary to gain enough strength to support the rear passengers. Usually the rear seat panel has to be raised to provide sufficient clearance for the deflection of the rear axle differential housing. The front edge of the rear seat panel is stiffened by the rear seat heelboard, and the rear edge of the seat panel is stiffened by the rear squab panel. The rear squab panel completes this unit and provides the rear bulkhead across the car. It seals off the boot or luggage compartment from the main body or passenger compartment.

The boot floor (7), which extends from the back of the rear squab panel to the extreme back of the body, completes the floor unit. In addition to the luggage, the spare wheel has to be accommodated here. The front edge of the boot floor is reinforced by the rear squab panel and the rear end by a cross member of some form (8). The sides of the floor are stiffened by vertical boot side panels at the rear, while the wheel arch panels complete the floor structure by joining the rear end of the main floor and its side members. The wheel arch panels (9) themselves seal the rear road wheels from the body.

In general the floor unit is made up from a series of panels with suitable cross members or reinforcements. The edge of the panels are stiffened either by flanging reinforcing members, or by joining to the adjacent panels. The boot framing is joined at the back to the rear end of the boot floor, at the sides to the boot side panels and at the top to the shelf panel behind the rear squab (10). It has to be sufficiently strong at the point where the boot lid hinges are fitted to carry the weight of the boot lid when this is opened. Surrounding the boot lid opening there is a gutter to carry away rain and water to prevent it entering the boot; opposite the hinges, provision is made for the boot lid lock striking plate (11) to be fixed. From the forward edge of the boot, the next unit is the back light and roof structure (12), and this extends to the top of the windscreen or canopy rail (13). The roof is usually connected to the body side frames, which comprise longitudinal rails or stringers and a pair of cantrails which form the door openings (14). Provision in the roof should be made for the interior lights and wiring and also the fixing of the interior trimming. The scuttle and windscreen unit, including the front standing pillar or A-post (15), provides the front bulkhead and seals the engine from the passenger compartment.

Accommodation has to be made for the instrumentation of the car, the wiring, radio, windscreen wipers and driving cable, demisters and ducting, steering column support, handbrake support and pedals. The scuttle (16) is a complicated structure which needs to be very strong. When the front door is hinged at the

---

(Facing page)

Figure 2.38 Body constructional details of Austin Rover Maestro (*Austin Rover*)

1 Main floor unit
2 Main centre floor panel
3 Access holes
4 Bottom sills
5 Rear seat heelboard
6 Rear seat panel
7 Boot floor
8 Cross member
9 Wheel arch panel
10 Rear squabs
11 Boot lid lock striking plate
12 Roof structure
13 Windscreen or canopy rail
14 Cantrails
15 Front standing pillar (A-post)
16 Scuttle
17 Centre standing pillar (BC-post)
18 Rear standing pillar (D-post)
19 Quarter panels

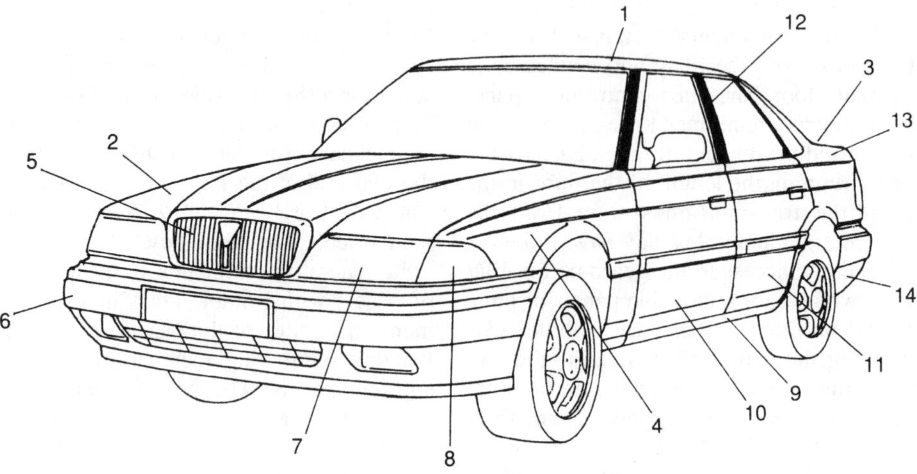

Figure 2.39 Major body panels

1 Roof panel
2 Bonnet panel
3 Boot lit
4 Front wing
5 Radiator grille
6 Front bumper bar
7 Headlamps
8 Sidelamps
9 Sill panel
10 Front door
11 Rear door
12 Centre pillar
13 Rear quarter panel
14 Rear bumper bar

forward edge, provision has to be made in the front pillar for the door hinges, door check and courtesy light switches.

The centre standing pillar or BC-post (17) is fixed to the side members of the main floor unit and supports the cantrails of the roof unit. It provides a shut face for the front door, a position for the door lock striking plate and buffers or dovetail, and also a hinge face for the rear door; as with the front standing pillar, provision is made for the door hinges and door check. The rear standing pillar or D-post (18) provides the shut face for the rear end of the floor side members at the bottom, while the top is fixed to the roof cantrails and forms the front of the quarters.

The quarters (19) are the areas of the body sides between the rear standing pillars and the back light and boot. If the body is a six-light saloon there will be a quarter window here with its necessary surrounding framing, but in the case of a four-light saloon this portion will be more simply constructed. Apart from the doors, bonnet, boot lid and front wings this completes the structure of the average body shell.

2.5 IDENTIFICATION OF MAJOR BODY PRESSINGS

The passenger-carrying compartment of a car is called the body, and to it is attached all the doors, wings and such parts required to form a complete body shell assembly (Figure 2.39).

2.5.1 Outer construction

This can be likened to the skin of the body, and is usually considered as that portion of a panel or panels which is visible from the outside of the car.

2.5.2 Inner construction

This is considered as all the brackets, braces and panel assemblies that are used to give the car strength (Figure 2.40). In some cases the entire panels are inner construction on one make of car and a combination of inner and outer on another.

2.5.3 Front-end assembly including cowl or dash panel

The front-end assembly (Figure 2.41) is made up from the two front side member assemblies which are designed to carry the weight of the engine, suspension, steering gear and radiator. The suspension system used will affect the design of the panels, but whatever system is used the loads must be transmitted to the wing valances and on to the body panels. The front cross member assembly braces the front of the car and carries the radiator and headlamp units. The side valance assemblies form a housing for the wheels, a mating edge for the bonnet and a strong box section for attachment of front wings. Both the side frames and valance assemblies are connected to the cowl or dash panel. The front-end assembly is attached to the main floor at the toe panel.

The cowl or dash panel forms the front bulkhead of the body (Figure 2.41) and is usually formed by joining smaller panels (the cowl upper panel and the cowl side panel) by welds to form an integral unit. In some cases the windscreen frame is integral with the cowl panel. The cowl extends upwards around the entire windscreen opening so that the upper edge of the cowl panel forms the front edge of the roof panel. In this case the windscreen pillars, i.e. the narrow sloping construction at either side of the windscreen opening, are merely part of the cowl panel. In other constructions, only a portion of the windscreen pillar is formed as part of the cowl. The cowl is sometimes called the fire wall because it is the partition between the passenger and engine compartments, and openings in the cowl accommodate the necessary controls, wiring and tubing that extend from one compartment to the other. The instrument panel, which is usually considered as part of the cowl panel although it is a complex panel in itself, provides a mounting for the instruments necessary to check the performance of the vehicle during operation. Cowl panels usually have both inner and outer construction, but in certain constructions only the upper portion of the cowl around the windscreen is visible. On many vehicles the front door hinge pillar is also an integral part of the cowl.

2.5.4 Front side member assembly

This is an integral part of the front-end assembly; it connects the front wing valances to the cowl or dash assembly. It is designed to strengthen the front end; it is part of the crumple zone, giving lateral strength on impact and absorbing energy by deformation during a collision. It also helps to support the engine and suspension units (see Figure 2.41; key figure references 13, 15, 16, 20, 22).

2.5.5 A-post assembly

This is an integral part of the body side frame. It is connected to the front end assembly and forms the front door pillar or hinge post. It is designed to carry the weight of the front door and helps to strengthen the front bulkhead assembly (Figure 2.41).

2.5.6 Main floor assembly

This is the passenger-carrying section of the main floor. It runs backwards from the toe panel to the heelboard or back seat assembly. It is strengthened to carry the two front seats, and in some cases may have a transmission tunnel running through its centre. Strength is built into the floor by the transmission tunnel acting like an inverted channel section. The body sill panels provide extra reinforcement in the form of lateral strength. Transverse strength is provided by box sections at right angles to the transmission tunnel, generally in the areas of the front seat and in front of the rear seat. The remaining areas of flat floor are ribbed below the seats and in the foot wells to add stiffness (Figure 2.42).

2.5.7 Boot floor assembly

This is a section of the floor between the seat panel and the extreme back of the boot. It is strengthened by the use of cross members to carry the rear seat passengers. This area forms the rear bulkhead between the two rear wheel arches, forming the rear seat panel or heelboard, and in a saloon body shell can incorporate back seat supports and parcel shelf. The boot floor is also strengthened to become the luggage compartment, carrying the spare wheel and petrol tank. At the extreme back it becomes the panel on to which the door or tailgate closes (Figure 2.42).

2.5.8 Complete underbody assembly

This is commonly called the floor pan assembly, and is usually composed of several smaller panels welded together to form a single floor unit. All floor panels are reinforced on the underside by stiffening members or cross members. Most floor pans are irregular in shape for several reasons. They are formed with indentations or heavily swaged areas to strengthen the floor sections between the cross members, and foot room for the passengers is often provided by these recessed areas in the floor. Figure 2.42 shows a complete underbody assembly.

2.5.9 Body side frame assembly

On a four-door saloon this incorporates the A-post, the BC-post, the D-post and the rear quarter section. The side frames reinforce the floor pan along the sill sections. The hinge pillar or A-post extends forward to meet the dash panel and front bulkhead to provide strength at this point. The centre pillars or BC-posts connect the body sills to the roof cantrails. They are

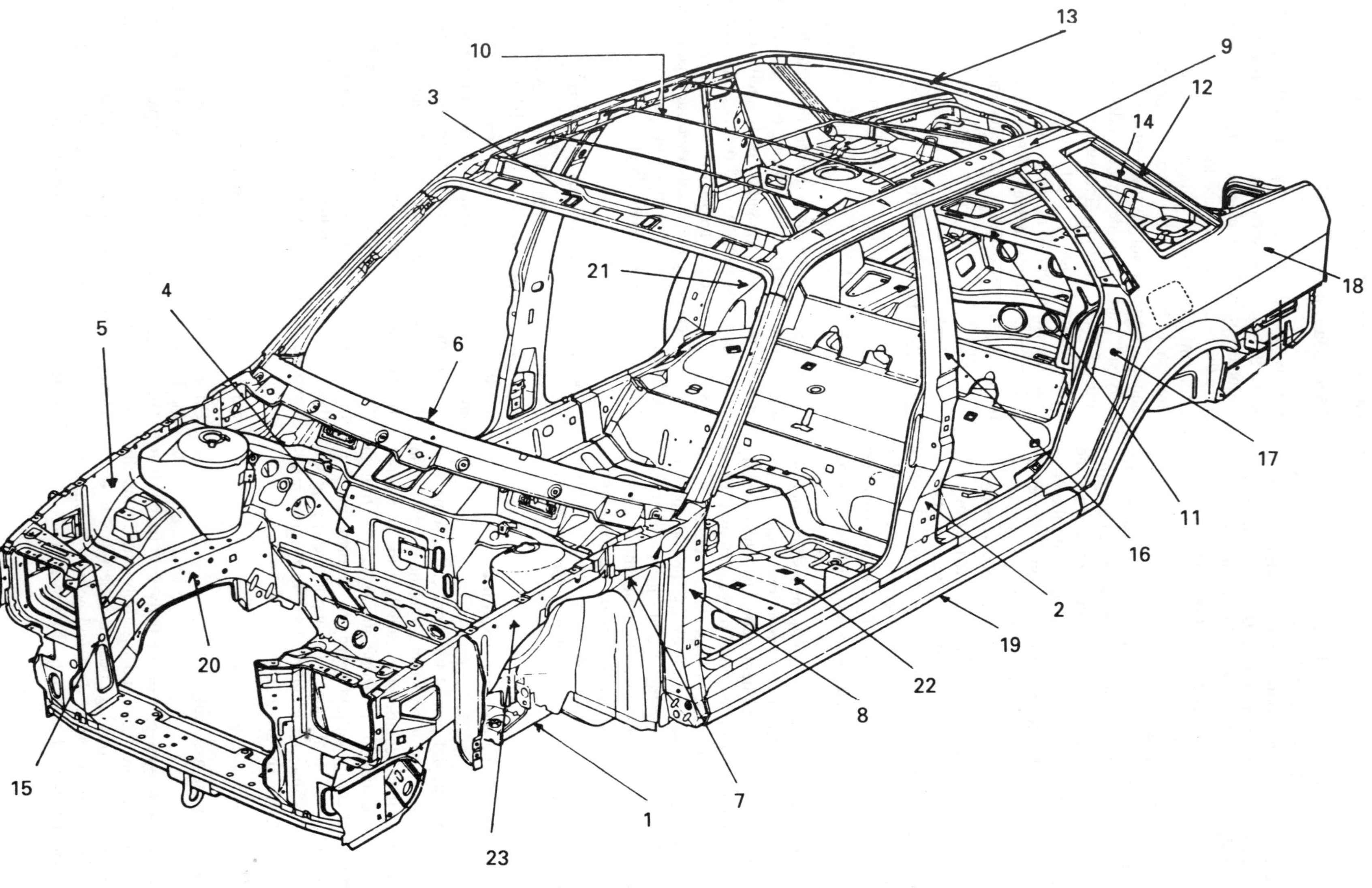

13
10
9
3
12
14
21
18
4
5
6
17
11
16
2
20
19
22
8
15
7
1
23

usually assembled as box sections using a top-hat section and flat plate. These are the flanges which form the attachments for the door weather seals and provide the four door openings. The D-post and rear quarter section is integral with the rear wheel arch and can include a rear quarter window (Figure 2.43).

2.5.10 Roof panel

The roof panel is one of the largest of all major body panels, and it is also one of the simplest in construction. The area which the roof covers varies between different makes and models of cars. On some cars, the roof panel ends at the windscreen. On others it extends downwards around the windscreen so that the windscreen opening is actually in the roof. On some cars the roof ends above the rear window, while on others it extends downwards so that the rear window opening is in the lower rear roof. When this is the case the roof panel forms the top panel around the rear boot opening. Some special body designs incorporate different methods of rear window construction, which affect the roof panel; this is particularly true for estate cars, hatchbacks and hardtop convertibles. Alternatively the top is joined to the rear quarter panel by another smaller panel which is part of the roof assembly.

The stiffness of the roof is built in by the curvature given to it by the forming presses, while the reinforcements, consisting of small metal strips placed crosswise to the roof at intervals along the inside surface, serve to stiffen the front and rear edges of the windscreen and rear window frames. In some designs the roof panel may have a sliding roof built in (Figure 2.43) or a flip-up detachable sunroof incorporated.

2.5.11 Rear quarter panel or tonneau assembly

This is integral with the side frame assembly and has both inner and outer construction. The inner construction comprises the rear wheel arch and the rear seat heelboard assembly. This provides the support for the rear seat squab in a saloon car; if the vehicle is a hatchback or estate car, the two back seats will fold flat and the seat squabs will not need support. This area is known as the rear bulkhead of the car; it gives additional transverse strength between the wheel arch sections and provides support for the rear seat. The rear bulkhead also acts as a partition between the luggage and passenger compartments (Figure 2.43).

2.5.12 Rear wheel arch assembly

This assembly is constructed as an integral part of the inner construction of the rear quarter panel. It is usually a two-piece construction comprising the wheel arch and the quarter panel, which are welded together (Figure 2.42).

2.5.13 Wings

A wing is a part of the body which covers the wheel. Apart from covering the suspension construction, the wing prevents water and mud from being thrown up on to the body by the wheels. The front wings (or the fender assembly) are usually attached to the wing valance of the front end assembly (see Figure 2.41) by means of a flange the length of the wing, which is turned inwards from the outer surface and secured by either welding or bolts. Adjustment for the front wing is usually provided for by slotting the bolt holes so that the wing can be moved either forwards or backwards by loosening the attaching bolts. This adjustment cannot be made if the wing is welded to the main body structure.

In some models the headlights and sidelights are recessed into the front wing and fastened in place by flanges and reinforcement rims on the wing. Any trim or chrome which appears on the side of the wing is

(Facing page)
Figure 2.40 Body shell assembly (*Austin Rover Group Ltd*)

1 Underbody assembly
2 Body side frame assembly
3 Windscreen upper rail assembly
4 Cowl and dash panel assembly
5 Front wheel house complete panel
6 Instrument panel assembly
7 Cowl side lower brace
8 Front body hinge pillar (A-post)
9 Roof panel assembly
10 Roof bow assembly
11 Bulkhead brace assembly
12 Rear quarter centre panel assembly (back window)
13 Back window upper rail panel assembly
14 Rear-end upper panel assembly
15 Radiator panel complete assembly
16 Centre pillar (BC-post)
17 D-post
18 Rear quarter assembly
19 Sill panel
20 Front side member assembly
21 Rear wheel arch assembly
22 Main floor assembly
23 Front valance complete assembly

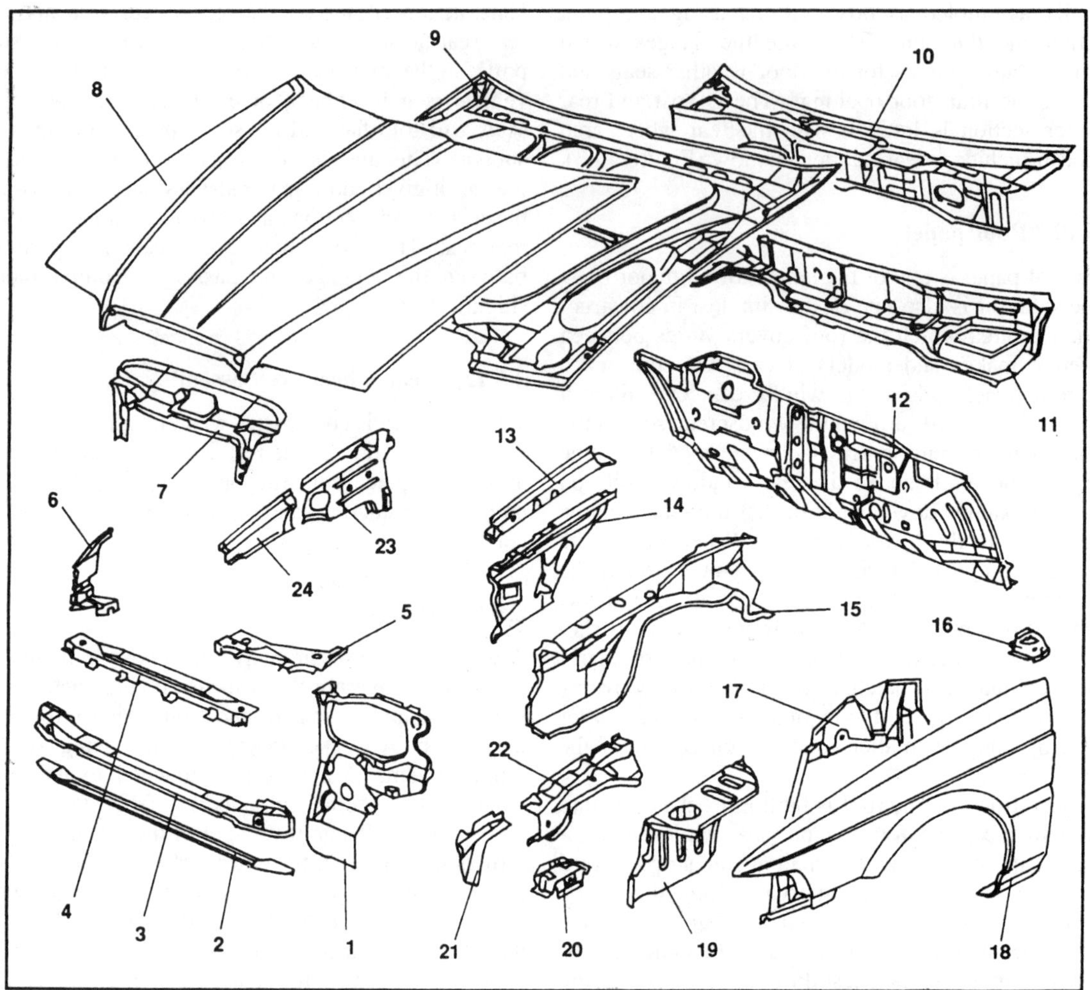

Figure 2.41 Complete front-end assemblies (*Rover Group Ltd*)

1 Headlamp panel RH and LH
2 Front cross member closing panel
3 Front cross member
4 Bonnet lock panel
5 Headlamp panel reinforcement RH and LH
6 Front wing corner piece RH and LH
7 Bonnet frame extension
8 Bonnet skin
9 Bonnet frame
10 Dash panel
11 Scuttle panel
12 Front bulkhead
13 Chassis leg reinforcement RH and LH
14 Front inner wing RH and LH
15 Front chassis leg RH and LH
16 Subframe mounting RH and LH
17 Front wheel arch RH and LH
18 Front wing RH and LH
19 Battery tray
20 Chassis leg gusset RH and LH
21 Bumper mounting reinforcement RH and LH
22 Chassis leg extension RH and LH
23 A-post rear reinforcement RH and LH
24 A-post front reinforcement RH and LH

usually held in place by special clips or fasteners which allows easy removal of the trim.

The unsupported edges of the wing are swaged edges known as beads. The bead is merely a flange which is turned inwards on some cars and then up to form a U-section with a rounded bottom. It not only gives strength but prevents cracks developing in the edges of the wing due to vibration, and it provides a smooth finished appearance to the edge of the wing.

In general the rear wing is an integral part of the body side frame assembly and rear quarter panel. When the wing forms an integral part of the quarter

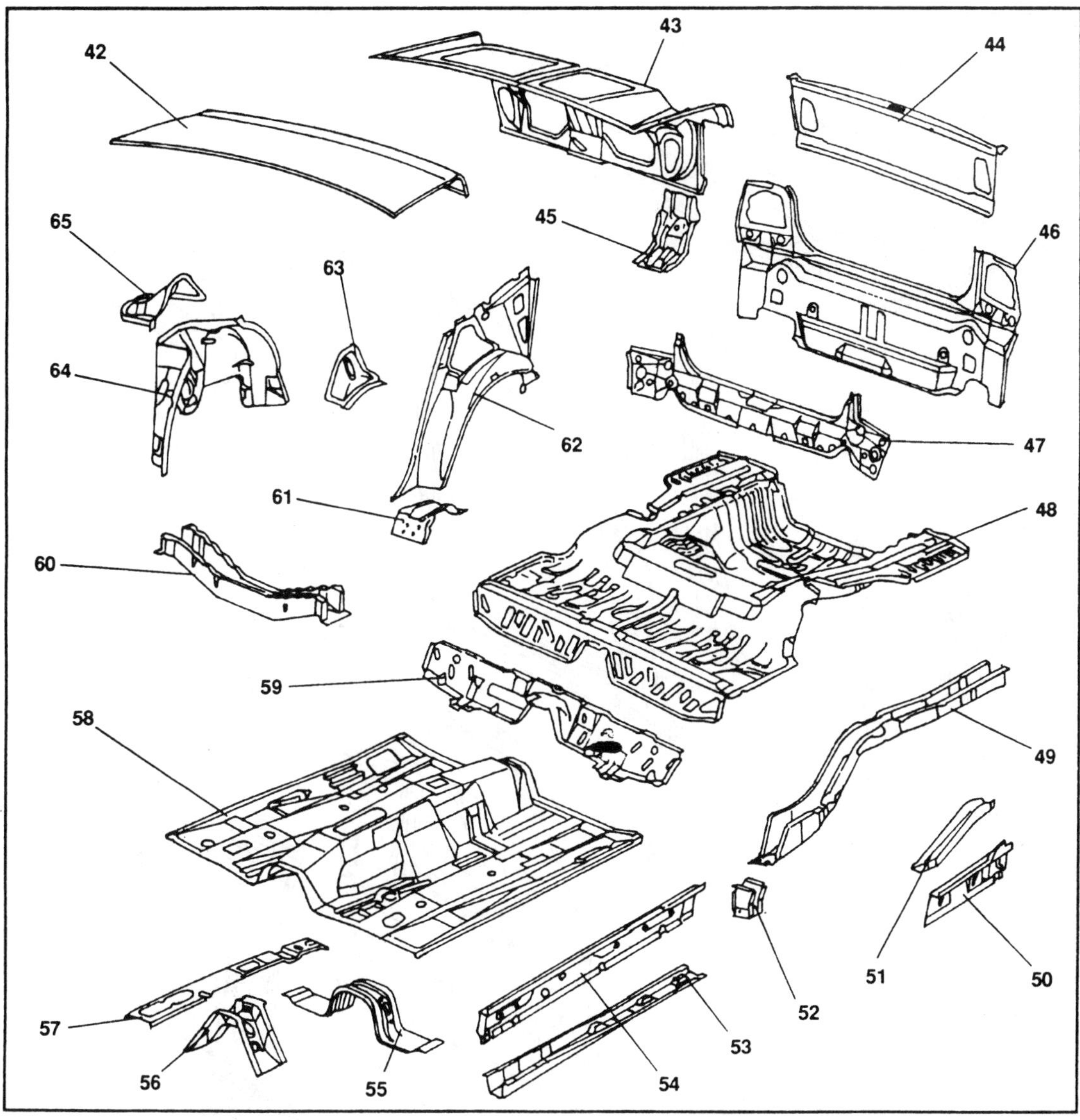

Figure 2.42 Main floor assemblies and boot floor assemblies (*Rover Group Ltd*)

42 Boot lid skin
43 Boot lid frame
44 Boot lid lower skin
45 Boot lid striker reinforcement
46 Outer rear panel
47 Rear panel reinforcement
48 Boot floor
49 Rear chassis leg RH and LH
50 Inner sill rear extension RH and LH
51 Chassis leg reinforcement RH and LH
52 Rear suspension mounting
53 Centre chassis leg RH and LH
54 Inner sill RH and LH
55 Centre cross member
56 Main floor centre reinforcement
57 Centre tunnel reinforcement
58 Main floor
59 Heelboard
60 Rear suspension cross member
61 Rear seatbelt upper mounting RH and LH
62 Outer rear wheel arch RH and LH
63 Wheel arch gusset RH and LH
64 Inner rear wheel arch RH and LH
65 Rear suspension turret capping RH and LH

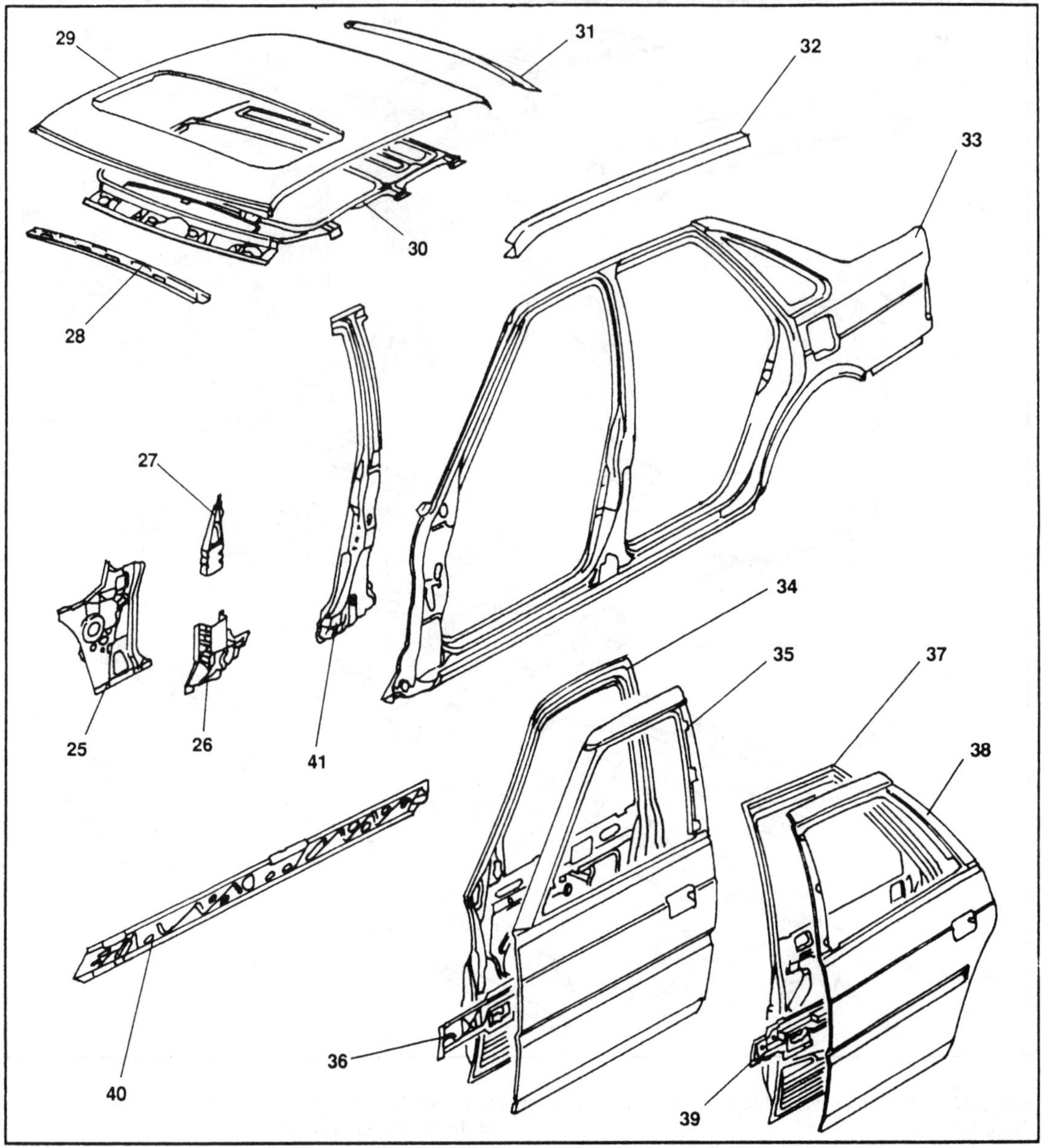

Figure 2.43 Body side assemblies, roof, BC-post, front and rear doors (*Rover Group Ltd*)

25 A-post RH and LH
26 A-post lower reinforcement RH and LH
27 A-post upper reinforcement RH and LH
28 Header Rail
29 Roof panel
30 Sunroof frame
31 Roof rear cross member
32 Cantrail RH and LH
33 Body side assembly RH and LH
34 Front door frame RH and LH
35 Front door skin RH and LH
36 Front door crash rail RH and LH
37 Rear door frame RH and LH
38 Rear door skin RH and LH
39 Rear door crash rail RH and LH
40 Intermediate sill RH and LH
41 B-post reinforcement RH and LH

panel, the inner construction is used to form part of the housing around the wheel arch. The wheel arch is welded to the rear floor section and is totally concealed by the rear quarter panel, while the outer side of the wheel arch is usually attached to the quarter panel around the wheel opening. This assembly prevents road dirt being thrown upwards between the outer panel and inner panel construction.

2.5.14 Doors

Several types of door are used on each vehicle built, although the construction of the various doors is similar regardless of the location of the door on the vehicle, as indicated on Figure 2.43. The door is composed of two main panels, an outer and an inner panel, both being of all-steel construction. The door derives most of its strength from the inner panel since this is constructed mainly to act as a frame for the door. The outer panel flanges over the inner panel around all its edges to form a single unit, which is then spot welded or, in some cases, bonded with adhesives to the frame.

The inner panel has holes or apertures for the attachment of door trim. This trim consists of the window regulator assembly and the door locking mechanism. These assemblies are installed through the large apertures in the middle of the inner panel. Most of the thickness of the door is due to the depth of the inner panel which is necessary to accommodate the door catch and window mechanism. The inner panel forms the lock pillar and also the hinge pillar section of the door. Small reinforcement angles are usually used between the outer and inner panel, both where the lock is inserted through the door and where the hinges are attached to the door. The outer panel is either provided with an opening through which the outside door handle protrudes, or is recessed to give a more streamlined effect and so creating better aerodynamics.

The upper portion of the door has a large opening which is closed by glass. The glass is held rigidly by the window regulator assembly, and when raised it slides in a channel in the opening between the outer and inner panels in the upper portion of the door. When fully closed the window seats tightly in this channel, effectively sealing out the weather.

2.5.15 Boot lid or tailgate

This is really another door which allows access to the luggage compartment in the rear of the car (Figure 2.42). A boot lid is composed of an outer and an inner panel. These panels are spot welded along their flanged edges to form a single unit in the same manner as an ordinary door. Hatchback and estate cars have a rear window built into the boot lid, which is then known as a tailgate. Some manufacturers use external hinges, while others use concealed hinges attached to the inner panel only. A catch is provided at the lower rear edge of the boot lid or tailgate and is controlled by an external handle or locking mechanism. This mechanism may be concealed from the eye under a moulding or some type of trim. In some models there is no handle or external locking mechanism; instead the hinges are spring loaded or use gas-filled piston supports, so that when the lid is unlocked internally it automatically rises and is held in the open position by these mechanisms.

2.5.16 Bonnet

The bonnet (Figure 2.41) is the panel which covers the engine compartment where this is situated at the front of the vehicle, or the boot compartment of a rear-engined vehicle. Several kinds of bonnets are in use on different makes of cars. The bonnet consists of an outer panel and an inner reinforcement constructed in the H or cruciform pattern, which is spot welded to the outer skin panel at the flanged edges of the panels. The reinforcement is basically a top-hat section, to give rigidity to the bonnet. In some cases the outer panel is bonded to the inner panel using epoxy resins. This system avoids the dimpling effect on the outer surface of the bonnet skin which occurs in spot welding.

Early models used a jointed type of bonnet which was held in place by bolts through the centre section of the top of the bonnet into the body of the cowl and into the radiator. A piano-type hinge was used where the bonnet hinged both at the centre and at the side.

The most commonly used bonnet on later constructions is known as the mono or one-piece type, and can be opened by a variety of methods. On some types it is hinged at the front so that the rear end swings up when the bonnet is open. Others are designed so that they can be opened from either side, or unlatched from both sides and removed altogether. Most bonnets, however, are of the alligator pattern, which is hinged at the rear so that the front end swings up when opened.

The type of bonnet catch mechanism depends on the type of bonnet used. When a bonnet opens from the rear the catch mechanism is also at the rear. When it opens from either side the combination hinge and catch are provided at each side. The alligator bonnets have their catches at the front, and in most cases the catches are controlled from inside the car.

Bonnets are quite large, and to make opening easier the hinges are usually counterbalanced by means of

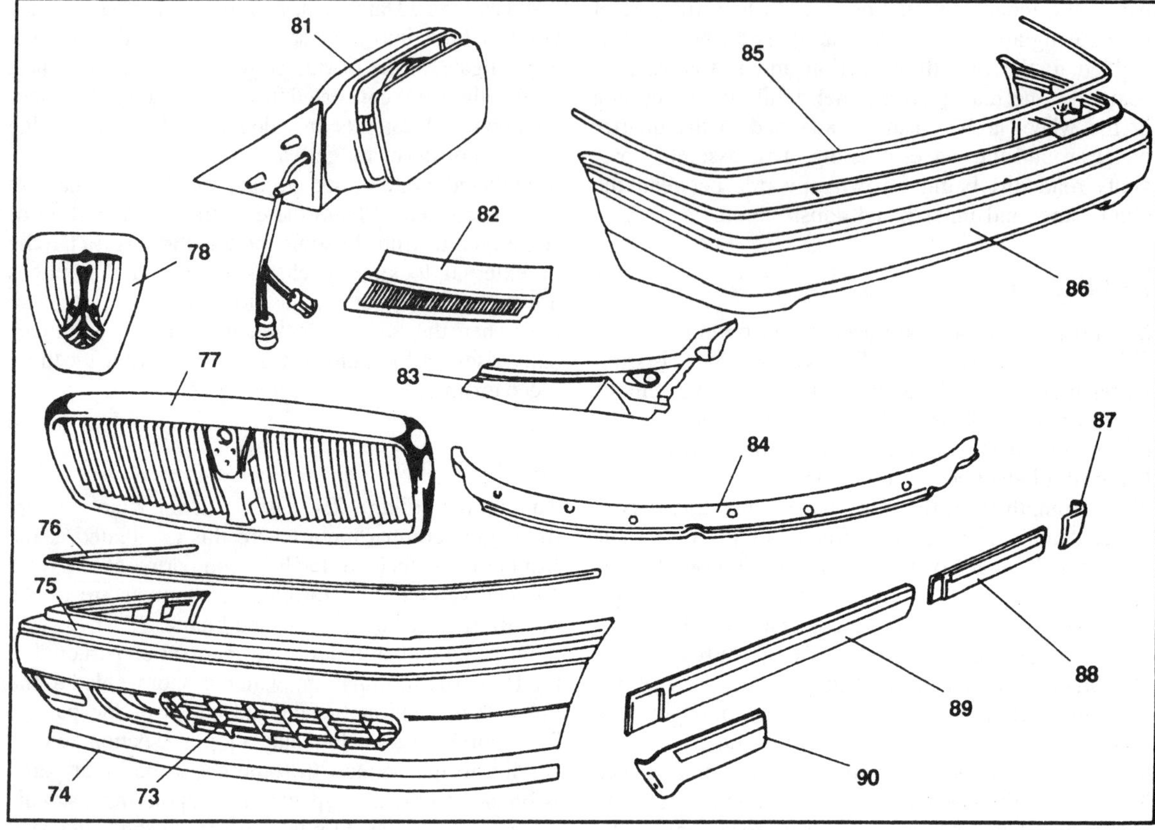

Figure 2.44 Exterior trim (*Rover Group Ltd*)
73 Lower front grille
74 Front spoiler
75 Front bumper
76 Front bumper insert
77 Front grille
78 Motif
81 Door mirror assembly
82 Scuttle grille
83 Scuttle moulding
84 Lower screen moulding
85 Rear bumper insert
86 Rear bumper
87 Rear wing waist moulding
88 Rear door waist moulding
89 Front door waist moulding
90 Front door waist moulding

tension or torsion springs. Where smaller bonnets are used the hinges are not counterbalanced and the bonnet is held in place by a bonnet stay from the side of the wing to the bonnet. Adjustment of the bonnet position is sometimes possible by moving the hinges.

2.5.17 Trims

Some details of exterior and interior trims are shown in Figures 2.44 and 2.45.

2.5.18 Complete body shell

A contemporary vehicle embracing all the techniques of panel assembly is shown in Figures 2.16 and 2.47. Figure 2.46 illustrates the completed structure with all panel assemblies in place. Figure 2.47 shows the completely finished vehicle ready for the road.

2.5.19 Comparative terms in common use by British, American and European car manufacturers

As manufacturers use differing terms for the various body panel assemblies and individual panels, difficulties may arise when identifying specific panels. The following are the terms in most common use:

Bonnet, hood
Boot lid, deck lid, trunk lid, tailgate

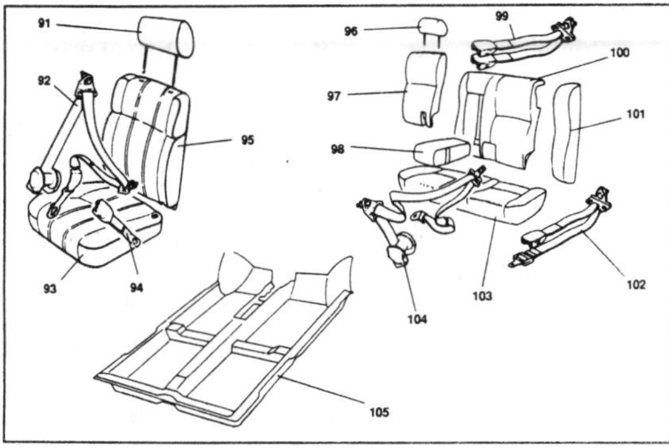

Figure 2.45 Interior trim (*Rover Group Ltd*)
91 Front seat headrest
92 Front seatbelt assembly
93 Front seat cushion
94 Front seatbelt centre stalk
95 Front seat back rest
96 Front seat head rest
97 Front seat back rest small section
98 Front seat centre arm rest
99 Front seatbelt buckle assembly
100 Front seat back rest large section
101 Front seat side bolster
102 Front seatbelt lap assembly
103 Rear seat cushion
104 Rear seatbelt assembly
105 Main floor carpet

Cantrail, roof side rail, drip rail
Centre pillar, BC-post
Courtesy light, interior light
Cowl, scuttle, bulkhead, fire wall
Dash panel, facia panel
Door opening plates, scuff plates
Door skin, outside door panel
Face bar, bumper bar
Front pillar, A-post, windscreen pillar
Light, window
Quarter panel, tonneau assembly
Roof, turret
Roof lining, headlining
Sill panel, rocker panel
Squab, seat back
Underbody, floor pan assembly
Valance of front wing, fender side shield
Vent window, flipper window
Waist rail, belt rail
Wheel arch, wheel house
Windscreen, windshield
Wing, fender

2.5.20 Vehicle identification numbers

The vehicle identification number (VIN) is stamped on a plate located typically inside the engine compartment or on a door pillar. A VIN system is shown in Figure 2.48. The figure also shows the paint and trim codes which are usually included on the VIN plate.

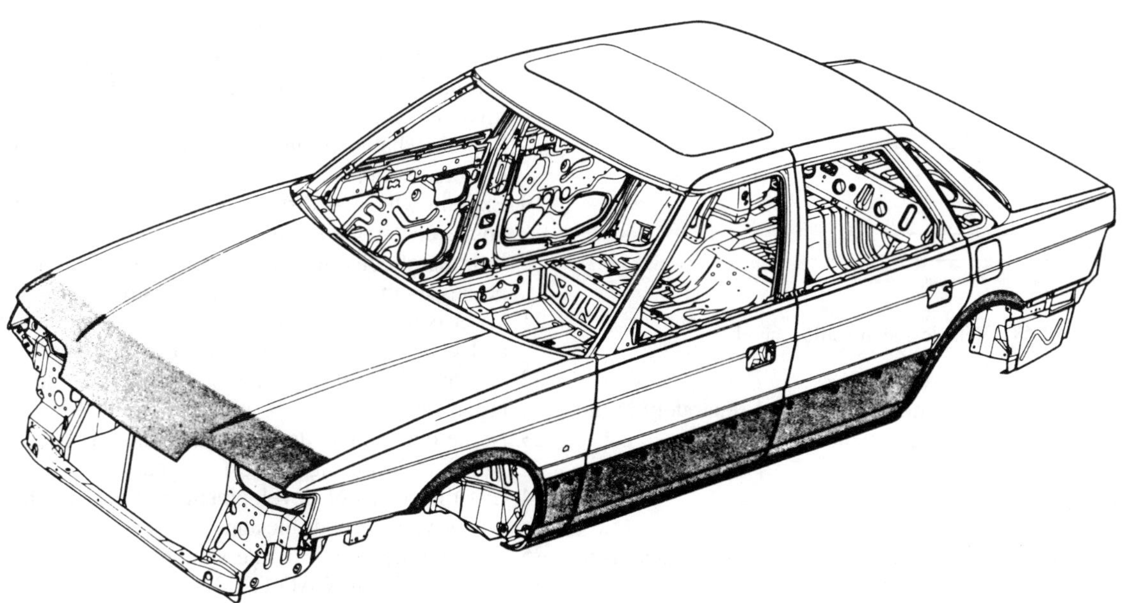

Figure 2.46 Complete body shell (*Rover Group Ltd*)

Figure 2.47 Rover 800 series (*Rover Group Ltd*)

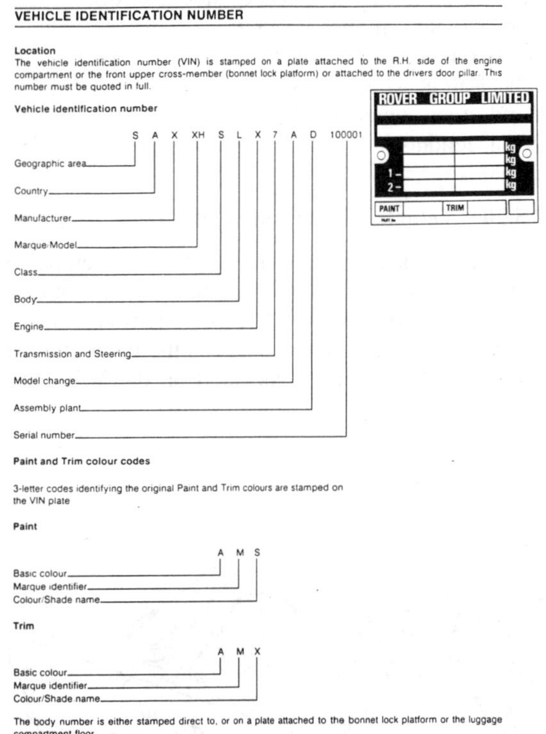

Figure 2.48 Vehicle identification number

The car body number is provided separately in the engine or boot compartments.

QUESTIONS

1. Why were the earliest motor vehicle bodies made almost entirely of wood?
2. When and why did manufacturers commence to use metal for the construction of vehicle bodies?
3. Give a brief history of the development of the vehicle body style, illustrating the significant changes which have taken place.
4. What is meant by monocoque construction, and why has it become so popular in motor vehicle manufacture?
5. With the aid of sketches, describe the general principles of monocoque construction.
6. Describe, with the aid of sketches, the general principles of composite and integral methods of body construction.
7. Draw a sketch of a vehicle body shell and name all the major body panels.
8. State the location and function on a vehicle body of the following sections: (a) BC-post (b) quarter panel (c) wheel arch (d) bonnet.
9. What is the most common form of vehicle body construction?
10. What are the alternatives to integral construction?
11. What is a load-bearing stressed panel assembly? Give examples.
12. What is a non-load-bearing panel assembly? Give examples.
13. Explain how rigidity and strength are achieved in mono construction.
14. Describe the location and function of the front and rear bulkheads.
15. Give a brief description of the following early vehicle body styles: coupé, cabriolet, limousine, saloon.
16. What is meant by a veteran vehicle? Name and describe three such vehicles.
17. Name two people associated with the early development of the motor vehicle, and state their involvement.
18. Explain what is meant by the semi-integral method of construction.
19. Explain why it is difficult to mass produce composite constructed vehicles.
20. In integral construction, what section of the body possesses the greatest amount of strength?
21. What is the front section of the body shell called, and what are its principal panel assemblies?
22. Explain the role of the stylist in the design organization.
23. Name one vehicle design stylist who has become well known during the last 25 years.
24. List the stages of development in the creation of a new vehicle body design.
25. State the definition of the symbol C_d.
26. Define the term CAD-CAM.
27. Explain the role of the clay modeller in the structure of the styling department.

28 With the aid of a sketch, explain what is meant by profile aerodynamic drag.
29 Explain the necessity for prototype testing.
30 Explain the use of dummies in safety research and testing.
31 Explain the difference in manufacture between a medium-bodied mass-produced vehicle and a high-quality coach-built limousine.
32 Describe the bodywork styling of a sports or GT vehicle.
33 What is the difference in design between a saloon and a hatchback vehicle?
34 With the aid of a sketch, explain the body styling of a coupé vehicle.
35 Explain ABS as an active safety feature on a vehicle.
36 How are vehicles made safe against side impact involvement?
37 Explain how the airbag system works in a vehicle.
38 Explain the VIN number and why it is used on a vehicle.
39 Name the two main types of seatbelt arrangement which are fitted to a standard saloon vehicle.
40 State the letters used in design to identify the body pillars on a four-door saloon.
41 State the main purpose of a vehicle subframe.
42 Explain why seatbelt anchorages must be reinforced on a vehicle body.
43 State why GRP bodywork is normally associated with separate body construction.
44 List the design features that characterize a vehicle body as a limousine.
45 Explain the necessity for a hydraulic damper in the suspension of a motor vehicle.
46 Why is GRP not used in the mass production of vehicle body shells on an assembly line?
47 Name one of the persons associated with the early development of the mass production of the motor vehicle and state his involvement.
48 State the purposes of the inner reinforcement members of a bonnet panel and say how they are held in place.
49 State the reasons for swaging certain areas of a vehicle floor pan.
50 Explain the importance of the use of scale models in vehicle design.

3

Materials used in vehicle bodies

3.1 MANUFACTURE OF STEEL COIL AND SHEET FOR THE AUTOMOBILE INDUSTRY

In the manufacture of steel coils, the raw material iron ore is fed into a blast furnace, together with limestone and coke; the coke is used as a source of heat, while the limestone acts as a flux and separates impurities from the ore. The ore is quickly reduced to molten iron, known as pig iron, which contains approximately 3–4 per cent carbon. In the next stage of manufacture, the iron is changed into steel by reheating it in a steel-making furnace and blowing oxygen either into the surface of the iron or through the liquid iron, which causes oxidation of the molten metal. This process burns out impurities and reduces the carbon content from 4 per cent to between 0.08 and 0.20 per cent.

3.1.1 Casting

The steel is cast into ingots; these are either heated in a furnace and rolled down to a slab, or more commonly continuously cast into a slab. Slabs by either casting process are typically 8–10 in (200–250 mm) thick, ready for further rolling. These slabs are reheated prior to rolling in a computer-controlled continuous hot strip mill to a strip around twice the thickness required for body panels. The strip is closely wound into coil ready for further processing.

3.1.2 Pickling

Before cold rolling, the surfaces of the coils must be cleaned of oxide or black scale formed during the hot rolling process and which would otherwise ruin the surface texture. This is done by pickling the coils in either dilute hydrochloric acid or dilute sulphuric acid and then washing them in hot water to remove the acid. The acid removes both the oxide scale and any dirt or grit which might also be sticking to the surface of the coil.

3.1.3 Cold rolling

In the cold rolling process the coil is rolled either in a single-stand reversing mill (narrow mills using either narrow hot mill product or slit wide mill product) or in a multiple-stand tandem mill to the required thickness. Most mills are computer controlled to ensure close thickness control, and employ specially prepared work rolls to ensure that the right surface standard is achieved on the rolled strip. The cold rolling process hardens the metal, because mild steel quickly work hardens. The cold rolled coils are suitable for applications such as panelling where no bending or very little deformation is needed. At this stage the coil is still not suitable for the manufacture of the all-steel body shell and it must undergo a further process to soften it; this is known as annealing.

3.1.4 Annealing

Coils used for the manufacture of a car body must not only have a bright smooth surface but must also be soft enough for bending, rolling, shaping and pressing operations, and so the hardness of cold rolled coils to be used for car bodies must be reduced by annealing. If annealing were carried out in an open furnace this would destroy the bright surface of the coil and therefore oxygen must be excluded or prevented from attacking the metal during the period of heating the coils.

The normal method of annealing the coils is box annealing. The coils are stacked on a furnace base, covered by an inner hood and sealed. The atmosphere is purged with nitrogen and hydrogen to eliminate oxygen. A furnace is then placed over the stack and fired to heat the steel coils to a temperature of about 650°C for around 24 hours, depending on charge size and steel grade.

3.1.5 Temper rolling

During the process of annealing the heat causes a certain amount of buckling and distortion, and a further operation is necessary to produce flat coils. The annealed coil is decoiled and passed through a single-stand temper mill using a specially textured work roll surface, where it is given a light skin pass, typically of 0.75–1.25 per cent extension. This is necessary to remove buckles formed during annealing, to impart the appropriate surface texture to the strip, and to control the metallurgical properties of the strip. The strip is then rewound ready for dispatch or finishing as appropriate.

3.1.6 Finishing

The temper rolled coils can be slit to narrow coil, cut to sheet, reinspected for surface critical applications, or flattened for flatness critical applications as appropriate. Material can be supplied with a protective coating oil, and packed to prevent damage or rusting during transit and storage.

3.2 SPECIFICATIONS OF STEELS USED IN THE AUTOMOBILE INDUSTRY

The motor body industry uses many different types of steel. Low-carbon steel is used for general constructional members. High-tensile steels are used for bolts and nuts which will be subjected to a heavy load. Specially produced deep-drawn steel including micro-alloyed steel is used for large body panels which require complex forming. Zinc-coated steel sheets are increasingly being specified for automobile production, both for body and chassis parts, as improved corrosion protection is sought. Stainless steel is used for its non-rusting, hard wearing and decorative qualities. The many different types of springs used in the various body fittings are produced from spring steel, while specially hardened steels make the tools of production. Drills, chisels, saws, hack-saws and guillotine blades are all produced from special alloy steels, which are made from an appropriate mixture of metals and elements.

Steel varies from iron chiefly in carbon content; iron contains 3–4 per cent carbon while carbon steels may contain from 0.08 per cent to 1.00 per cent carbon. The chemical composition and mechanical properties of these carbon steels, especially when alloyed with other elements such as nickel, chromium and tungsten, have been gradually standardized over the years, and now the different types of steels used are produced to specifications laid down by the British Standards Institution. A British Standard specification defines the chemical composition and mechanical properties of the steel, and also the method and apparatus to be used when testing samples to prove that the mechanical properties are correct. The tensile strength, and in the case of sheet and strip steel the bend test, are the properties of most interest, but the British Standard specification also defines the elongation, the yield point and the hardness of the steel.

The steels used in the motor trade may be grouped as follows:

1 Cold forming steels.
2 Carbon steels.
3 Alloy steels.
4 Free cutting steels.
5 Spring steels.
6 Rust-resisting and stainless steels.

As each group may contain many different specifications, some idea of the variety of steels may be gained. However, in the motor body industry the specifications which apply are those pertaining to cold forming steels, namely BS 1449: Part 1: 1983.

The greatest percentage of steel used in motor bodies is in the form of coil, strip, sheet or plate. Sheet steel is a rolled product produced from a wide rolling mill (600 mm or wider); to come under the heading of sheet steel, the steel must be less than 3 mm thick. Steel 3–16 mm thick comes under the heading of plate.

Tables 3.1, 3.2 and 3.3 are the specifications for steel sheet strip and coil for the manufacture of motor body shells in the automobile industry.

3.3 CARBON STEEL

Carbon steels can be classified as follows (Table 3.3):

Low-carbon steel
Carbon-manganese steel
Micro-alloyed steel
Medium-carbon steel
High-carbon steel

The properties of plain carbon steel are determined principally by carbon content and micro-structure, but it may be modified by residual elements other than carbon, silicon, manganese, sulphur and phosphorus, which are already present. As the carbon content increases so does the strength and hardness, but at the expense of ductility and malleability.

3.3.1 Low-carbon steel

For many years low-carbon steel (sometimes referred to as mild steel) has been the predominant autobody material. Low-carbon coil, strip and sheet steel have been used in the manufacture of car bodies and chassis members. This material has proved an excellent general-purpose steel offering an acceptable combination of strength with good forming and welding properties. It is ideally suited for cold pressings of thin steel sheet and is used for wire drawing and tube manufacture because of its ductile properties.

Table 3.1 Symbols for material conditions: BS 1449: Part 1: 1983

Condition	Symbol	Description
Rimmed steel	R	Low-carbon steel in which deoxidation has been controlled to produce an ingot having a rim or skin almost free from carbon and impurities, within which is a core where the impurities are concentrated
Balanced steel	B	A steel in which processing has been controlled to produce an ingot with a structure between that of a rimmed and a killed steel. It is sometimes referred to as semi-killed steel
Killed steel	K	Steel that has been fully deoxidized
Hot rolled on wide mills	HR	Material produced by hot rolling. This will have an oxide scale coating, unless an alternative finish is specified (see Table 3.2)
narrow mills	HS	
Cold rolled on wide mills	CR	Material produced by cold rolling to the final thickness
Narrow mills	CS	
Normalized	N	Material that has been normalized as a separate operation
Annealed	A	Material in the annealed last condition (i.e. which has not been subjected to final light cold rolling)
Skin passed	SP	Material that has been subjected to a final light cold rolling
Temper rolled		Material rolled to the specified temper and qualified as follows:
	H1	Eighth hard
	H2	Quarter hard
	H3	Half hard
	H4	Three-quarters hard
	H5	Hard
	H6	Extra hard
Hardened and tempered	HT	Material that has been continuously hardened and tempered in order to give the specified mechanical properties

Low-carbon steel is soft and ductile and cannot be hardened by heating and quenching, but can be case hardened and work hardened. It is used extensively for body panels, where its high ductility and malleability allows easy forming without the danger of cracking. In general low-carbon steel is used for all parts not requiring great strength or resistance to wear and not subject to high temperature or exposed to corrosion.

However, factors such as a worldwide requirement for fuel conservation for lighter-weight body structures, and safety legislation requiring greater protection of occupants through improved impact resistance, have brought about a change in materials and production technology. This has resulted in the range of micro-alloyed steels known as high-strength steels (HSSs) or high-strength low-alloy steels (HSLAs).

3.3.2 Micro-alloyed steel

This steel is basically a carbon-manganese steel having a low carbon content, but with the addition of micro-alloying elements such as niobium and titanium. Therefore it is classed as a low-alloy high-strength steel within the carbon range. As a result of its strength, toughness, formability and weldability, car body manufacturers are using this material to produce stronger, lighter-weight body structures.

A typical composition utilized for a micro-alloyed high-strength steel (HSS) is as follows:

	Percentage
Carbon (C)	0.05–0.08
Manganese (Mn)	0.80–1.00
Niobium (Nb)	0.015–0.065

The percentage of niobium used depends on the minimum strength required.

Formable HSSs were developed to allow the automotive industry to design weight out of the car in support of fuel economy targets. A range of high-strength formable steels with good welding and painting characteristics have been developed. The steels are hot rolled for chassis and structural components,

Table 3.2 Symbols for surface finishes and surface inspection: BS 1449: Part 1: 1983

Finish	Symbol	Description
Pickled	P	A hot rolled surface from which the oxide has been removed by chemical means
Mechanically descaled	D	A hot rolled surface from which the oxide has been removed by mechanical means
Fully finish	FF	A cold rolled skin passed material having one surface free from blemishes liable to impair the appearance of a high-class paint finish
General-purpose finish	GP	A cold rolled material free from gross defects, but of a lower standard than FF
Matt finish	M	A surface finish obtained when material is cold rolled on specially prepared rolls as a last operation
Bright finish	BR	A surface finish obtained when material is cold rolled on rolls having a moderately high finish. It is suitable for most requirements, but is not recommended for decorative electroplating
Plating finish	PL	A surface finish obtained when material is cold rolled on specially prepared rolls to give one surface which is superior to a BR finish and is particularly suitable for decorative electroplating
Mirror finish	MF	A surface finish having a high lustre and reflectivity. Usually available only in narrow widths in cold rolled material
Unpolished finish	UP	A blue/black oxide finish; applicable to hardened and tempered strip
Polished finish	PF	A bright finish having the appearance of a surface obtained by fine grinding or abrasive brushing; applicable to hardened and tempered strip
Polished and coloured blue	PB	A polished finish oxidized to a controlled blue colour by further heat treatment; applicable to hardened and tempered strip
Polished and coloured yellow	PY	A polished finish oxidized to a controlled yellow colour by further heat treatment; applicable to hardened and tempered strip
Vitreous enamel	VE	A surface finish for vitreous enamelling of material of specially selected chemical composition
Special finish	SF	Other finishes by agreement between the manufacturer and the purchaser

and cold reduced for body panels. Through carefully controlled composition and processing conditions, these steels achieve high strength in combination with good ductility to allow thinner gauges to be used: a reduction from 0.90 mm down to 0.70 mm is a general requirement.

There was a danger that the new HSS, thinner but just as strong, would lack the ductility which allows it to be press formed into shape. This problem was overcome by the use of micro-alloying. In a metal crystal the atoms are in layers; when the crystals are stretched (as in forming), one layer of atoms slides over another. The layers of atoms slide like playing cards in a pack, and in doing so are changing shape in a ductile way. The sliding can be controlled by adding elements to the steel such as niobium or titanium. The element reacts with the carbon to produce fine particles which spread through the steel. The element controls the ductility and strengthens the steel, thereby improving the properties of the material.

These steels are low carbon, employing solution strengthening (cold reduced rephosphorized) or precipitation hardening (hot rolled and cold reduced micro-alloyed) elements to produce fine-grained steels which are suitable for welding by spot and MIG processes only.

3.3.3 Medium-carbon steel

This can be hardened by quenching, and the amount it can be hardened increases with its carbon content. This type of steel can be used for moving parts such as connecting rods, gear shafts and transmission shafts, which require a combination of toughness and strength, but it is being replaced in the car industry by high-alloy steels.

Table 3.3 Summary of material grades, chemical compositions and types of steel available: BS 1449: Part 1: 1983

Material grade	Rolled condition (see Table 3.1)	Chemical composition C min.	C max.	Si min.	Si max.	Mn min.	Mn max.	S max.	P max.	
Materials having specific requirements based on formability										
		%	%	%	%	%	%	%	%	
1	HR, HS, –, –,	–	0.08	–	–	–	0.45	0.030	0.025	Extra deep drawing aluminium-killed steel
1	–, –, CR, CS	–	0.08	–	–	–	0.45	0.030	0.025	Extra deep drawing aluminium-killed stabilized steel
2	HR, HS, CR, CS	–	0.08	–	–	–	0.45	0.035	0.030	Extra deep drawing
3	HR, HS, CR, CS	–	0.10	–	–	–	0.50	0.040	0.040	Deep drawing
4	HR, HS, CR, CS	–	0.12	–	–	–	0.60	0.050	0.050	Drawing or forming
14	HR, HS, –, –,	–	0.15	–	–	–	0.60	0.050	0.050	Flanging
15	HR, HS, –, –,	–	0.20	–	–	–	0.90	0.050	0.060	Commercial
Materials having specific requirements based on minimum strengths										
Carbon-manganese steels										
34/20	HR, HS, CR, CS	–	0.15	–	–	–	1.20	0.050	0.050	Available as rimmed (R), balanced (B) or killed (K) steels
37/23	HR, HS, CR, CS	–	0.20	–	–	–	1.20	0.050	0.050	
43/25	HR, HS, –, –,	–	0.25	–	–	–	1.20	0.050	0.050	
50/35	HR, HS, –, –,	–	0.20	–	–	–	1.50	0.050	0.050	Grain-refined balanced (B) or killed (K) steel
Micro-alloyed steels										
40/30	HR, HS, –, CS	–	0.15	–	–	–	1.20	0.040	0.040	Grain-refined niobium- or titanium-treated fully killed steels having high yield strength and good formability
43/35	HR, HS, –, CS	–	0.15	–	–	–	1.20	0.040	0.040	
46/40	HR, HS, –, CS	–	0.15	–	–	–	1.50	0.040	0.040	
50/45	HR, HS, –, CS	–	0.20	–	–	–	1.50	0.040	0.040	
60/55	–, HS, –, CS	–	0.20	–	–	–	1.50	0.040	0.040	
40F30	HR, HS, –, CS	–	0.12	–	–	–	1.20	0.035	0.030	The steels including F in their designations in place of the oblique line offer superior formability for the same strength levels
43F35	HR, HS, –, CS	–	0.12	–	–	–	1.20	0.035	0.030	
46F40	HR, HS, –, CS	–	0.12	–	–	–	1.20	0.035	0.030	
50F45	HR, HS, –, CS	–	0.12	–	–	–	1.20	0.035	0.030	
60F55	–, HS, –, CS	–	0.12	–	–	–	1.20	0.035	0.030	
68F62	–, HS, –, –	–	0.12	–	–	–	1.50	0.035	0.030	
75F70	–, HS, –, –	–	0.12	–	–	–	1.50	0.035	0.030	
Narrow strip supplied in a range of conditions for heat treatment and general engineering purposes										
4	–, HS, –, CS	–	0.12	–	–	–	0.60	0.050	0.050	Low-carbon steel available hot rolled, annealed, skin passed or cold rolled to controlled hardness ranges H1 to H6 inclusive
10	–, HS, –, CS	0.08	0.15	0.10	0.35	0.60	0.90	0.045	0,045	For case hardening
12	–, HS, –, CS	0.10	0.15	–	–	0.40	0.60	0.050	0.050	A range of carbon steels available in the hot rolled or annealed condition
17	–, HS, –, CS	0.15	0.20	–	–	0.40	0.60	0.050	0.050	
20	–, HS, –, CS	0.15	0.25	0.05	0.35	1.30	1.70	0.045	0.045	
22	–, HS, –, CS	0.20	0.25	–	–	0.40	0.60	0.050	0.050	
30	–, HS, –, CS	0.25	0.35	0.05	0.35	0.50	0.90	0.045	0.045	
40	–, HS, –, CS	0.35	0.45	0.05	0.35	0.50	0.90	0.045	0.045	A range of carbon steels for use in the hot rolled, normalized, annealed and (except for grade 95) in the temper rolled (half hard) conditions. Grades 40 and 50 may be induction or flame hardened and grades 60, 70, 80, and 95 may be supplied in the hardened and tempered condition
50	–, HS, –, CS	0.45	0.55	0.05	0.35	0.50	0.90	0.045	0.045	
60	–, HS, –, CS	0.55	0.65	0.05	0.35	0.50	0.90	0.045	0.045	
70	–, HS, –, CS	0.65	0.75	0.05	0.35	0.50	0.90	0.045	0.045	
80	–, HS, –, CS	0.75	0.85	0.05	0.35	0.50	0.90	0.045	0.045	
95	–, HS, –, CS	0.90	1.00	0.05	0.35	0.30	0.60	0.040	0.040	

3.3.4 High-carbon steel

This can be hardened to give a very fine cutting edge, but with some loss of its ductile and malleable properties. It is used for metal and wood cutting tools, turning tools, taps and dies, and forging and press dies because of its hardness and toughness, but is seldom used now for motor vehicle parts because of the introduction of high-alloy steels.

3.3.5 Zinc-coated steels

The automotive industry, in seeking to provide extended warranties, is turning increasingly to the use of zinc-coated steels. Modern automobiles must be not only of high quality but also durable and economical, as perceived by their purchasers. These vehicles are expected to exceed an average of seven years without structural or cosmetic deterioration due to corrosion. Increasingly aggressive environmental influences tend to shorten the life of the car, whereas ever more specialized steel sheets are being incorporated into vehicle construction in the battle against corrosion.

The use of zinc-coated steels has dramatically increased to meet these challenges. Different areas of a vehicle require different zinc coatings and coating weights to meet appearance and performance criteria. These are available in both hot dipped (BS 2989: 1982) and electrolytically deposited (BS 6687: 1986) versions in a range of coating weights or thicknesses. Both types offer barrier and sacrificial corrosion protection, and the choice of product depends on the particular application and requirements. The hot dip product (available as plain zinc, or iron-zinc alloy) is generally used for underbody parts. The electrolytic product is used for exposed body panels, where a full-finish surface quality is available to ensure that a showroom paint finish is achieved. The electrolytic product is available in single-sided and double-sided coating (see Figure 3.1).

Single-sided zinc-coated steel Free zinc is applied to one side of a steel sheet by either the hot dip or the electrolytic process for this material. Its uncoated side provides a good surface for paint appearance, so it is used mainly for outer body panels. Since free zinc is towards the inside of the car, it protects against perforation corrosion.

One-and-a-half-sided zinc-coated steel In this case, one side of the sheet is coated with free zinc, and a thin layer of zinc-iron alloy is formed on the other side. This product is produced mainly by the hot dip process. It is primarily used for exposed panels, where the zinc-iron layer is on the outside for cosmetic protection and the free zinc side provides perforation protection.

Double-sided zinc-coated steel This product is manufactured by applying free zinc to both sides of the sheet with equal or differential coating weight.

All types are readily paintable and weldable. However, care should be taken to ensure that welding conditions are comparable with the material used: for example, higher weld current ratings may be necessary on the heavier coatings.

3.4 ALLOY STEELS

Alloy steel is a general name for steels that owe their distinctive properties to elements other than carbon. They are generally classified into two major categories:

Low-alloy steel possesses similar microstructures to and requires similar heat treatments to plain carbon steels (see Section 3.3.2 on micro-alloyed steel).

High-alloy steel may be defined as a steel having enhanced properties owing to the presence of one or more special elements or a larger proportion of element than is normally present in carbon steel.

This section is concerned primarily with high-alloy steels.

Alloy steels usually take the name of the element or elements, in varying percentages, having the characteristics the alloy.

Chromium	Increased hardness and resistance to corrosion.
Cobalt	Increased hardness, especially at high temperatures.
Manganese	High tensile strength, toughness and resistance to wear.
Molybdenum	Increased hardness and strength, at high temperatures.
Nickel	Increased tensile strength, toughness, hardness and resistance to fatigue.
Niobium	Strong carbide forming effect; increases tensile strength and improves ductility.
Silicon	Used as a deoxidizing agent, and has the slight effect of improving hardness.
Titanium	Strong carbide forming element.
Tungsten	Greater hardness, especially at high temperatures; improved tensile strength and resistance to wear.
Vanadium	Increased toughness and resistance to fatigue.

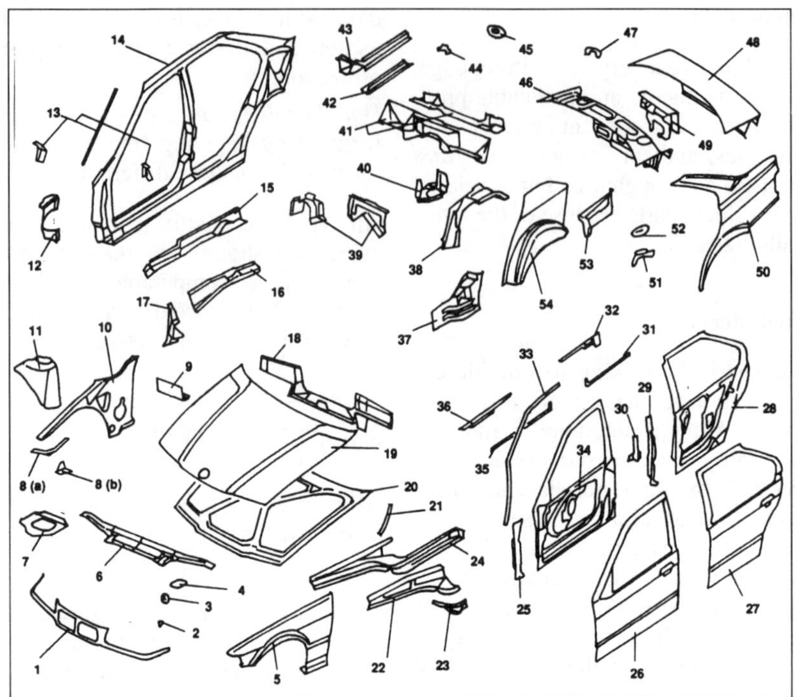

Figure 3.1 Body shell panels showing galvanized protection (*Motor Insurance Repair Research Centre*)

No.	Panel		No.	Panel	
1	Front grille panel	B	27	Rear door skin RH and LH	A
2	Reinforcement	B	28	Rear door frame RH and LH	B
3	Front bumper mounting reinforcement RH and LH	B	29	Rear door frame stiffener RH and LH	A
4	Bonnet lock reinforcement RH and LH	B	30	Rear door frame gusset RH and LH	B
5	Front wing RH and LH	B	31	Rear door skin stiffener RH and LH	A
6	Bonnet lock panel	B	32	Rear door lock stiffener RH and LH	A
7	Front suspension turret stiffener RH and LH	A	33	Front window frame stiffener RH and LH	A
8a/8b	Front wheel arch gussets RH and LH	B(a)/C(b)	34	Front door frame RH and LH	B
9	Front bulkhead stiffener RH and LH	A	35	Front door skin stiffener RH and LH	A
10	Front inner wing RH and LH	B	36	Front door lock stiffener RH and LH	A
11	Front suspension turret RH and LH	A	37	Sill rear gusset RH and LH	A
12	A-post reinforcement RH and LH	B	38	Rear chassis leg RH and LH	A
13	Body side gussets RH and LH	B	39	Floor/heelboard gussets	A
14	Body side RH and LH	A	40	Rear chassis leg gusset RH and LH	B
15	Inner sill RH and LH	A	41	Boot floor cross members	A
16	Inner sill reinforcement RH and LH	A	42	RH boot floor brace closing panel	A
17	B-post gusset RH and LH	A	43	Boot floor brace RH and LH	A
18	Upper dash panel	B	44	Boot floor bracket RH and LH	A
19	Bonnet skin	B	45	Rear bumper mounting gusset RH and LH	B
20	Bonnet frame	B	46	Boot lid frame	B
21	Dash stiffener RH and LH	B	47	Exhaust rear hanger stiffener	B
22	Front chassis leg closing panel RH and LH	A	48	Boot lid skin	A
23	Front chassis leg gusset RH and LH	B	49	Boot lid lock mounting stiffener	B
24	Font chassis leg RH and LH	A	50	Rear wing RH and LH	A
25	Front door frame stiffener RH and LH	A	51	Rear suspension turret RH and LH	A
26	Front door skin RH and LH	A	52	Suspension turret capping RH and LH	B
			53	Boot floor side extension RH and LH	A
			54	Outer rear wheel arch RH and LH	A

A = Galvanized one side only
B = Galvanized both sides
C = Galvanized layer applied individually

Correct heat treatment is essential to develop the properties provided by alloying elements.

There are many alloy steels containing different combinations and percentages of alloying elements, of which some of the most popular are as follows:

High-tensile steel Used whenever there is an essential need for an exceptionally strong and tough steel capable of withstanding high stresses. The main alloying metals used in its manufacture are nickel, chromium and molybdenum, and such steels are often referred to as nickel-chrome steels. The exact percentage of these metals used varies according to the hardening processes to be used and the properties desired. Such steels are used for gear shafts, engine parts and all other parts subject to high stress.

High-speed steels These are mostly used for cutting tools because they will withstand intense heat generated by friction and still retain their hardness at high temperatures. It has been found that by adding tungsten to carbon steel, an alloy steel is formed which will retain a hard cutting edge at high temperatures. High-speed steels are based on tungsten or molybdenum or both as the primary heat-resisting alloying element; chromium gives deep hardening and strength, and vanadium adds hardness and improves the cutting edge.

Manganese steel An addition of manganese to steel produces an alloy steel which is extremely tough and resistant to wear. It is used extensively in the manufacture of chains, couplings and hooks.

Chrome-vanadium steel This contains a small amount of vanadium which has the effect of intensifying the action of the chromium and the manganese in the steel. It also aids in the formation of carbides, hardening the alloy and increasing its ductility. These steels are valuable where a combination of strength and ductility is desired. They are often used for axle half-shafts, connecting rods, springs, torsion bars, and in some cases hand tools.

Silicon-manganese steel This is a spring steel using the two elements of manganese and silicon. These steels have a high strength and impact resistance and are used for road-springs and valve springs.

3.5 STAINLESS STEEL

The discovery of stainless steel was made in 1913 by Harry Brearley of Sheffield, while experimenting with alloy steels. Among the samples which he threw aside as unsuitable was one containing about 14 per cent chromium. Some months later he saw the pile of scrap test pieces and noticed that most of the steels had rusted but the chromium steel was still bright. This led to the development of stainless steels. The classic Rolls-Royce radiator was one of the first examples of the use of stainless steel.

The designer, engineer or fabricator of a particular component may think that stainless steel is going to be both difficult to work and expensive. This is quite wrong, and perhaps stems from the fact that many people tend to fall into the trap of the generic term 'stainless steels'. In fact, this is the title for a wide range of alloys. Therefore if such materials are to be used effectively and maximum advantage is to be taken of the many benefits they have to offer, there should be very close collaboration and consultation over which grade of stainless steel is best for the particular job in hand.

There are over 25 standard grades of stainless steel specified by BS 1449: Part 2. Each provides a particular combination of properties, some being designed for corrosion resistance, some for heat resistance and others for high-temperature creep resistance. Many, of course, are multi-purpose alloys and can be considered for more than one of these functions. In terms of composition, there is one element common to all the different grades of stainless steel. This is chromium, which is present to at least 10 per cent. It is this element which provides the basis of the resistance to corrosion by forming what is known as a 'passive film' on the surface of the metal. This film is thin, tenacious and invisible and is essentially a layer of chromium oxide formed by the chromium in the steel combining with the oxygen in the atmosphere. The strength of the passive film, in terms of resistance to corrosion, increases within limits with the chromium content and with the addition of other elements such as nitrogen and molybdenum. The formation of the passive film, therefore, is a natural characteristic of this family of steels and requires no artificial aid. Consequently, if stainless steels are scratched or cut or drilled, the passive film is automatically and instantaneously repaired by the oxygen in the atmosphere.

Stainless steels can be conveniently divided into the following four main groups:

Austenitic Generally containing 16.5–26 per cent chromium and 4–22 per cent nickel.

Ferritic Usually containing 12–18 per cent chromium.

Austenitic/ferritic duplex Usually containing 22 per cent chromium, 5.5 per cent nickel, 3 per cent molybdenum, and 0.15 per cent nitrogen.

Table 3.4 Typical stainless steels used in vehicle: BS 1449: Part 2

Steel grade	Typical alloying elements (%)	Characteristics	Typical applications
Austenitic			
301 S21	17Cr 7Ni	Good corrosion resistance	Riveted body panels, wheel covers, hubcaps, rocker panel mouldings
304 S16	18Cr 11Ni 0.06C	Good corrosion resistance	Mild corrosive tankers
305 S19	18Cr 11Ni 0.10C	Low work hardening rate	Rivets
316 S31	17Cr 11Ni 2.25Mo 0.07C	Highest corrosion resistance of the commercial grades	Road tankers for widest cargo flexibility
Ferritic			
409 S19	11Cr 0.08C + Ti	Reasonable combination of weldability, formability and corrosion resistance	Exhaust systems and catalytic converter components and freight container cladding
430 S17	17Cr	Good corrosion resistance. Can be drawn and formed	Tanker jackets, interior trim, body mouldings, windshield wiper arms
Martensitic			
410 S21	13Cr 0.12C	Corrosion resistant. Heat-treatable composition capable of high hardness	Titanium modified for silencer components

Martensitic Based on a chromium content of 11–14 per cent, although some grades may have a small amount of nickel.

Of the above groups, the austenitic steels are by far the most widely used because of the excellent combination of forming, welding and corrosion resisting properties that they offer. Providing that the correct grade is selected as appropriate to the service environment, and that the design and production engineering aspects are understood and intelligently applied, long lives with low maintenance costs can be achieved with these steels.

HyResist 22/5 duplex is a highly alloyed austenitic/ferritic stainless steel. It has more than twice the proof strength of normal austenitic stainless steels while providing improved resistance to stress corrosion cracking and to pitting attacks. It possesses good weldability and can be welded by conventional methods for stainless steel. The high joint integrity achievable combined with good strength and toughness permit fabrications to be made to a high standard. It is being increasingly used in offshore and energy applications.

Table 3.4 shows typical stainless steels used in motor vehicles.

3.6 ALUMINIUM

In the present-day search for greater economy in the running of motor vehicles, whether private, public or commercial, the tendency is for manufacturers to produce bodies which, while still maintaining their size and strength, are lighter in weight. Aluminium is approximately one-third of the weight of steel, and aluminium alloys can be produced which have an ultimate tensile strength of 340–620 MN/m^2. In the early 1920s the pioneers of aluminium construction were developing its use for both private and commercial bodies; indeed, the 1922 40 hp Lanchester limousine body had an aluminium alloy construction for the bulkhead and bottom frame, and aluminium was used for all the body panels. Before the Second World War aluminium was used mainly for body panels, but since the war aluminium alloys have been and are now being used for body structures. Although aluminium is more expensive than steel, it is easy to work and manipulate and cleaner to handle. It also has the advantage of not rusting, and, provided that the right treatment is adopted for welding, corrosion is almost non-existent. In recent years the use of aluminium and aluminium alloys for motor bodies, especially in the commercial field, has developed enormously.

In the modern motor body the saving of weight is its most important advantage, and although on average the panel thickness used is approximately double that of steel, a considerable weight saving can be achieved. One square metre of 1.6 mm thick aluminium weighs 4.63 kg while one square metre of 1.00 mm thick steel weighs 7.35 kg; the use of aluminium results in a saving in weight of just under 40 per cent.

The non-rusting qualities of the aluminium group are well known and are another reason for their use in bodywork. An extremely thin film of oxide forms on all surfaces exposed to the atmosphere, and even if this film is broken by a scratch or chip it will re-form, providing complete protection for the metal. The oxide film, which is only 0.0002 cm thick, is transparent, but certain impurities in the atmosphere will turn it to various shades of grey.

3.6.1 Production

The metal itself has only been known about 130 years, and the industrial history of aluminium did not begin until 1886 when Paul Heroult in France discovered the basis of the present-day method of producing aluminium. Aluminium is now produced in such quantities that in terms of volume it ranks second to steel among the industrial metals. Aluminium of commercial purity contains at least 99 per cent aluminium, while higher grades contain 99.5–99.8 per cent of pure metal.

In the production of aluminium, the ore bauxite is crushed and screened, then washed and pumped under pressure into tanks and filtered into rotating drums, which are then heated. This separates the aluminium oxide from the ore. In the next stage the aluminium oxide is reduced to metal aluminium by means of an electrolytic reduction cell. This cell uses powdered cryolite and a very heavy current of electricity to reduce the aluminium oxide to liquid metal, which passes to the bottom of the cell and is tapped off into pigs of aluminium of about 225 kg each.

3.6.2 Types of sheet

Sheet, strip and circle blanks are sold in hard and soft tempers possessing different degrees of ductility and tensile strength. Sheet is supplied in gauges down to 0.3 mm, but it is generally more economical to order strip for gauges less than 1.6 mm.

3.6.3 Manufacturing process

Sheet products are first cast by the semicontinuous casting process, then scalped to remove surface roughness and preheated in readiness for hot rolling. They are first reduced to the thickness of plate, and then to sheet if this is required. Hot rolling is followed by cold rolling, which imparts finish and temper in bringing the metal to the gauge required. Material is supplied in the annealed (soft) condition and in at least three degrees of hardness, H1, H2 and H3 (in ascending order of hardness).

3.7 ALUMINIUM ALLOYS

From the reduction centre the pigs of aluminium are remelted and cast into ingots of commercial purity. Aluminium alloys are made by adding specified amounts of alloying elements to molten aluminium. Some alloys, such as magnesium and zinc, can be added directly to the melt, but higher-melting-point elements such as copper and manganese have to be introduced in stages. Aluminium and aluminium alloys are produced for industry in two broad groups:

1 Materials suitable for casting.
2 Materials for the further mechanical production of plate sheet and strip, bars, tubes and extruded sections.

In addition both cast and wrought materials can be subdivided according to the method by which their mechanical properties are improved:

Non-heat-treatable alloys Wrought alloys, including pure aluminium, gain in strength by cold working such as rolling, pressing, beating and any similar type of process.

Heat-treatable alloys These are strengthened by controlled heating and cooling followed by ageing at either room temperature or at 100–200°C.

The most commonly used elements in aluminium alloys are copper, manganese, silicon, magnesium and zinc. Manufacturers can supply these materials in a variety of conditions. The non-heat-treatable alloys can be supplied either as fabricated (F), annealed (O) or strain hardened (H1, H2, H3). The heat-treatable alloys can be supplied as fabricated (F) or annealed (O), or, depending on the alloy, in variations of the heat treatment processes (T3, T4, T5, T6, T8).

3.7.1 Wrought light aluminium alloys: BS specifications 1470–75

Material designations
Unalloyed aluminium plate, sheet and strip:

1080A commercial pure aluminium 99.8 per cent
1050A commercial pure aluminium 99.5 per cent
1200 commercial pure aluminium 99.0 per cent

Non-heat-treatable aluminium alloy plate, sheet and strip:

3103 AlMn
3105 AlMnMg
5005 AlMg
5083 AlMgMn
5154A AlMg

5251 AlMg
5454 AlMgMn

Heat-treatable aluminium alloy plate, sheet and strip:

2014A	AlCuSiMg
Clad 2014A	AlCuSiMg clad with pure aluminium
2024	AlCuMg
Clad 2024	AlCuMg clad with pure aluminium
6082	AlSiMgMn

Abbreviations for basic temper

As fabricated: F The temper designation F applies to the products of shaping processes in which no special control over thermal conditions or strain hardening is employed. For wrought products there are no specified requirements for mechanical properties.

Annealed: O The temper designation O applies to wrought products which are annealed to obtain the lowest strength condition.

Abbreviations for strain hardened materials

The temper designation H for strain hardened products (wrought products only) applies to products subjected to the application of cold work after annealing (hot forming) and partial annealing or stabilizing, in order to achieve the specified mechanical properties. The H is always followed by two or more digits, indicating the final degree of strain hardening. The first digit (1, 2 or 3) indicates the following:

H1 strain hardened only
H2 strain hardened and partially annealed
H3 strain hardened and stabilized

The second digit (2, 4, 6 or 8) indicates the degree of strain hardening, as follows:

HX2 tensile strength approximately midway between O temper and HX4 temper

HX4 tensile strength approximately midway between O temper and HX8 temper

HX6 tensile strength approximately midway between HX4 temper and HX8 temper

HX8 full hard temper

Abbreviations for heat-treated materials

The temper designation T applies to products which are thermally treated, with or without supplementary strain hardening, to produce stable tempers. The T is always followed by one or more digits, indicating the specific sequence of treatments as follows:

T3 solution heat treated, cold worked, and naturally aged to a substantially stable condition

T4 solution heat treated and naturally aged to a substantially stable condition

T5 cooled from an elevated temperature shaping process and then artificially aged

T6 solution heat treated and then artificially aged

T8 solution heat treated, cold worked and then artificially aged

The additional digits for the T tempers TX51 indicate that the products have been stress relieved by controlled stretching.

Tables 3.5, 3.6, 3.7 and 3.8 show the characteristics and properties of aluminium alloys.

3.7.2 Aluminium alloys used in bodywork

The choice of material and the condition in which it is required must depend largely upon design requirements and the manufacturing processes within the factory. The alloys most commonly used in vehicle bodywork are as shown in Table 3.9.

Alloy 5154A is suitable for use in car panels which are to be pressed into shape; it is supplied in either annealed condition or H2 condition, which are the most suitable for press work on vehicle bodies. Of the other materials, 1200 is a commercial purity sheet, and is widely used for exterior and interior panelling where no great strength is required. Types 3103, 5251, 5154A and 5056A are non-heat-treatable alloys of the aluminium-magnesium range with a strength of 90–325 N/mm^2. They come in sheet form, and provide a range of mechanical properties to suit different applications. They are used extensively in panel work, and also for forming, pressing and machining, and can be welded without much difficulty. The plate material 5083 is a medium-strength non-heat-treatable alloy particularly suitable for welding. It can be used for parts carrying fairly high stress loads and is often used in the form of patterned tread plate for floor sections.

For internal structure members which need to be stronger than the outer panels, the heat-treatable alloys usually used are 6063 and 6082, and in odd cases 2014A. Type 6082 is a heat-treatable medium-strength alloy which combines good mechanical properties with high corrosion resistance. Permissible stresses in this alloy can be as high as 200 N/mm^2 under static loading conditions, although some reduction below this would normally be made for transport applications where there is a considerable element of dynamic loading. The alloy is weldable by the inert gas arc process, but there is a considerable loss of strength near the weld owing to the annealing effect of the

Table 3.5 General characteristics of all wrought forms of aluminium alloys

		General characteristics						
Purity or alloy	Temper or condition	Cold forming	Machining	Durability	Inert-gas shielded arc (MIG or TIG)	Welding resistance (spot, seam, flash butt or stub)[b,c]	Oxy-acetylene[d]	Metal arc
5251	F	V	G	V	V[e]	E	G[e]	F[e]
	O	V	G					
	H22	G	G					
	H24	G	V					
	H28	G	V					
	H39X	F	V	F	V[e]	E	G[e]	G[e]
5454	F	V	V	V	V[e]	E	F[e]	N
	O	V		V				
	H22	G		G				
	H24	G		G				
5154	F	V	V	V[f]	E	E	F	N
	O	V		V[f]				
	H22	G		G[f]				
	H24	F		G[f]				
	H28							
5083	F	G	E	V[f]	E	E	F	N
	O	G		V[f]				
	H22	F		F[f]				
	H24	G		F[f]				
6082	F	V	G	G	V[e]	V	F[e]	G[e]
	O	E	G	G				
	T4	G	V	V				
	T6	F	E	G				
	T451							
	T651							
1200	F	E	F	V	E	V	V	G
	O	E	F					
	H12	V	F					
	H14	V	F					
	H16	G	G					
	H18	F	G					
3103	F	E	F	V	E	E	V	G
	O	E	F					
	H12	V	F					
	H14	V	F					
	H16	G	G					
	H18	F	G					

(a) Materials are graded thus: E excellent; V very good; F fair; P poor; N not recommended.
(b) The mechanical properties of work hardened or heat-treated materials will be reduced in the vicinity of the weld.
(c) The weld zone is generally discernible after anodic treatment, the degree depending on the material and welding process.
(d) The oxy-acetylene process is normally recommended only for material thinner than 6–4 mm.
(e) Filler or electrode of other than parent metal composition is recommended.
(f) Applicable at temperature of 70°C and less.

Table 3.6 Chemical composition limits[1] and mechanical properties of unalloyed aluminium plate, sheet and strip (BS 1470)

												Others[2]				Thickness		Tensile strength		Elongation on 50 mm: Materials thicker than					Elongation on 5.65 $\sqrt{S_o}$ over 12.5 mm thick
Material designation	Tolerance category	Silicon (%)	Iron (%)	Copper (%)	Manganese (%)	Magnesium (%)	Chromium (%)	Nickel (%)	Zinc (%)	Gallium (%)	Titanium (%)	Each (%)	Total (%)	Aluminium min. (%)	Temper[4]	> (mm)	≤ (mm)	Min. (N/mm^2)	Max. (N/mm^2)	0.5 mm min. (%)	0.8 mm min. (%)	1.3 mm min. (%)	2.6 mm min. (%)	3.0 mm min. (%)	(min.) (%)
1080A	A	0.15	0.15	0.03	0.02	0.02	–	–	0.06	0.03	0.02	0.02	–	99.80[3]	F	3.0	25.0	–	–	–	–	–	–	–	–
															O	0.2	6.0	–	90	29	29	29	35	35	–
															H14	0.2	12.5	90	125	5	6	7	8	8	–
															H18	0.2	3.0	125	–	3	4	4	5	–	–
1050A	A	0.25	0.40	0.05	0.05	0.05	–	–	0.07	–	0.05	0.03	–	99.50[3]	F	3.0	25.0	–	–	–	–	–	–	–	–
															O	0.2	6.0	55	95	22	25	30	32	32	–
															H12	0.2	6.0	80	115	4	6	8	9	9	–
															H14	0.2	12.5	100	135	4	5	6	6	8	–
															H18	0.2	3.0	135	–	3	3	4	4	–	–
1200	A	1.0 Si + Fe		0.05	0.05	–	–	–	0.10	–	0.05	0.05	0.15	99.00[3]	F	3.0	25.0	–	–	–	–	–	–	–	–
															O	0.2	6.0	70	105	20	25	30	30	30	–
															H12	0.2	6.0	90	125	4	6	8	9	9	–
															H14	0.2	12.5	105	140	3	4	5	5	6	–
															H16	0.2	6.0	125	160	2	3	4	4	4	–
															H18	0.2	3.0	140	–	2	3	4	4	–	–

1 Composition in per cent (m/m) maximum unless shown as a range or a minimum.

2 Analysis is regularly made only for the elements for which specific limits are shown. If, however, the presence of other elements is suspected to be, or in the case of routine analysis is indicated to be, in excess of the specified limits, further analysis should be made to determine that these other elements are not in excess of the amount specified.

3 The aluminium content for unalloyed aluminium not made by a refining process is the difference between 100.00% and the sum of all other metallic elements in amounts of 0.010% or more each, expressed to the second decimal before determining the sum.

4 An alternative method of production, designated H2, may be used instead of the H1 routes, subject to agreement between supplier and purchaser and providing that the same specified properties are achieved.

Table 3.7 Chemical composition limits[1] and mechanical properties of aluminium alloy plate, sheet and strip (non-heat-treatable) (BS 1470)

Material designation	Tolerance category	Silicon (%)	Iron (%)	Copper (%)	Manganese (%)	Magnesium (%)	Chromium (%)	Nickel (%)	Zinc (%)	Other restrictions (%)	Titanium (%)	Others[2] Each (%)	Others[2] Total (%)	Aluminium (%)	Temper[3]	Thickness > (mm)	Thickness ≤ (mm)	0.2% proof stress min. (N/mm^2)	Tensile strength Min. (N/mm^2)	Tensile strength Max. (N/mm^2)	Elongation on 50 mm, Materials thicker than 0.5 mm min. (%)	0.8 mm min. (%)	1.3 mm min. (%)	2.6 mm min. (%)	3.0 mm min. (%)	Elongation on 5.65 $\sqrt{S_o}$ over 12.5 mm thick (min.)
3103	A	0.50	0.7	0.10	0.9–1.5	0.30	0.10	–	0.20	0.10 Zr + Ti	–	0.05	0.15	Rem.*	F	0.2	25.0	–	–	–	–	–	–	–	–	–
															O	0.2	6.0	–	90	130	20	23	24	24	25	–
															H12	0.2	6.0	–	120	155	5	6	7	9	9	–
															H14	0.2	12.5	–	140	175	3	4	5	6	7	–
															H16	0.2	6.0	–	160	195	2	3	4	4	4	–
															H18	0.2	3.0	–	175	–	2	3	4	4	–	–
3105	A	0.6	0.7	0.30	0.30–0.8	0.20–0.8	0.20	–	0.40	–	0.10	0.05	0.15	Rem.	O	0.2	3.0	–	110	155	16	18	20	20	–	–
															H12	0.2	3.0	115	130	175	2	3	4	5	–	–
															H14	0.2	3.0	145	160	205	2	2	3	4	–	–
															H16	0.2	3.0	170	185	230	1	1	2	3	–	–
															H18	0.2	3.0	190	215	–	1	1	1	2	–	–
5005	A	0.30	0.7	0.20	0.20	0.50–1.1	0.10	–	0.25	–	–	0.05	0.15	Rem.	O	0.2	3.0	–	95	145	18	20	21	22	–	–
															H12	0.2	3.0	80	125	170	4	5	6	8	–	–
															H14	0.2	3.0	100	145	185	3	3	5	6	–	–
															H18	0.2	3.0	165	185	–	1	2	3	3	–	–
5083	B	0.40	0.40	0.10	0.40–1.0	4.0–4.9	0.05–0.25	–	0.25	–	0.15	0.05	0.15	Rem.	F	3.0	25.0	–	–	–	–	–	–	–	–	–
															O	0.2	80.0	125	275	350	12	14	16	16	16	14
															H22	0.2	6.0	235	310	375	5	6	8	10	8	–
															H24	0.2	6.0	270	345	405	4	5	6	8	6	–
5154A	B	0.50	0.50	1.10	0.50	3.1–3.9	0.25	–	0.20	0.10–0.50 Mn + Cr	0.20	0.05	0.15	Rem.	O	0.2	6.0	85	215	275	12	14	16	18	18	–
															H22	0.2	6.0	165	245	295	5	6	7	8	8	–
															H24	0.2	6.0	225	275	325	4	4	4	6	6	–
5251	A	0.40	0.50	0.15	0.10–0.50	1.7–2.4	0.15	–	0.15	–	0.15	0.05	0.15	Rem.	F 3.0	25.0	–	–	–	–	–	–	–	–	–	
															O	0.2	6.0	60	160	200	18	18	18	20	20	–
															H22	0.2	6.0	130	200	240	4	5	6	8	8	–
															H24	0.2	6.0	175	225	275	3	4	5	5	5	–
															H28	0.2	3.0	215	255	285	2	3	3	4	4	–
5454	B	0.25	0.40	0.10	0.50–1.0	2.4–3.0	0.05–0.20	–	0.25	–	0.20	0.05	0.15	Rem.	F	3.0	25.0	–	–	–	–	–	–	–	–	–
															O	0.2	6.0	80	215	285	12	14	16	18	18	–
															H22	0.2	3.0	180	250	305	4	5	7	8	–	–
															H24	0.2	3.0	200	270	325	3	4	5	6	–	–

* Remainder

1 Composition in per cent (m/m) maximum unless shown as a range or a minimum.

2 Analysis is regularly made only for the elements for which specific limits are shown. If, however, the presence of other elements is suspected to be, or in the case of routine analysis is indicated to be, in excess of the specified limits, further analysis should be made to determine that these other elements are not in excess of the amount specified.

3 An alternative method of production, designated H2, may be used instead of the H1 routes, subject to agreement between supplier and purchaser and providing that the same specified properties are achieved.

Table 3.8 Chemical composition limits and mechanical properties of aluminium alloy plate, sheet and strip (heat-treatable) (BS 1470)

Material designation	Tolerance category	Silicon (%)	Iron (%)	Copper (%)	Manganese (%)	Magnesium (%)	Chromium (%)	Nickel (%)	Zinc (%)	Other Restrictions (%)	Titanium (%)	Others[2]		Aluminium (%)	Temper	Thickness		0.2% proof stress min. (N/mm^2)	Tensile strength		Elongation on 50 mm: Materials thicker than					Elongation on 5.65 $\sqrt{S_o}$ over 12.5 mm thick (min.) (%)
												Each (%)	Total (%)			> (mm)	≤ (mm)		Min. (N/mm^2)	Max. (N/mm^2)	0.5 mm min. (%)	0.8 mm min. (%)	1.3 mm min. (%)	2.6 mm min. (%)	3.0 mm min. (%)	
2014A	B	0.50–0.9	0.50	3.9–5.0	0.40–1.2	0.20–0.8	0.10	0.10	0.25	0.20 Zr + Ti	0.15	0.05	0.15	Rem.*	O	0.2	6.0	110	–	235	14	14	16	16	16	–
															T4	0.2	6.0	225	400	–	13	14	14	14	14	–
															T6	0.2	6.0	380	440	–	6	6	7	7	8	–
															T451	6.0	25.0	250	400	–	–	–	–	–	14	12
																25.0	40.0	250	400	–	–	–	–	–	–	10
																40.0	80.0	250	395	–	–	–	–	–	–	7
Clad 2014A	B	0.50–0.9	0.50	3.9–5.0	0.40–1.2	0.20–0.8	0.10	0.10	0.25	0.20 Zr + Ti	0.15	0.05	0.15	Rem.	T651	6.0	25.0	410	460	–	–	–	–	–	–	6
																25.0	40.0	400	450	–	–	–	–	–	–	5
																40.0	60.0	390	430	–	–	–	–	–	–	5
																60.0	90.0	390	430	–	–	–	–	–	–	4
																90.0	115.0	370	420	–	–	–	–	–	–	4
																115.0	140.0	350	410	–	–	–	–	–	–	4
2024	B	0.50	0.50	3.8–4.9	0.30–0.9	1.2–1.8	0.10	–	0.25	–	0.15	0.05	0.15	Rem.	O	0.2	6.0	100	–	220	14	14	16	16	16	–
															T4	0.2	1.6	240	385	–	13	14	14	–	–	–
																1.6	6.0	245	395	–	–	–	–	14	14	–
															T6	0.2	1.6	345	420	–	7	7	8	9	–	–
																1.6	6.0	355	420	–	–	–	–	9	9	–
															O	0.2	6.0	110	–	235	12	12	14	14	14	–
															T3	0.2	1.6	290	440	–	11	11	11	–	–	–
																1.6	6.0	290	440	–	–	–	–	12	12	–
															T351	6.0	25.0	280	430	–	–	–	–	–	10	10
																25.0	40.0	280	420	–	–	–	–	–	–	9
																40.0	60.0	270	410	–	–	–	–	–	–	9
																60.0	90.0	270	410	–	–	–	–	–	–	8
																90.0	115.0	270	400	–	–	–	–	–	–	8
																115.0	140.0	260	390	–	–	–	–	–	–	7

* Remainder

Table 3.9 Standard aluminium alloys: availability, physical properties, and applications

Purity or alloy (new nomenclature)	Related BS/GE specification (BS 1470–1475) (old BS alloy designation)	Nominal composition: % alloying elements (remainder aluminium and normal impurities) Mg	Si	Mn	Standard forms[a] Sheet	Plate	Extrusions	Hollow extrusions	Tube	Physical properties Density g/cm³	lb/in³	Melting range °C (approx.)	Coefficient of linear expansion per °C (20–100°C)	Typical road transport applications
Non-heat treatable alloys														
1200	IC	–	–	–	×	×	×	×	×	2.71	0.098	660	0.000 023 5	Vehicle panelling where panel beating is required; mouldings and trim. Tank cladding
3103	N3	–	–	1.2	×	×	×[b]	×[b]	×[b]	2.73	0.099	645–655	0.000 023 5	Flat panelling of vehicles and general sheet metal work. Tank cladding.
5251	N4	2.25	–	0.4	×	×	×	–	×	2.69	0.097	595–650	0.000 024	Body panelling. Head boards and drop sides. Truss panels in buses. Cab panelling. Tanker shells and divisions
5154A	N5	3.5	–	0.4	×	–	×	–	×	2.67	0.096	600–640	0.000 024 5	Welded body construction. Tipper body panelling. Truss panels in buses.
5454	N51	2.7	–	0.8										Tanker shells and divisions
5083	N8	4.5	–	0.5	×	×	×	–	×	2.65	0.096	580–635	0.000 024 5	Pressurized bulk transport at ambient or low temperature
Heat-treatable alloys														
6082	H30	0.7	1.0	0.5	×	×	×	×	×	2.70	0.098	570–660	0.000 023 5–0.000 024[c]	All structural sections in riveted vehicle bodies. Tipper body panelling. Highly stressed underframe gussets, truss panels

(a) Other forms may be available by special arrangement.
(b) Not covered by British Standard (general engineering) specification.
(c) Depending on condition.

welding process. Type 6063 is also a heat-treatable alloy but of somewhat lower strength, and is used mainly in applications requiring good surface finish or where the parts are required to be anodized. Alloy 2014A contains a greater percentage of copper than the others, is more expensive, is more difficult to form and is less resistant to corrosion, but has the advantage of a greater tensile strength.

Fastenings and solid rivets can be of commercial purity material or of aluminium alloy 5154A, and for smaller sizes 6082 is sometimes used. Rivets are also available in 5056A material, but should not be used in cases where high temperatures occur in service. Bolts used in bodywork are normally of the 6082 alloy.

The condition in which heat-treatable alloys are supplied should be related to their application or use in bodywork. For example, if a section is to remain straight and is part of a framework which is to be bolted, riveted or welded in place, it is obvious that the material used should already be fully heat treated so that maximum strength is provided to support the framework or structure of the body. On the other hand, if the section has to be shaped, bent or formed in any way the material should be used in the annealed condition and then heat treated after the shaping operations have all been carried out.

Aluminium alloys are now being accepted by the automobile manufacturers as a standard material for exterior and interior trim, and are used for all normal bright trim applications such as radiator grilles, headlamp bezels, wheel trim, instrument panels, body mouldings and window and windscreen surrounds. Alloys used for trimming can be divided into two groups: high-purity alloys bright finished on one side only, in which the majority of the trim components are made; and super-purity alloys for use when maximum specular reflectivity is an advantage, such as would be required by light units.

3.8 RUBBER

The value of rubber lies in the fact that it can be readily moulded or extruded to any desired shape, and its elastic quality makes it capable of filling unavoidable and irregular gaps and clearances. It is an ideal material in door shuts and as the gasket for window glass, and in both instances it provides the means for excluding dust and water, although with windscreens and backlights additional use has generally to be made of a sealing material. Rubber specifications have been built on the basis of the properties of material which has given satisfaction. One major difficulty has been to ensure and measure resistance to weathering; rubber is subject to oxidation by ozone in the atmosphere, and this results in cracking. In addition to natural rubber, a variety of types of synthetic rubbers are used by the motor industry; these vary in price and characteristics, and all are more expensive than natural rubber. For complete ozone resistance it is necessary to use either butyl or neoprene rubber: both satisfy atmospheric and ozone ageing tests. Butyl rubber, however, is 'dead' to handle and contains no wax, and so while neoprene is costly its use is essential for some parts.

Sponge sealing rubbers can be provided with built-in ozone resistance by giving them a live skin of neoprene, and a further way of providing ozone resistance is to coat the rubber components with Hypalon; more recently, continuously extruded neoprene sponge has been adopted. Apart from weather resistance, the important requirement of door and boot lid seals is that their compression characteristics should ensure that they are capable of accommodating wide variations in clearance, without giving undue resistance to door closing.

Various types of foam rubber have been evolved to suit the different parts of the car seating, and the designer's choice of material is governed by cost, comfort, durability, the type of base, the type of car, and whether it is a cushion or a squab, a rear or front seat. When considering the foams available, it is apparent that the number of permutations is large. The types of foam available today fall into seven broad categories. These are:

Moulded latex foam
Low-grade fabricated polyether
Fabricated polyether
Moulded polyether
Fabricated polyester
Polyvinyl chloride foam
Reconstituted polyether

Latex foam today utilizes a mixture of natural and synthetic latexes to obtain the best qualities of both. After being stabilized with ammonia, natural latex is shipped in liquid form to this country from Malaysia, Indonesia and other rubber producing countries. Synthetic latex, styrene butadiene rubber, is made as a byproduct of the oil cracker plants. Polyether foam can now be made in different grades, and the physical properties of the best grades approach those of latex foam. As a general guide, service life and physical properties improve as the density increases for any given hardness. As the cost is proportional to weight, it follows that the higher-performance foams are more costly.

Flexible polyurethane foam seats are replacing heavy and complicated padded metal spring structures. Moulded seats simplify assembly, reduce weight and give good long-term performance. A major innovation here has been the cold cure systems. These produce foams of superb quality, particularly in terms of strength, comfort and long-term ageing. The systems are particularly suitable for the newer seat technologies such as dual hardness, where the wings are firm to give lateral support, leaving the seat pad softer and more comfortable.

3.9 SEALERS

The history of sealers is longer than that of the motor car. Mastic, bitumen compounds and putties of various kinds have been used since the invention of the horseless carriage. It is likely that early coach-builders used putties of some kind – possibly paint fillers – to bridge joints in various applications on motor bodies, but it is generally conceded that the first use of specialized sealers on a large scale was in the early 1920s when, in America, the pressed steel body became popular. In this country it was 1927 before one of the first truly effective sealers was introduced. It was known as Dum Dum and is still in use. It was a modified roof sealer, and proved to have many applications in body production. It was not until the late 1930s, when all-steel bodies and unitary construction became a common feature of mass-produced cars, that more thought was given to the points that required the use of sealing compounds, and to the nature of these products. Among the first developed was interweld sealing compound primarily to prevent corrosion. Since then, particularly in the post-war years, there have been remarkable developments, probably accelerated by criticisms from overseas markets that British cars were susceptible to dust and water entry. Companies specializing in the manufacture of mastic compounds have developed a range of materials which are now used not only for welding and for general putty application but also for floor pans, drip rails, body joints, exterior trim and many other points, leading to well-sealed car bodies equal to any produced elsewhere in the world (Figure 3.2, Table 3.10).

The term 'sealer' covers a wide variety of materials used in the motor industry for sealing against water and dust, from products which remain virtually mastic throughout their life to others which harden up but still retain some measure of elasticity. They range from mixtures of inert fillers and semi-drying oils to heat curing plastisols which may be applied in a thin paste form as an interweld sealer or as extruded beads. Sealing compounds can be categorized into the following general groups: oil-based compositions, rubber-based compositions and synthetic-resin-based compositions. The choice of each of the types will be dependent on the site for application, on the eventual conditions of exposure and often on price. These categories can be subdivided further into the various physical forms in which they can be made available, which include mastic putties for hand application, extruded sections for placing in precise locations, gun grade compositions which have the advantage of speed and economy of application, and pouring and spraying grades.

The properties of sealers will obviously vary according to their type and to their application. Thus preformed strip or putty sealers must adhere to the surfaces to which they are applied, and must not harden or crumble in service. Glazing sealers must be capable of being readily applied from a gun, with the ability to harden off on the surface, but must remain mastic in the assembly so that they are capable of maintaining a leaktight joint whatever deflection the body undergoes. Heat gelling sealers must be capable of being readily applied by extrusion or possibly by spraying, and then must set up when cured but still retain a degree of flexibility.

As a result of soaring energy costs together with the need for car aerodynamic design, direct glazing of windscreens and fixed body glass was introduced and an adhesive was required to bond glass windscreens to the metal aperture. The material used is polyurethane adhesive sealant. It possesses a combination of adhesion, sealant and gap filling qualities; it is a one-component adhesive and sealing compound of permanent elasticity. This dual-purpose material is based on a special moisture cured polyurethane with an accelerated setting time. The curing time is dependent on the humidity levels prevailing, as well as the temperature. For example, at 20°C with a relative humidity level of 65 per cent, a 6 mm diameter bead will be tack free within 1 hour and fully cured in 24 hours.

Table 3.11 indicates the uses of various sealant materials.

3.10 SOUND DEADENING, THERMAL INSULATING AND UNDERSEALING MATERIALS

The type of material used for sound damping or deadening depends on whether or not it is also required to provide undersealing. A material required for sound deadening only will normally be applied to the interior of a vehicle, whereas one required to provide

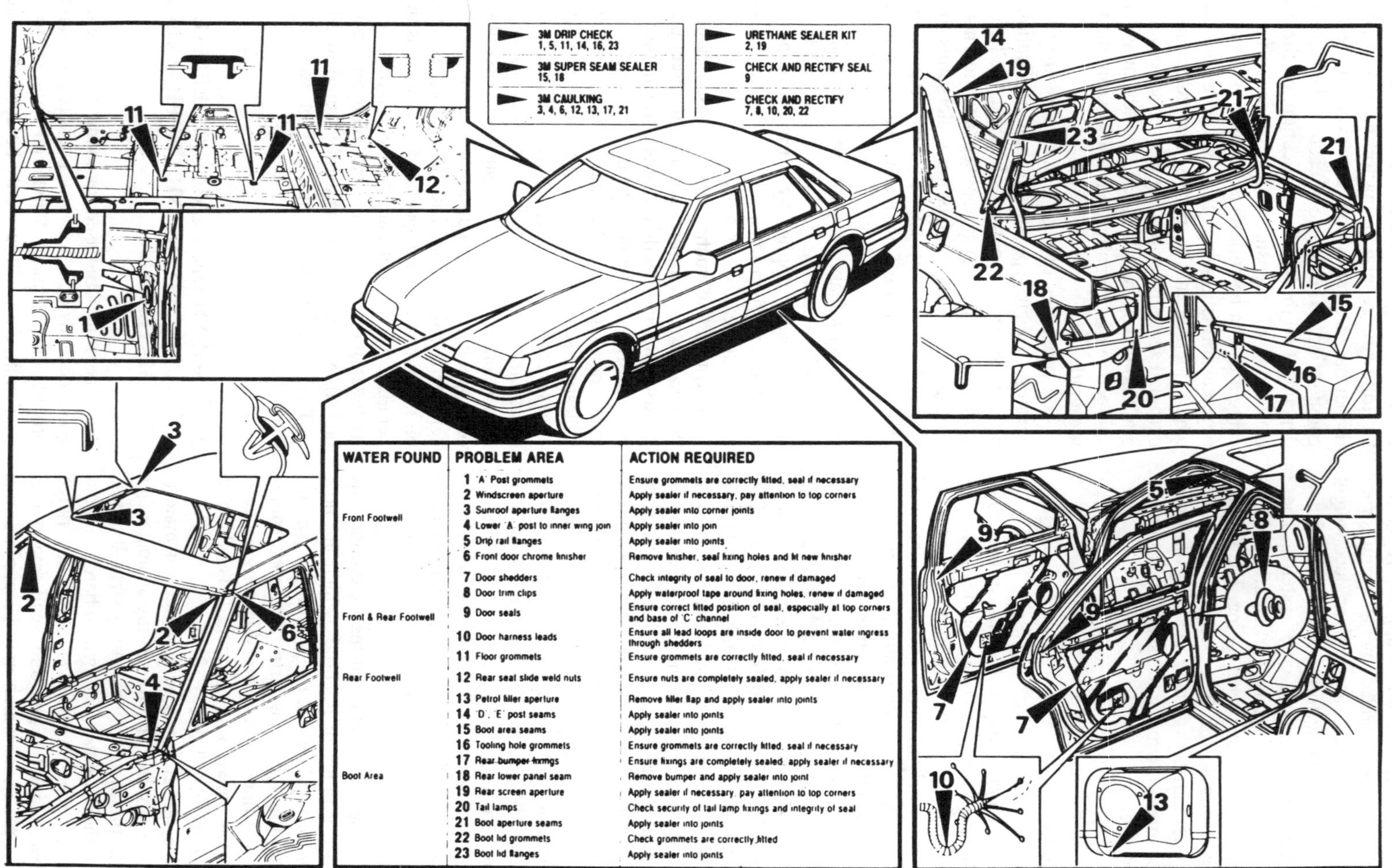

WATER FOUND	PROBLEM AREA	ACTION REQUIRED
Front Footwell	1 'A' Post grommets	Ensure grommets are correctly fitted, seal if necessary
	2 Windscreen aperture	Apply sealer if necessary, pay attention to top corners
	3 Sunroof aperture flanges	Apply sealer into corner joints
	4 Lower 'A' post to inner wing join	Apply sealer into join
	5 Drip rail flanges	Apply sealer into joints
	6 Front door chrome finisher	Remove finisher, seal fixing holes and fit new finisher
Front & Rear Footwell	7 Door shedders	Check integrity of seal to door, renew if damaged
	8 Door trim clips	Apply waterproof tape around fixing holes, renew if damaged
	9 Door seals	Ensure correct fitted position of seal, especially at top corners and base of 'C' channel
	10 Door harness leads	Ensure all lead loops are inside door to prevent water ingress through shedders
	11 Floor grommets	Ensure grommets are correctly fitted, seal if necessary
Rear Footwell	12 Rear seat slide weld nuts	Ensure nuts are completely sealed, apply sealer if necessary
Boot Area	13 Petrol filler aperture	Remove filler flap and apply sealer into joints
	14 'D', 'E' post seams	Apply sealer into joints
	15 Boot area seams	Apply sealer into joints
	16 Tooling hole grommets	Ensure grommets are correctly fitted, seal if necessary
	17 Rear bumper fixings	Ensure fixings are completely sealed, apply sealer if necessary
	18 Rear lower panel seam	Remove bumper and apply sealer into joint
	19 Rear screen aperture	Apply sealer if necessary, pay attention to top corners
	20 Tail lamps	Check security of tail lamp fixings and integrity of seal
	21 Boot aperture seams	Apply sealer into joints
	22 Boot lid grommets	Check grommets are correctly fitted
	23 Boot lid flanges	Apply sealer into joints

Figure 3.2 Problem areas requiring body sealing (*Rover Group Ltd*)

Table 3.10 Types of sealed joints used in vehicle bodywork (*Rover Group Ltd*)

Application		Material description	Application equipment
Bolted joints	Between panels	Preformed strip	Hand or palette/putty knife
		Zinc-rich primer	Brush or spray
	Panel edges	Seam sealer light	Applicator gun (hand)
Spot-welded joints	Between panels	Zinc-rich primer	Brush or spray
		Structural adhesive or seam sealer	
	Panel edges	Seam sealer light	Applicator gun (hand)
Bonded joints	Between panels	Metal-to-metal adhesive semi-structural	Caulking gun

sound deadening and undersealing properties will be applied to the underside of the vehicle. Thus the former need not be fully water resistant, whereas the latter must be water resistant in addition to many other necessary requirements.

The sound deadening properties of a material are related to its ability to damp out panel vibrations, and this in turn is related to some extent (but not solely) to its weight per unit volume. Thus the cheapest sound deadening materials are based on mixtures of sand and bitumen, although these tend to be brittle. A better material is bitumen filled with asbestos; although this is probably less effective as a panel damper than sand-filled bitumen, it is nevertheless more suitable owing to its better ductility. In general, those sound deadeners applied to the interior of the vehicle are water-based bitumen emulsions with fillers, whereas sound deadener/sealers applied externally should be solvent-based materials. A more effective sound deadener than asbestos-filled bitumen, now being phased out is a clay-filled water dispersed polyvinyl acetate (PVA) resin emulsion; this has damping characteristics approximately three times better than the bitumen-based material, but naturally it is more expensive.

Other sound absorbing materials are now used for insulation in the automotive industry. Needle felts are blends of natural and manmade fibres locked together by needle punching. These are used in die-cut flat sheet forms for attachment to moulded carpets, floor boot mats and as anti-rattle pads. Bonded, fully cured felts are similar blends of fibres bonded together with synthetic resins. All the binder is cured during the felt making process. As with needled materials, these are mainly used for flat products, especially where low density is required such as in sound absorption pads and floor mats. In moulded felts and moulded glass wool the binder is only partially cured during the felt making process. The curing sequence is completed under the action of temperature and pressure in matched die compression moulded tools to produce components which have three dimensional form and a controllable degree of rigidity. Fully cured and needled products can be given various surface treatments including abrasion-resistant and waterproof coatings such as latex, PVC or rubber, or they can be combined with bitumen to improve the sound insulation properties. Moulded felt can be supplied covered with a range of woven textile covers or with various grades of PVC and heavy-layer bitumen EVA products.

Polyurethane flexible moulded foam can be modified or filled to meet different insulating requirements in the vehicle. The foam can also be moulded directly on to the hard layer, allowing simple tailoring of insulation thickness.

Table 3.12 indicates the use of various undersealing materials.

Table 3.11 Sealers used in vehicle body repair work

Type	Base material	Application
Visible seams	Polyurethane	Extremely adherent sealant used on front and rear aprons, rear panel, engine compartment, bottom of boot, passenger compartment, side panels, wheel arches, vehicle underbody, tank filler caps and wings. Can be painted over with primer and fillers after curing, is non-shrinking, can be brushed and smoothed with a spatula. Cures by means of air moisture
	Synthetic rubber	Particularly suitable for all automotive problem areas where cleaning is difficult. Extremely adherent to raw, degreased, bonded, primed and painted sheet steel. Can be painted over with lacquers after thoroughly drying. Following application, can be passed through drying ovens at a maximum temperature of 90°C
	Acrylic dispersion (water based)	Particularly suitable for sealing joints, welded seams and butt joints on vehicle bodies. Substrate must be primed and can be readily painted over. When cured, is resistant to water
Structural seams	MS polymer	Applied by means of an air pressure pistol. All structural seams sprayed by the manufacturer can be re-created with this sealant, so that the original finish can be restored after repair. Also can be painted over immediately wet-on-wet.
	Nitrile-butadiene rubber	Special brushable sealant used for front and rear aprons, bottom of boot, inside floor, side walls, wheel arches. Has excellent adhesion to raw, primed and painted sheet metal, and can be painted over after drying
Underbody-seams	Bitumen rubber	Specially used for the underbody area of the vehicle. Resistant to water, salt spray, alcohol and dilute sulphuric acid
Overlapping joints	Synthetic rubber	Used for sealing bolted wings, headlight units, rear light housings and cable inlet holes
Sealing tape	Synthetic rubber	Suitable for all overlapping and screwable joints on vehicle bodies, metal to metal, metal to wood, metal to plastic, wood to wood or plastic to plastic
Rubber profiled windows	Synthetic resin, synthetic rubber	Particularly suitable for sealing rubber profiled front, rear and side window units between rubber and glass or rubber and the vehicle body

3.11 INTERIOR FURNISHINGS

3.11.1 Carpets and floor coverings

The body engineer has a choice of materials, ranging from carpeting of the Wilton type for prestige vehicles, through polyamide and polypropylene moulded needle felts, to rubber flooring for economy versions. The main requirements of flooring are wear resistance, colour fastness to light and water, and adequate strength to enable the customer to remove the flooring from the car without it suffering damage. The method of manufacture of pile carpets varies: in some cases the pile is bonded to hessian backing; in other cases it is woven simultaneously into the backing and then anchored in position with either a rubber coating on the back surface or a vinyl coating. Quality is normally controlled by characteristics such as number of rows of tufts of pile per unit length, height and weight of the free pile, overall weight of the carpet, strength as determined by a tensile test in both the warp and weft directions, together with adhesion of pile if applicable. Rubber flooring generally has a vinyl coating to provide colour.

3.11.2 Leather (hide)

Large numbers of motor vehicle users all over the world continue to specify hide upholstery when the option is available, and will gladly pay the extra cost involved for a material which defies complete simulation. Great advances have been made in the development of suitable substitutes, and the best of

Table 3.12 Undersealing and protection materials used in vehicle body repair work

Type	Base material	Application
Coatings for underbodies, spray type	Bitumen/rubber	Coating for underbodies, wheel arches, new and repaired parts. Also a corrosion protection for vehicle underbodies against elements such as moisture, road salt, road stone chippings. Good adhesion, and durable at extreme temperatures. Applied with a spray pistol
	Rubber/resin	Suitable for underbodies, wheel arches, front and rear aprons, sills, new parts, repair sheets. Can be painted over and has high abrasion resistance. Applied with a spray pistol
	Wax	Suitable for underbodies, touching up and subsequent treatment of all protective coatings
	Polymer wax	Long-term corrosion protection even when thinly applied. Has good flowing properties
Small-scale repair application (brushable)	Rubber/bitumen	Brushable coating, suitable for underbody and wheel arches. High abrasion resistance and good sound deadening properties
Road stone chip repair material	Synthetic/dispersion	Good protection against stone chips and corrosion. Particularly suitable for front and rear aprons, sills, spoilers, wheel arches

the plasticized materials are to many people quite indistinguishable by eye from leather. The unique character of leather lies in its micro-structure, the like of which is not obtained in any manmade material. Under a microscope leather can be seen to consist of the hairy epidermis and under that the corium, or bulk of the hide, this being the basis of the leather as we know it. By virtue of the millions of minute air spaces between the fibres and bundles of fibres, leather is able to 'breathe'. To the motorist this means that leather does not get hot and uncomfortable in warm weather or cold and inflexible in winter, and although permeable to water vapour it offers sufficient resistance if it is exposed to normal liquids. It is also strongly resistant to soiling, and when it does get dirty the dirt can usually be removed fairly easily without special materials. Unlike some plasticized materials, leather does not appear to attract dirt and dust as a result of static electricity. Many people, moreover, regard the distinctive smell of leather as an asset, and this defies imitation by manufacturers of substitute materials. However, natural hide has to go through many complex processes before it attains the form familiar to the upholsterer trimmer or motorist.

3.11.3 Fabrics for interior trim

Vinyl coated fabrics are now well established as trim material. Their vast superiority over the linseed oil coatings, and later the nitrocellulose coatings, of yesteryear are almost forgotten in the march of progress. Vinyl coatings are now sufficiently familiar for their merits to be taken for granted; nevertheless they continue to provide a material which for durability, uniformity and appearance at a reasonable cost so far remains unsurpassed. The resin polyvinyl chloride, the main ingredient of the coating, became available in commercial quantities in the early 1930s, and now a coating based on polyvinyl chloride (PVC) is used for seating material in this country. Although by tradition PVC is produced with a simulated leather appearance, on the Continent and particularly in the USA it is widely used with fancy embosses and patterns. An extremely wide range of qualities is available, and in recent years there has been an effort to achieve some degree of rationalization. Additional qualities are necessary for tilt covers, headlinings and hoods for convertibles; the material for hooding convertibles must be resistant to mildew, to shrinkage and to wicking, this last term relating to the absorption of water on the inside surface of the cloth from the bottom edge of the hood. The characteristics necessary to provide serviceability over the life of the vehicle are the strength of the material under tension and under tearing conditions; adhesion of the coating to the backing cloth; resistance to flexing, and resistance to cracking at low temperatures; low friction, to enable the owner to slide on the seat; colour fastness, soiling resistance and, of course, wear resistance. In the case of the breathable leathercloth, the air permeability of the fabric has to be controlled.

3.11.4 Modern trends

The interior furnishing of a car is gaining in importance within the automotive trade. In response stylists are endeavouring to upgrade and soften the interior, using fabrics with the appearance and feel of textiles to appeal visually and functionally. This has manifested itself in all areas of the car, including the boot, seating, carpets, door trims and headliner cover fabrics. In all four areas of fabric use in car interiors (seating, door and side panels, bolsters and headliners) fabrics are gaining ground against exposed plastics. Both polyamide and polyester are giving designers new scope for attractive colours and variation in seating upholstery and in panels, while fully meeting light fastness and other performance standards.

3.12 PLASTICS

3.12.1 Development

Celluloid might well qualify for the honour of being the first plastic, though its inventor, Alexander Parkes, was certainly not aware of that fact. He made it around 1860 and patented his method for making it in 1865. An American, John Hyatt, found a way of solving the technical problems which plagued Parkes, and he set up business in 1870 to sell the same sort of material. He called it celluloid to indicate its raw material, cellulose.

In 1920 a German chemist, Hermann Staudinger, put forward a theory about the chemical nature of a whole group of substances, natural and synthetic. He called them macromolecules; today we call them polymers. His theory not only explained the nature of plastics, but also indicated the ways in which they could be made. It provided the foundation for the world of plastics as we know it.

3.12.2 Polymerization

The raw materials for plastics production are natural products such as cellulose, coal, oil, natural gas and salt. In every case they are compounds of carbon (C) and hydrogen (H). Oxygen (O), nitrogen (N), chlorine (Cl) and sulphur (S) may also be present. Oil, together with natural gas, is the most important raw material for plastics production.

The term plastics in the broadest sense encompasses (a) organic materials which are based on (b) polymers which are produced by (c) the conversion of natural products or by synthesis from primary chemicals coming from oil, natural gas or coal.

The basic building blocks of plastics are monomers. These are simple chemicals that can link together to form long chains or polymers. The type of monomer used and the way it polymerizes, or links together, give a plastic its individual characteristics. Some monomers form simple linear chains. In polyethylene, for example, a typical chain of 50 000 ethylene links is only about 0.02 mm long. Other monomers form chains with side branches. Under certain circumstances, the individual chains can link up with each other to form a three-dimensional or cross-linked structure with even greater strength and stability. Cross-linking can be caused either chemically or by irradiating the polymer.

To get the advantages of two different plastics, two different monomers can be combined in a copolymer. By combining the monomers in different proportions and by different methods, a vast range of different properties can be achieved (see Tables 3.13 and 3.14). The properties of plastics can also be enhanced by mixing in other materials, such as graphite or molybdenum disulphide (for lubrication), glass fibre or carbon fibre (for stiffness), plasticizers (to increase flexibility) and a range of other additives (to make them resistant to heat and light).

3.12.3 Thermoplastics and thermosetting plastics

The simplest way of classifying plastics is by their reaction to heat. This gives a ready subdivision into two basic groups: thermoplastics and thermosetting plastics. *Thermoplastic* materials soften to become plastic when heated, no chemical change taking place during this process. When cooled they again become hard and will assume any shape into which they were moulded when soft. *Thermosetting* materials, as the name implies, will soften only once. During heating a chemical change takes place and the material cures; thereafter the only effect of heating is to char or burn the material.

As far as performance is concerned, these plastics can be divided into three groups:

General-purpose thermoplastics
Polyethylene
Polypropylene
Polystyrene
SAN (styrene/acrylonitrile copolymer)
Impact polystyrene
ABS (acrylonitrile butadiene styrene)
Polyvinyl chloride
Poly(vinylidene chloride)
Poly(methyl methacrylate)
Poly(ethylene terephthalate)

Engineering thermoplastics
Polyesters (thermoplastic)
Polyamides
Polyacetals
Polyphenylene sulphide

Table 3.13 Physical properties of polymers

Material	P or S	Density (kg/m³)	Melting (softening) range (°C)	Specific-heat capacity (J/kg/K × 10³)	Thermal conductivity (W/m/K)	Coefficient of linear expansion (K × 10⁻⁶)
LD polyethylene	P	0.01–0.93	80	2.3	0.13	120–140
HD polyethylene	P	0.04–0.97	90–100	2.1–2.3	0.42–0.45	120
Polypropylene	P	0.90	100–120	1.9	0.09	120
GFR polypropylene	P	1.00–1.16	110–120	3.5	–	55–85
Polyvinylchloride	P	1.16–1.35	56–85	0.8–2.5	0.16–0.27	50–60
Polystyrene	P	1.04–1.11	82–102	1.3–1.45	0.09–0.21	60–80
Polystyrene copolymer (ABS)	P	0.99–1.10	85	1.4–1.5	0.04–0.30	60–130
Nylon 66	P	1.14	250–265	1.67	0.24	80
Nylon 11	P	1.04	185	2.42	0.23	150
PTFE (Teflon)	P	2.14–2.20	260–270	1.05	0.25	100
Acrylic (Perspex)	P	1.10–1.20	70–90	1.45	0.17–0.25	50–90
Polyacetals	P	1.40–1.42	175	1.45	0.81	80
Polycarbonates	P	1.20	215–225	1.25	0.19	65
Phenol formaldehyde	S	1.25–1.30	–	1.5–1.75	0.12–0.25	25–60
Urea formaldehyde	S	1.40–1.50	–	1.65	0.25–0.38	35–45
Melamine formaldehyde	S	1.50	–	1.65	0.25–0.40	35–45
Epoxies	S	1.20	–	1.65	0.17–0.21	50–90
Polyurethanes R	S	3.2–6.0	150–185	1.25	0.02–0.025	20–70
Polyurethanes F	S	4–8	150–185	1.25	0.035	50–70
Polyesters	S	1.10–1.40	–	1.26	0.17–0.19	100–150
Silicones	S	1.15–1.8	200–250	–	0.17	24–30

GFR glass fibre reinforced; P thermoplastic; S thermosetting; R rigid; F flexible; LD low density; HD high density

Table 3.14 Typical mechanical properties of representative plastics

Material	Modulus of elasticity *E* (MN/m²)	Tensile strength (MN/m²)	Compressive strength	Elongation (%)
LD polyethylene	120–240	7–13	9–10	300–700
HD polyethylene	550–1050	20–30	20–25	300–800
Polypropylene	900–140	32–35	35	20–300
GFR polypropylene	1500+	34–54	40–60	5–20
Flexible PVC	3500–4800	10–25	7–12	200–450
Rigid PVC	2000–2800	40	90	60
Polystyrene	2400–4200	35–62	90–110	1–3
ABS copolymer	1380–3400	17–58	17–85	10–140
Perspex	2700–3500	55–75	80–130	2–3
PTFE	350–620	15–35	10–15	200–400
Nylon 11	1250–1300	52–54	55–56	180–400
GFR nylon 6	7800–800	170–172	200–210	3–4

Polycarbonates
Polysulphone
Modified polyphenylene ether
Polyimides
Cellulosics
RIM/polyurethane
Polyurethane foam

Thermosetting plastics
Phenolic
Epoxy resins
Unsaturated polyesters
Alkyd resins
Diallyl phthalate
Amino resins

3.12.4 Amorphous and crystalline plastics

An alternative classification of plastics is by their shape. They may be crystalline (with shape) or amorphous (shapeless).

Amorphous plastics
Amorphous plastics basically are of three major types:

ABS: acrylonitrile butadiene styrene
ABS/PC blend
PC: polycarbonate.

Amorphous engineering plastics have the following properties:

High stiffness
Good impact strength
Temperature resistance
Excellent dimensional stability
Good surface finish
Electrical properties
Flame retardance (when required)
Excellent transparency (polycarbonate only)

In the automotive industry use is made of the good mechanical properties (even at low temperatures), the thermal resistance and the surface finish. The applications are:

1 Body embellishment.
2 Interior cladding.
3 Lighting where, apart from existing applications of back lamp clusters, polycarbonate has replaced glass for headlamp lenses.

Semi-crystalline plastics
Semi-crystalline plastics are in two basic types:

Polyamide 6 and 66 types
Polybutylene terephthalate (PBT)

Semi-crystalline plastics have the following properties:

High rigidity
Hardness
High heat resistance
Impact resistance
Abrasion, chemical and stress crack resistance

The semi-crystalline products find major application in the automotive sector, where full use is made of the mechanical and thermal properties, together with abrasion and chemical resistance. Examples include:

1 Underbonnet components.
2 Mechanical applications.
3 Bumpers, using elastomeric PBT for paint on-line.
4 Body embellishment (wheel trims, handles, mirrors).
5 Lighting, headlamp reflectors.

Blended plastics
Blended plastics have been developed to overcome inherent specific disadvantages of individual plastics. For large-area body panels, the automotive industry demands the following properties:

Temperature resistance
Low-temperature impact resistance
Toughness (no splintering)
Petrol resistance
Stiffness

Neither polycarbonate nor polyester could fulfil totally these requirements. This led to the combination of PC and PBT to form Macroblend PC/PBT, which is used for injection moulded bumpers.

3.12.5 Plastics applications

Plastic products can be decorated by vacuum metallizing and electroplating. They have replaced metals in a lot of automotive applications, such as mirror housings, control knobs and winder handles as well as decorative metallic trim. It is a field which uses their advantages to the full without relying on properties they lack.

Thin parts must be tough and resistant to the occasional impact. They must be impervious to attack by weather, road salts, extremes of temperatures and all the other hazards that reduce older forms of body embellishments to pitted, rusted, dull, crumbling metal. They do not need high tensile strength or flexural strength as they do not have to carry heavy stresses. They must be cheap and capable of being formed into highly individual and complex shapes. All these requirements are satisfied by thermoplastics and thermosetting resins. They can be pressed, stamped, blow moulded, vacuum formed and injection moulded into any decorative shape required.

Apart from their decorative properties, the mechanical properties of acrylic resins are among the highest of the thermoplastics. Typical values are a tensile strength of 35–75 MN/m^2 and a modulus of elasticity of 1550–3250 MN/m^2. These properties apply to relatively short-term loadings, and when long-term service is envisaged tensile stresses in acrylics must be limited to 10 MN/m^2 to avoid surface cracking or crazing. Chemical properties are also good, the acrylics being inert to most common chemicals. A particular advantage to the automotive industry is their complete stability against petroleum products and salts.

Acetal resins are mostly used for mechanical parts such as cams, sprockets and small leaf springs, but also find application for housings, cover plates, knobs and levers. They have the highest fatigue endurance limits of any of the commercial thermoplastics, and these properties, coupled with those of reduced friction and noise, admirably qualify the acetal resins for small gearing applications within the vehicle.

Plastics can be self-coloured so that painting costs are eliminated and accidental scratching remains inconspicuous, and they can be given a simulated metal finish. For large-scale assemblies, such as automobile bodies, painting is necessary to obtain uniformity of colour, especially when different types of plastics are used for different components. Plastics can also be chrome plated, either over a special undercoating which helps to protect and fix the finish, or by metal spraying or by vacuum deposition in which the plastic part is made to attract metal particles in a high-vacuum chamber. The use of a plastic instead of a metal base for chrome plating eliminates the possibility of the base corroding and damaging the finish before the chromium plating itself would have deteriorated. The chrome coating can be made much thinner and yet have a longer effective life, with a consequent saving in cost.

Originally polymer materials were joined only by means of adhesives. Now the thermoplastic types can be welded by using various forms of equipment, in particular by hot gas welding, hot plate machines which include pipe welding plant, ultrasonic and vibration methods, spin or friction welding machines, and induction, resistance and microwave processes. Lasers have been used experimentally for cutting and welding. The joining of metals to both thermoplastic and thermosetting materials is possible by some welding operations and by using adhesives.

3.12.6 Future of plastics in the automotive industry

The automotive industry has grown to appreciate the potential of plastics as replacements for metal components within their products. The realization that plastics are, in their own right, engineering materials of high merit has led to rapid advancement of material and application technology, with the end result that plastics have gained a firm and increasing footing in the motor vehicle. Many factors have aided the adoption of plastics by the automotive industry, which uses them in the following areas: body, chassis, engine, electrical system, interior and vehicle accessories. Lower costs of plastics parts must, of course, be the major contributing factor in the replacement of existing parts, and this is closely followed by the ease with which modern plastics can be formed by comparatively inexpensive tooling. The inert properties of synthetic materials also contribute greatly; properties like corrosion resistance, low friction coefficients and light weight are of prime importance.

The use of plastics in the automotive industry continues to accelerate at a phenomenal rate as research into plastic technology results in new developments and applications. The future growth of plastics in the automotive industry will be controlled by two factors: the growth of the industry itself, and the greater penetration of plastic per car (see Figure 3.3). A key constituent in world growth, therefore, is the developing nations which are involved in the assembly and production of motor vehicles. They will consequently favour the use of plastics as a first choice, rather than as a replacement for metal.

Over the past years, the natural applications for plastics in automobiles (interior fittings, cushioning and upholstery, trim, tail lights, electrical components) have become saturated. The growth for the future can be expected to come from the use of plastic for bodywork and some mechanical components. There is a widespread use of plastics for front and rear bumpers. We now see bonnets, boot lids and front wings in plastics.

All the major volume producers of cars are engaged in long-term development work towards the all-plastic car. Whether or not such targets can be realized remains to be seen. Factors such as energy costs and availability of resources may play a greater part in the total picture than simple objects like vehicle weight reduction.

3.13 SAFETY GLASS

More and more glass is being used on modern cars. Pillars are becoming slimmer and glass areas are increasing as manufacturers approach the ideal of almost complete all-round vision and the virtual elimination of blind spots. Windscreens have become deeper and wider. They may be gently curved, semi-wrapped round, or fully wrapped. With few exceptions they are of one-piece construction, sometimes swept back as much as 65° from the vertical. Styling trends, together with a growing knowledge of stress design in metal structures, have resulted in a significant increase in the glazed areas of modern car body designs. As a result of this move towards a more open style, the massive increase in the cost of energy has brought growing pressure on vehicle designers to achieve more economic operations, principally in respect to lower fuel consumption through better power/weight ratios. An outcome of these two lines of development has been a situation in which, although the areas of glass has increased, the total weight of glass has remained constant or even decreased.

Figure 3.3 Applications of plastics in automobiles (*Motor Insurance Repair Research Centre*)

1 Front bumper (Pocan S7931)
2 Front spoiler (Santoprene grade 123-50 and 121 with aluminium inert)
3 Fog lamp blanking plate (Xenoy EPX500)
4 Lower front grille (Xenoy CL100)
5 Front number plate plinth (Xenoy CL100)
6 Front bumper insert (PVC and EB-type Nylar)
7 Front grille (moulding, ABS; Bezel, MS Chrome)
8 Bonnet/boot lid/tailgate badges (ABS, aluminium and PU skin)
9 Underbonnet felt (moulded felt)
10 Door mirror casing RH and LH (polymide, 15% glass reinforced)
11 Door mirror mounting RH and LH (polymide, 15% glass reinforced)
12 Front/rear wheel trims RH and LH (*cap*, Noryl 731; *moulding*, Bayer Duretan BM30X, ICI Maranyl TB570)
13 Front/rear mudflaps RH and LH (front, rubber to BLS.22 RD.27 Ref. 421; rear, EPDM mix 4080)
14 Scuttle grille/mouldings (ABS)
15 Front/rear screen upper and side mouldings (PVC with stainless steel co-extrusion)
16 Front/rear wing splashguards RH and LH (PP)
17 Front wing waist moulding RH and LH (Noryl)
18 Front door waist moulding RH and LH (Noryl)
19 Rear door waist moulding RH and LH (Noryl)
20 Rear wing waist moulding RH and LH (Noryl)
21 Front/rear door outer handles RH and LH (*body*, Xenoy; *flap*, Glass-filled nylon)

Broadly speaking, motor vehicle regulations specify that windscreens must be of safety glass. To quote one section: 'On passenger vehicles and dual-purpose vehicles first registered on or after 1 January 1959, the glass of all outside windows, including the windscreen, must be of safety glass.' The British Standards Institution defines safety glass indirectly as follows: 'All glass, including windscreen glass, shall be such that, in the event of shattering, the danger of personal injury is reduced to a minimum. The glass shall be sufficiently resistant to conditions to be expected in normal traffic, and to atmospheric and heat conditions, chemical action and abrasion. Windscreens shall, in addition, be sufficiently transparent, and not cause any confusion between the signalling colours normally used. In the event of the windscreen shattering, the driver shall still be able to see the road clearly so that he can brake and stop his vehicle safely.'

Two types of windscreen fulfil these requirements – those made from heat-treated (or toughened) glass, and those of laminated glass. In addition there are plastic coated laminated or annealed safety glasses. Most windscreens and some rear windows fitted in motor vehicles are of ordinary laminated glass. For the main part, toughened glass is confined to door glass, quarter lights and rear windows where the use of more expensive laminated products has yet to be justified. However, laminated glass is being increasingly used on locations other than windscreens for reasons of vehicle security and also for passenger safety (containment in an accident), especially in estates with seating in the rear. Note that the applicable EEC Directive (see later) has effectively banned the fitment of toughened windscreens from the end of 1992.

Ordinary laminated safety glass is the older of the two types and is the result of a basic process discovered in 1909. Some years earlier a French chemist, Edouard Benedictus, had accidentally knocked down a flask which held a solution of celluloid. Although the flask cracked it did not fall into pieces, and he found that it was held together by a film of celluloid adhering to its inner surface. This accident led to the invention of laminated safety glass, made from two pieces of glass with a celluloid interlayer. An adhesive, usually gelatine, was used to hold them together and the edges had to be sealed to prevent delaminating. However, despite the edge sealing, the celluloid (cellulose nitrate plastic) discoloured and blistered; hence celluloid was replaced by cellulose acetate plastic, but this, although a more stable product than celluloid, still needed edge sealing. Nowadays a polyvinyl butyral (PVB) self-bonding plastic interlayer is used; no adhesive is necessary and the edges do not need sealing, making it quite practical to cut to size after laminating. When producing glasses to a particular size, however, the glass and vinyl interlayer are usually cut to size first. In the process the vinyl plastic interlayer is placed between two, clean, dry pieces of glass and the assembly is heated and passed between rubber-covered rollers to obtain a preliminary adhesion. The sandwich of glass and interlayer is then heated under pressure for a specified period in an autoclave. This gives the necessary adhesion and clarity to the interlayer, which is not transparent until bonded to the glass. If a piece of laminated glass is broken, the interlayer will hold the splinters of glass in place and prevent them flying.

Plastic coated laminated safety glass is an ordinary laminated glass which has soft elastic polyurethane films bonded on to the inner surface to provide improved passenger protection if fragmentation occurs. There is some interest in the use of bilayer construction which uses 3 mm or 4 mm annealed glass bonded with a load-bearing surface layer of self-healing polyurethane.

Uniformly toughened glass is produced by a completely different process, involving heating of the glass followed by rapid cooling. Although patents were taken out in 1874 covering a method of increasing the strength of flat glass sheet by heating and cooling it in oil, toughened glass was not in common use until the 1930s. Modern toughened glass is produced by heating the glass in a furnace to just below its softening point. At this temperature it is withdrawn from the furnace and chilled by blasts of cold air. The rapid cooling hardens and shrinks the outside of the glass; the inside cools more slowly. This produces

Figure 3.3 (Continued)
22 Rear quarterlight moulding RH and LH (*4-door*, PVC/Stainless steel extrusion; *5-door and coupé*, PU with stainless steel moulding)
23 Boot lid moulding (ABS)
24 Rear spoiler (PU core and polyester skin)
25 Rear number plate plinth (ABS)
26 Rear bumper insert (PVC and EB-type Nylar)
27 Rear bumper (Pocan S7913)
28 Front/rear door upper mouldings RH and LH (PVC with stainless steel moulding)
29 Front/rear door outer weatherseals RH and LH (PVC with stainless steel co-extrusion)
30 Fog lamp bezel (PP)

compressional strain on the surfaces with a compensating state of tension inside, and has the effect of making the glass far stronger mechanically than ordinary glass. If, however, the glass does fracture in use, it disintegrates into a large number of small and harmless pieces with blunted edges. The size of these particles can be predetermined by an exact temperature control and time cycle in the toughening process, and manufacturers now produce a uniformly toughened safety glass which will, when broken, produce not less than 40 or more than 350 particles within a 50 mm square of glass. This conforms to the British Standard specification.

The main standards for the UK are now:

BS 857

ECE R43 (UN regulation)

EEC Directive (AUE/178) (a common market regulation)

BS 857 glazing is still valid but is seldom used because ECE R43 is accepted throughout Europe, Japan and Australia.

There are other types of safety glass – mostly crossbreeds of the pure toughened glass screen which are designed to combine vision with safety. These are modified zone-toughened glasses, having three zones with varying fragmentation characteristics. The inner zone is a rectangular area directly in front of the driver, not more than 200 mm high and 500 mm long. This is surrounded by two other zones, the outer one of which is 70 mm wide all round the edge of the windscreen. This type of windscreen has been fitted to various vehicles since 1962. As a result of ECE Regulation 43 this type of windscreen has been superseded by the fully zebra-zoned windscreen. Many countries, including the USA but with the exception of the UK, legislate against toughened windscreens.

Although sheet and plate glass are manufactured satisfactorily for use in doors, rear lights and windscreens, float glass has now largely superseded their use for reasons of economy and improved flatness.

Most laminated windscreens used in the motor vehicle trade are 4.4 mm, 5 mm, 5.8 mm or 6.8 mm in overall thickness, with an 0.76 mm PVB interlayer. However, 4.4 mm is the thinnest laminated glass available, and as this has to be made from two pieces of glass it needs very careful handling during manufacture and is therefore expensive. Windscreens made from float glass should be a maximum of 6.8 mm thick, whether toughened or laminated. However, some large coaches and lorry windscreens are 7.8 mm thick (4 mm glass + 0.76 mm PVB interlayer + 3 mm glass). This gives immense strength and robustness against stone impact. Other body glasses, because they can be made from sheet glass and also can be toughened safety glass, are usually between 3 mm and 4 mm thick.

From our brief look at the history of glass manufacture it is obvious that the curving of glass presents no problems; in fact the problem has been to produce flat, optically perfect glass. However, to curve safety glass and still retain its optical and safety qualities requires careful control. Glass has no definite melting point, but when it is heated to approximately 600°C it will soften and can be curved. Curved glasses should be specified as 6.8 mm thick, as it is more difficult to control the curving of 4.4 mm glass. Even 6.8 mm thick glasses will have slight variations of curvature. To accommodate this tolerance, all curved glasses should be glazed in a rubber glazing channel, of which there are many different sections available. Glazed edges of glasses should be finished with a small chamfer known as an arrissed edge, while edges of glasses that are visible or which run in a felt channel should be finished with a polished, rounded edge. Should a glass be required for glazing in a frame, a notch will usually be required to clear the plate used to join the two halves of the frame together. The line of this notch must not have sharp corners because of the possibilities of cracking. Although laminated safety glass can be cut or ground to size after laminating, toughened safety glass must be cut to size and edge finished before the heat treating process.

Nearly all fixed glazing is now glazed using adhesive systems. Shapes are becoming more complex, needing very good angles of entry control to meet bonding requirements. The trend is towards aerodynamical designs involving flush glazing and the removal of sudden changes in vehicle shape; therefore corners must be rounded rather than angular as in older vehicle designs. Glass is often supplied with moulded-on finisher (encapsulation). Consequently bending processes are becoming very sophisticated. Adhesive glazing (polyurethane is the adhesive normally used) has added considerably to the complexity of vehicle glazing in a scientific sense. It has many advantages, however, if carried out correctly: it will reduce water leaks, it suits modern car construction, it results in a load-bearing glazing member, and it lends itself to robotic assembly in mass production. As a consequence of adhesive glazing, all the associated glazing is now printed with a ceramic fired-in black band to protect the polyurethane adhesive from ultraviolet degradation, and also for cosmetic reasons so that the adhesive cannot be seen.

By a Ministry of Transport regulation, safety glass was made compulsory in 1937 for windscreens and

Table 3.15 Physical properties of metals and alloys

Metal	Melting temperature range (°C)	Density (kg/m³)	Specific heat capacity/ (J/kg/K × 10³)	Thermal conductivity (W/m/K)	Electrical conductivity (% IACS)	Coefficient of linear expansion/ (K × 10⁻⁶)
Aluminium	660	2.69	0.22	218	63	23
Al–3.5 magnesium	550–620	2.66	0.22	125	25	23
Duralumin type	530–610	2.80	0.21	115–140	20–36	23
Copper	1085	8.92	0.39	393	101	17
70/30 brass	920/950	8.53	0.09	120	17	19
95/5 tin bronze	980/990	8.74	0.09	80	12	17
Lead	327	11.34	0.13	35	8	29
Magnesium	650	1.73	1.04	146	35	30
Nickel	1455	8.90	0.51	83	21	13
Monel	1330–1360	8.80	0.43	26	3	10
Tin	232	7.30	0.22	64	13	20
Titanium	1665	4.50	0.58	17	3	8.5
Zinc	419	7.13	0.39	113	26	37
Iron	1535	7.86	0.46	71	7	12
Mild steel	1400	7.86	0.12	45	31	11

Table 3.16 Typical mechanical properties of metals and alloys

Material	Modulus of elasticity E (kN/mm²)	Tensile and compressive strength (N/mm²)	Elongation (%)	Hardness (HV)
Pure aluminium	68–70	62–102	45–7	15.30
Aluminium alloys	68–72	90–500	20–5	20–80
Magnesium	44	170–310	5–8	30–60
Cast irons (grey)	75–145	150–410	0.5–1.0	160–300
SG cast iron	170–172	370–730	17–2	150–450
Copper	122–132	155–345	60–5	40–100
Copper alloys	125–135	200–950	70–5	70–250
Mild steel	190	420–510	22–24	130
Structural steels	190	480–700	20–24	130
Stainless steel	190	420–950	40–20	300–170
Titanium	100–108	300–750+	5–35	55–90
Zinc	90	200–500	25–30	45–50

other front windows. As already indicated, with effect from 1 January 1959 the Road Traffic and Vehicle Order 359 has demanded that for passenger vehicles and dual-purpose vehicles, all glass shall be safety glass. For goods vehicles, windscreens and all windows in front of, or at the side of, the driver's seat shall be safety glass.

3.14 PROPERTIES OF METALS

Metals and alloys possess certain properties which make them especially suitable for the processes involved in vehicle bodywork, particularly the forming and shaping of vehicle body parts either by press or by hand, and some of the jointing processes. These properties are described in the following sections, and some typical values of characteristics are shown in Tables 3.15 and 3.16.

3.14.1 Malleability

A malleable metal may be stretched in all directions without fracture occurring, and this property is essential in the processes of rolling, spinning, wheeling, raising, flanging, stretching and shrinking. In the operation of beating or hammering a metal on a steel block (such as planishing) an action takes place at each blow wherein the metal is squeezed under the blow of the hammer and is forced outwards around the centre-point of the blow. The thinner the metal can be rolled or hammered into sheet without fracture the more malleable is the metal.

After cold working, metals tend to lose their malleable properties and are said to be in a work hardened condition. This condition may be desirable for certain purposes, but if further work is to be carried out the malleability may be restored by annealing. Annealing, or softening, of the metal is usually carried out before or during curvature work such as raising and hollowing, provided the metal is not coated with a low-melting-point material. However, the quality of the modern sheet metal is such that many forming operations, such as deep drawing and pressing, may be carried out without the need for an application of heat.

The following are examples in which the properties of malleability are most evident:

Riveting Here the metal will be seen to have spread to a marked degree. If splitting occurs the metal is insufficiently malleable or has been overworked (work hardened).

Shaping The blank for a dome consists of a flat disc which has to be formed by stretching and shrinking into a double-curvature shape. The more malleable and ductile the material of the blank is, the more readily it can be formed; the less malleable and ductile, the more quickly does the metal work hardened thus need more frequent annealing.

The degree of malleability possessed by a metal is measured by the thinness of leaves that can be produced by hammering or rolling. Gold is extremely malleable and may be beaten into very thin leaf. Of the metals used for general work, aluminium and copper are outstanding for their properties. The property of malleability is used to advantage in the manufacture of mild steel sheets, which are rolled to a given size and gauge for the motor industry. It is also evident in the ability of mild steel and aluminium panels to be formed by mechanical presses into complicated contours for body shells. Malleability and ductility are the two essential properties needed in order to mass produce vehicle body shells by pressing.

The order of malleability of various metals by hammering is as follows: gold, silver, aluminium, copper, tin, lead, zinc, steel.

3.14.2 Ductility

Ductility depends on tenacity or strength in tension and the ease with which a metal is deformed, and is the property which enables a metal to be drawn out along its length, that is drawn into a wire. In wire drawing, metal rods are drawn through a hole in a steel die; the process is carried out with the metal cold, and the metal requires annealing when it becomes work hardened.

Ductile properties are also necessary in metals and alloys using the following processes:

Pressed components Special sheets which have extra deep drawing qualities are manufactured especially for press work such as that used in modern motor vehicle body production. These sheets undergo several deformations during the time they are being formed into components, yet because of their outstanding ductile properties they seldom fracture.

Welding electrodes and rods Ductility is an essential property in the production of electrodes, rods and wires. The wire drawing machines operate at exceptionally high speeds and the finished product conforms to close tolerances of measurement; frequent failure of the material during the various stages of drawing would be very costly.

The order of ductility of various metals is as follows: gold, aluminium, steel, copper, zinc, lead.

3.14.3 Tenacity

A very important property of metals is related to its strength in resistance to deformation; that property is tenacity, which may be defined as the property by which a metal resists the action of a pulling force. The ultimate tensile strength of a metal is a measure of the force which ultimately fractures or breaks the metal under a tensile pull. The ultimate tensile strength (UTS) of a material is normally expressed in tons per in^2 or MN/m^2 and may be calculated as follows:

$$\text{UTS} = \frac{\text{tensile force in N}}{\text{cross-sectional area in mm}}$$

In this case the load is the maximum required to fracture a specimen of the material under test, and the calculation is based on fracture taking place across the original cross-sectional area. In ductile materials a special allowance must be made for wasting or reduction of original cross-sectional area.

High-carbon steels possess a high degree of tenacity, evidence of which can be seen in the steel cables used to lift heavy loads. The mild steels used in general engineering possess a small amount of tenacity, yet a bar of metal of one inch square (6.5 cm^2) cross-section, made from low-carbon steel, is capable of supporting a load in excess of 20 tonnes.

Methods of increasing tensile strength
It is possible to increase the tensile strength of both sheet steel and pure aluminium sheets by cold rolling, but this has the added effect of reducing their workable qualities. In the manufacture of vessels to contain liquids or gases under pressure it is not always possible to use metals with a high tensile strength; for instance, copper is chosen to make domestic hot water storage cylinders because this metal has a high resistance to corrosion. In this case, the moderate strength of copper is increased by work hardening such as planishing, wheeling and cold rolling. Work hardening has the added effect of decreasing the malleability.

The order of tenacity of various metals in tons per in^2 (MN/m^2) is: steel 32 (494); copper 18 (278); aluminium 8 (124); zinc 3 (46); lead 1.5 (23).

3.14.4 Hardness

When referring to hardness, it should be carefully stated which kind of hardness is meant. For example, it may be correctly said that hardness is that property in a metal which imparts the ability to:

1 Indent, cut or scratch a metal of inferior hardness.
2 Resist abrasive wear.
3 Resist penetration.

A comparison of hardness can be made with the aid of material testing machines such as those used to carry out the Brinell or Vickers Diamond tests. Hardness may be increased by the following methods:

Planishing In addition to increasing the tensile strength of a metal, planishing also imparts hardness.

Heat treatment Medium- and high-carbon steels, such as those used in many body working tools, can be hardened by heating to a fixed temperature and then quenching.

The order of hardness of various metals is as follows: high-carbon steel, white cast iron, cast iron, mild steel, copper, aluminium, zinc, tin, lead.

3.14.5 Toughness

This property imparts to a metal the ability to resist fracture when subjected to impact, twisting or bending. A metal need not necessarily be hard to be tough; many hard metals are extremely brittle, a property which may be regarded as being opposite to toughness.

Toughness is an essential property in rivets. During the forming process the head of the rivet is subject to severe impact, and when in service rivets are frequently required to resist shear, twist and shock loads. Toughness is also a requisite for steel motor car bodies, which must be capable of withstanding heavy impacts and must often suffer severe denting or buckling without fracture occurring. Further, when repairs are to be made to damaged areas it is often necessary to apply force in the direction opposite to that of the original damaging force; and the metal must possess a high degree of toughness to undergo such treatment.

3.14.6 Compressibility

Compressibility may be defined as the property by which a metal resists the action of a compressing force. The *ultimate compressive strength* of a metal is a measure of the force which ultimately causes the metal to fail or yield under compression. Compressibility is related to malleability in so far as the latter refers to the degree to which a metal yields by spreading under the action of a compressing or pushing force, while the former represents the degree to which a metal opposes that action.

3.14.7 Elasticity

All metals possess some degree of elasticity; that is, a metal regains its original shape after a certain amount of distortion by an external force. The elastic limit of a metal is a measure of the maximum amount by which it may be distorted and yet return to its original form on removal of the force. Common metals vary considerably in elasticity. Lead is very soft yet possesses only a small amount of elasticity. Steel, on the other hand, may reveal a considerable degree of elasticity as, for example, in metal springs. The elasticity of mild steel is very useful in both the manufacture of highly curved articles by press work and in the repair of motor car bodies.

3.14.8 Fatigue

Most metals in service suffer from fatigue. Whether the metal ultimately fails by fracture or by breaking depends on a number of factors associated with the type and conditions of service. When metal structures in service are subjected to vibration over long periods, the rapid alternations of push and pull, i.e. compressive and tensile stresses, ultimately cause hardening of the metal with increased liability to fracture. Any weak points in the structure are most affected by the action and become the probable centres of failure, either by fracture or by breaking. Various methods have been devised for testing the capacity of metals to resist fatigue, all of which depend on subjecting the metal specimen to alternating vibratory stresses until failure occurs. The rate and extent of vibration over a specified time form the basis of most fatigue tests.

3.14.9 Weldability

This property is a measure of the ease with which a metal can be welded using one of the orthodox systems of welding. Certain metals are welded very easily by all recognized methods; some metals can only be welded by special welding processes; and other metals and alloys cannot be welded under any circumstances.

3.15 HEAT TREATMENT OF METALS AND METAL ALLOYS

Heat treatment can be defined as a process in which the metal in the solid state is subjected to one or more temperature cycles, to confer certain desired properties. The heat treatment of metals is of major importance in motor bodywork. All hand tools used by the body worker are made from a type of steel which is heat treatable so that the tools are strong, hard and have lasting qualities. The mechanical parts of a motor vehicle are also subject to some form of heat treatment, and in the constructional field heat-treatable aluminium alloys are being used extensively in commercial bodywork.

3.15.1 Work hardening and annealing

Most of the common metals cannot be hardened by heat treatment, but nearly all metals will harden to some extent as a result of hammering, rolling or bending. Annealing is a form of heat treatment for softening a metal which has become work hardened so that further cold working can be carried out. The common metals differ quite a lot in their degree and rate of work hardening. Copper hardens rather quickly under the hammer and, as this also reduces the malleability and ductility of the metal, it needs frequent annealing in order that it may be further processed without risk of fracture. Lead, on the other hand, may be beaten into almost any shape without annealing and without undue risk of fracture. It possesses a degree of softness which allows quite a lot of plastic deformation with very little work hardening. However, copper, though less soft than lead, is more malleable. Aluminium will withstand a fair amount of deformation by beating, rolling and wheeling before it becomes necessary to anneal it. The pure metal work hardens much less rapidly than copper, though some of the sheet aluminium alloys are too hard or brittle to allow very much cold working. Commercially pure iron may be cold worked to a fair extent before the metal becomes too hard for further working. Impurities in iron or steel impair the cold working properties to the extent that most steels cannot be worked cold (apart from very special low-carbon mild steel sheets used in the car industry), although nearly all steels may be worked at the red-heat condition.

The exact nature of the annealing process used depends to a large extent on the purpose for which the annealed metal is to be used. There is a vast difference in technique between annealing in a steel works where enormous quantities of sheet steel are produced, and annealing in a workshop where single articles may require treatment. Briefly, cold working causes deformation by crushing or distorting the grain structure within the metal. In annealing, a metal or alloy is heated to a temperature at which recrystallization occurs, and then allowed to cool at a predetermined rate. In other words, crystals or grains within the metal which have been displaced and deformed during cold working are allowed to rearrange themselves into their natural formation during the process of annealing. Iron and low-carbon steels should be heated to about 900°C and allowed to cool very slowly to ensure maximum softness, as far as possible out of contact with air to prevent oxidation of the surface; this can be done by cooling the metal in warm sand. High-carbon steels require similar treatment except that the temperature to which the steel needs to be heated is somewhat lower and is in the region of 800°C. Copper should be heated to a temperature of about 550°C or dull red, and either quenched in water or allowed to cool out slowly. The rate of cooling does not affect the resulting softness of this metal. The advantage of quenching is that the surface of the metal is cleaned of dirt and scale. Aluminium may be annealed by heating to a temperature of 350°C. This may be done in a suitable oven or salt bath. In the workshop aluminium is annealed by the use of a blowpipe, and a stick or splinter of dry wood is rubbed on the heated metal; when the wood leaves a charred black mark the metal is annealed. Sometimes a piece of soap is used instead of the wood; when the soap leaves a brown mark the heating should be stopped. The metal may then be quenched in water or allowed to cool out slowly in air. Zinc becomes malleable between 100 and 150°C, and so may be annealed by immersing it in boiling water. Zinc should be worked while still hot, as it loses much of its malleability when cold.

3.15.2 Heat treatment of carbon steel

Steel is an important engineering material because, although cheap, it can be given a wide range of mechanical properties by heat treatment. Heat treatment can both change the size and shape of the grains, and alter the microconstituents. The shape of the grains can be altered by heating the steel to a temperature above that of recrystallization. The size of the grains can be controlled by the temperature and the duration of the heating, and the speed at which the steel is cooled after the heating; the microconstituents can be altered by heating the steel to a temperature that is sufficiently high to produce the solid solution austenite so that the carbon is dispersed, and then cooling it at a rate which will produce the desired structure. The micrograin structure of carbon steel has the following constituents:

Ferrite Pure iron.

Cementite Carbon and iron mixed.

Pearlite A sandwich layer structure of ferrite and cementite.

All low-carbon steels of less than 0.83 per cent carbon content consist of a combination of ferrite and pearlite. Carbon steel containing 0.83 per cent carbon is called eutectoid and consists of pure pearlite structure. Steels over 0.83 per cent carbon up to 1.2 per cent carbon are a mixture of cementite and pearlite structures. If a piece of carbon steel is heated steadily its temperature will rise at a uniform rate until it reaches 700°C. At this point, even though the heating is continued, the temperature of the steel will first remain constant for a short period and then continue to rise at a slower rate until it reaches 775°C. The pause in the temperature rise and the slowing down of the rate indicate that energy is being absorbed to bring about a chemical and structural change in the steel. The carbon in the steel is changing into a solid solution with the iron and forming what is known as austenite. The temperature at which this change in the structure of the steel starts is 700°C, which is known as the lower critical point; the temperature at which the change ends is known as the upper critical point. The difference between these points is termed the critical range. The lower critical point is the same for all steels, but the upper critical point varies with the carbon content as shown in Figure 3.4. Briefly, steels undergo a chemical and structural change, forming austenite, when heated to a temperature above the upper critical point; if allowed to cool naturally they return to their normal composition.

Steel can be heat treated by normalizing, hardening, tempering and case hardening as well as by annealing, which has already been described in Section 3.15.1.

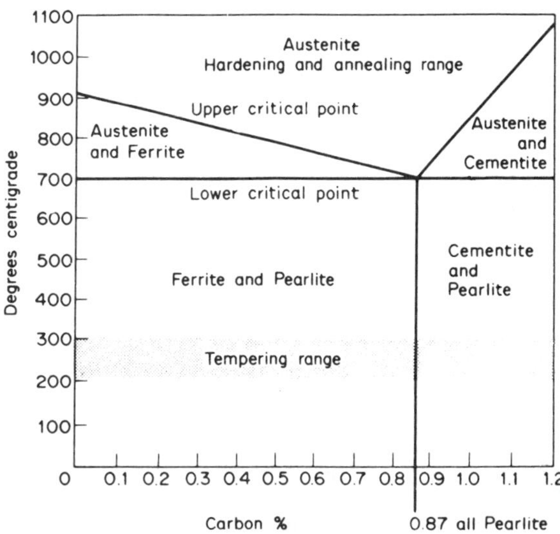

Figure 3.4 Changes in structure of carbon steel with temperature and carbon content

Normalizing
Normalizing is a process used to refine the grain structure of steel after it has been subjected to prolonged heating above the critical range (as in the case of forging) and to remove internal stresses caused by cold working. The process may appear to differ little from annealing, but as its name suggests the effect of normalizing is to bring the steel back to its normal condition and no attempt is made to soften the steel for further working. Normalizing is effected by slowly heating the steel to just above its upper critical range for just sufficient time to ensure that it is uniformly heated, and then allowing it to cool in still air.

Hardening
It has already been said that if a piece of steel is allowed to cool naturally after heating to above its upper critical point, it will change from austenite back to its original composition. If, however, the temperature of the heated steel is suddenly lowered by quenching it in clean cold water or oil, this change back from austenite does not take place, and instead of pearlite, a new, extremely hard and brittle constituent is formed, called martensite. This process makes steels containing 0.3 per cent or more carbon extremely hard, but steels having a carbon content of less than 0.3 per cent cannot be hardened in this way because the small amount of carbon produces too little martensite to have any noticeable hardening effect. The steel to be

Table 3.17 Temperature colours for steel

Colour	Temperature (°C)
Black	450–550
Very dark red	600–650
Dark red	700–750
Cherry red	800–850
Full red	850–900
Bright red	950–1000
Dark orange	1050–1100
Light orange	1150–1200
Yellow white	1270–1300
White (welding heat)	1400–1550

hardened should be quenched immediately it is uniformly heated to a temperature just above the upper critical point. It is also important not to overheat the steel and to allow it to cool to the quenching temperature. Whether water or oil is used for quenching depends on the use to which the steel is to be put. Water quenching produces an extremely hard steel but is liable to cause cracks and distortion. Oil quenching is less liable to cause these defects but produces a slightly softer steel. A more rapid and more even rate of cooling can be obtained if the steel is moved about in the cooling liquid, but only that part of the steel which is to be hardened should be moved up and down in the liquid in order to avoid a sharp boundary between the soft and hard portions.

A workshop method of hardening carbon tool steel is to heat the steel, using the forge or oxyacetylene blowtorch, to a dull red colour (see Table 3.17) and then quench it in water or oil. This would harden the article ready for tempering.

Tempering

Hardened steel is too brittle for most purposes, and the process of tempering is carried out to allow the steel to regain some of its normal toughness and ductility. This is done by heating the steel to a temperature below the lower critical point, usually between 200 and 300°C, thereby changing some of the martensite back to pearlite. The exact temperature will depend on the purpose for which the steel is intended; the higher the temperature, the softer and less brittle the tempered steel becomes. Methods used for controlling the tempering process depend on the size and class of the article to be tempered. One method is to heat the hardened steel in a bath of molten lead and tin, the melting points of various combinations of these two metals being used as an indication of the temperature. Another method of tempering small articles is to polish one face or edge and heat it with a flame. This polished surface will be seen to change colour as the heat is absorbed. The colour changes are caused by the formation at different temperatures of thin films of oxide, called tempering colours (Table 3.18). After tempering, the steel is either quenched or allowed to cool naturally.

Case hardening

Although mild steels having a carbon content of less than 0.3 per cent cannot be hardened, the surface of the mild steel can be changed to a high-carbon steel. Case hardening of mild steel can be divided into three main processes:

1 Carburizing.
2 Refining and toughening the core.
3 Hardening and tempering the outer case.

A method called pack carburizing is often used in small workshops. After thoroughly cleaning them, the steel parts to be carburized are packed in an iron box so that each part is entirely surrounded by 2.5 cm of carburizing compound. The box is sealed with fireclay, heated in a furnace to 950°C and kept at that temperature for two to 12 hours, during which time the carbon in the compound is absorbed by the surface of the steel parts. The steel parts are then allowed to cool slowly in the box. At this stage their cores will have a coarse grain structure due to prolonged heating, and the grain is refined by heating the steel parts to 900°C and quenching them in oil. Next the casing is hardened by reheating the steel to just under 800°C and then

Table 3.18 Tempering colours for plain carbon steel

Colour	Temperature (°C)	Type of article
Pale straw	220–230	Metal turning tools, scrapers, scribes
Dark straw	240–245	Taps, dies, reamers, drills
Yellow-brown	250–255	Large drills, wood turning tools
Brown	260–265	Wood working tools, chisels, axes
Purple	270–280	Cold chisels, press tools, small springs, punches, knives
Blue	290–300	Springs, screwdrivers, hand saws

quenching it in water or oil. Tempering of the casing may then be carried out in the normal way.

Small mild steel parts can be given a very thin casing by heating them in a forge to a bright red heat and coating them with carburizing powder. The parts are then returned to the forge and kept at a bright red heat for a short period to allow the carbon in the compound to penetrate the surface of the metal. Finally the parts are quenched and ready for use.

3.15.3 Heat treatment of aluminium alloys

An aluminium alloy is heat treated by heating it for a prescribed period at a prescribed temperature, and then cooling it rapidly, usually by quenching. The particular form of heat treatment which results in the alloy attaining its full strength is known as solution treatment. The alloy is raised to a temperature of 490°C by immersing it in a bath of molten salt. The bath is usually composed of equal parts of sodium nitrate and potassium nitrate contained in an iron tank. This tank is heated by gas burners and, except for its open top, is enclosed with the burners in a firebrick structure which conserves the heat. The temperature of the bath must be carefully regulated, as any deviation either above or below prescribed limits may result in the failure of the metal to reach the required strength. The alloy is soaked at 490°C for 15 minutes and then quenched immediately in cold water.

At the moment of quenching the alloy is reasonably soft, but hardening takes place fairly rapidly over the first few hours. Some alloys, chiefly the wrought materials, harden more rapidly and to a greater extent than others. Their full strength is attained gradually over four or five days (longer in cold weather); this process is known as natural age hardening. As age hardening reduces ductility, any appreciable cold working must be done while the metal is still soft. Working of the natural ageing aluminium alloys must be completed within two hours of quenching, or for severe forming within 30 minutes. Age hardening may be delayed by storing solution-treated material at low temperatures. Refrigerated storage, usually at 6–10°C, is used for strip sheet and rivets, and work may be kept for periods up to four days after heat treatment. If refrigerated storage is not used to prevent age hardening it may be necessary to repeat solution treatment of the metal before further work is possible.

Alloys of the hiduminium class may be artificially age hardened when the work is finished. Artificial ageing is often called precipitation treatment; this refers to the precipitation of the two intermetallic compounds responsible for the hardening, namely copper and manganese silicon. The process consists of heating the work in an automatically controlled oven to a temperature in the region of 170°C for a period of ten to 20 hours. Artificial ageing at this temperature does not distort the work. The temperature of the oven must be maintained to within a few degrees, and a careful check on the temperature is kept by a recording instrument. In order to ensure uniform distribution of temperature a fan is fitted inside the oven to keep the air in circulation. At the end of a period of treatment the oven is opened to allow the work to cool down. One of the chief advantages of this process is that work of a complicated character may be made and completed before ageing takes place. Moreover, numerous parts may be assembled or riveted together and will not suffer as a result of the ageing treatment.

3.16 DEVELOPMENT OF REINFORCED COMPOSITE MATERIALS

Glass fibres have been known in one form or another since 1500 BC, and some glass fabrics have been produced as far back as the beginning of the eighteenth century, but all of these were far too coarse to be of any industrial value. During the Iate 1930s glass fibre became available on a commercial scale and of a quality and smallness of diameter which enabled it to be manufactured into textile products. This was due to the development of the continuous filament process. It was soon apparent that the fibres made by this new process had many desirable properties such as strength, smallness of diameter, high elasticity, and the ability to withstand high temperatures.

Early attempts to use glass fibre as a reinforcement were disappointing. The resin then being used required high moulding pressures and this led to a crushing of the fibres with a resulting loss in strength. During the early 1940s an entirely new group of resins was introduced; these were known as contact resins as they could be used without pressure. Of the contact resins the polyesters are undoubtedly the best known, and they are the most widely used resins in the preparation of glass reinforced laminates because of their combination of low cost and ability to be moulded without pressure, and because their conditions of use fitted into normal workshop practice. With glass fibre as a reinforcement they could be used by relatively unskilled labour to produce strong, lightweight structures in complicated shapes which would be either too expensive to produce by panel beating or too complicated for pressing. The glass fibre reinforced plastic, having characteristics of high strength, low weight, resistance to corrosion, good electrical properties and

Table 3.19 Advantages of reinforced composite plastics (*Owens-Corning Fiberglas*)

Compared with metals	Compared with laminated thermoplastics	Compared with injection moulded thermoplastics	Compared with injection moulded thermosets	Compared with wood
1 Higher strength weight ratio 2 Easier and cheaper manufacture of complex shapes 3 Good corrosion resistance 4 Ability to incorporate self colours	1 Greater scope in mouldable shapes 2 Higher strength 3 Comparable or better electrical properties	1 Far higher strength 2 Improved dimensional stability 3 Higher temperature resistance	1 Far higher strength 2 Ability to be formed into thin flat sections or panels	1 Much higher strength 2 Greatly increased strength/weight ratio 3 Improved dimensional stability 4 Better weathering properties 5 Higher water resistance 6 Ease of fabricating complete structures

design flexibility, together with its suitability for a wider range of moulding methods, has resulted in a rapid growth of its use. A big advantage was the low tooling cost needed for limited production.

In the automobile field the availability of resins of various types and an extensive range of reinforcing materials has widened the scope of the designer. Increased production of polyester, although small compared with many other resins, has permitted raw material costs to be reduced sufficiently to make the polyester/glass combination competitive with traditional materials for the manufacture of many components.

Although glass reinforced laminates are characterized by high strength/weight ratio, it is well known that they do not offer the same flexural strength as steel and aluminium. Thus for bodywork the successful application of the materials is largely dependent on careful design and a full understanding of their properties. To the body designer one of the most outstanding characteristics of these materials is the ease with which complex curves can be produced as compared with methods of panel beating and wheeling, while from a production point of view the fact that complex structures can be produced as one-piece mouldings can greatly reduce assembly time and the need for complicated assembly fixtures.

Although this system would appear advantageous by reducing the large number of assembly operations required for building a steel body, it must be realized that in the layout of a steel body plant little or no allowance is made for processing time and that the speed of production is determined only by the time required for handling and transporting panels from one stage to the next. When time has to be allowed for curing, and with resins this may be considerable, production speed is greatly reduced and the floor area requirements may make the use of these materials prohibitive for quantity production.

The last few years have seen a rapidly increasing acceptance of the material as an ideal alternative to steel in the motor industry. Use of glass fibre as a replacement for steel in body panels is economically possible at comparatively low volume levels, say 1000 vehicles a year. The reason is that, unlike steel bodies, glass fibre panels do not require a large investment in presses and dyes, and moulds to produce the panels, as well as jigs and fixtures to manufacture the vehicle can also largely be produced from glass fibre.

From the manufacturer's viewpoint, glass fibre production means low tooling costs, comparatively small initial investment, flexibility of design and strength with lightness (Table 3.19). The user benefits by the complete resistance of glass fibre vehicles to corrosion, their sturdiness in a collision, and ease of repair.

3.17 BASIC PRINCIPLES OF REINFORCED COMPOSITE MATERIALS

The basic principle involved in reinforced plastic production is the combination of polyester resin and reinforcing fibres to form a solid structure (Figure 3.5). Glass reinforced plastics are essentially a family of structural materials which utilize a very wide range of thermoplastic and thermosetting resins. The incorporation of glass fibres in the resins changes them from relatively low-strength, brittle materials into strong and resilient structural materials. In many ways glass fibre reinforced plastic can be compared to concrete, with the glass fibres performing the same

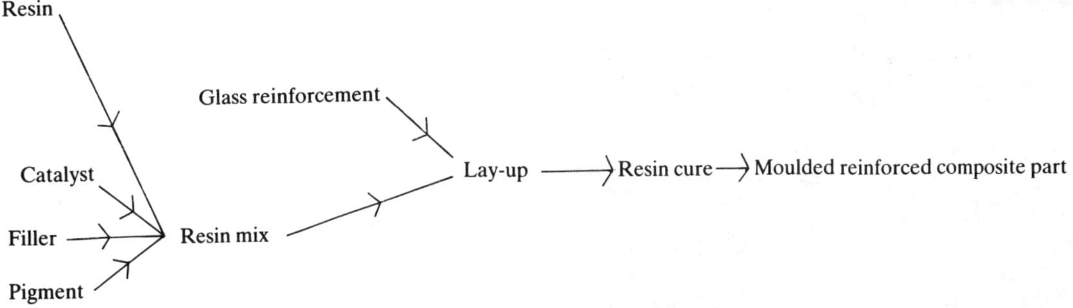

Figure 3.5 Flow chart showing the principles of reinforced composite materials

function as the steel reinforcement and the resin matrix acting as the concrete. Glass fibres have high strength and high modulus, and the resin has low strength and low modulus. Despite this the resin has the important task of transferring the stress from fibre to fibre, so enabling the glass fibre to develop its full strength.

Polyester resins are supplied as viscous liquids which solidify when the actuating agents, in the form of a catalyst and accelerator, are added. The proportions of this mixture, together with the existing workshop conditions, dictate whether it is cured at room temperature or at higher temperatures and also the length of time needed for curing. In common practice pre-accelerated resins are used, requiring only the addition of a catalyst to effect the cure at room temperature. Glass reinforcements are supplied in a number of forms, including chopped strand mats, needled mats, bidirectional materials such as woven rovings and glass fabrics, and rovings which are used for chopping into random lengths or as high-strength directional reinforcement. Other materials needed are the releasing agent, filler and pigment concentrates for the colouring of glass fibre reinforced plastic.

Among the methods of production, the most used method is that of contact moulding, or the wet laying-up technique as it is sometimes called. The mould itself can be made of any material which will remain rigid during the application of the resin and glass fibre, which will not be attacked by the chemicals involved, and which will also allow easy removal after the resin has set hard. Those in common use are wood, plaster, sheet metal and glass fibre itself, or a combination of these materials. The quality of the surface of the completed moulding will depend entirely upon the surface finish of the mould from which it is made. When the mould is ready the releasing agent is applied, followed by a thin coat of resin to form a gel coat. To this a fine surfacing tissue of fibreglass is often applied. Further resin is applied, usually by brush, and carefully cut-out pieces of mat or woven cloth are laid in position. The use of split washer rollers removes the air and compresses the glass fibres into the resin. Layers of resin and glass fibres are added until the required thickness is achieved. Curing takes place at room temperature but heat can be applied to speed up the curing time. Once the catalyst has caused the resin to set hard, the moulding can be taken from the mould.

3.18 MANUFACTURE OF REINFORCED COMPOSITE MATERIALS

When glass is drawn into fine filaments its strength greatly increases over that of bulk glass. Glass fibre is one of the strongest of all materials. The ultimate tensile strength of a single glass filament (diameter 9–15 micrometres) is about 3 447 000 kN/m^2. It is made from readily available raw materials, and is non-combustible and chemically resistant. Glass fibre is therefore the ideal reinforcing material for plastics. In Great Britain the type of glass which is principally used for glass fibre manufacture is E glass, which contains less than 1 per cent alkali borosilicate glass. E glass is essential for electrical applications and it is desirable to use this material where good weathering and water resistance properties are required. Therefore it is greatly used in the manufacture of composite vehicle body shells, both for private and for commercial vehicles.

Basically the glass is manufactured from sand or silica and the process by which it is made proceeds through the following stages:

1 Initially the raw materials, including sand, china clay and limestone, are mixed together as powders in the desired proportions.
2 The ‘glass powder’, or frit as it is termed, is then fed into a continuous melt furnace or tank.
3 The molten glass flows out of the furnace through a forehearth to a series of fiberizing units usually

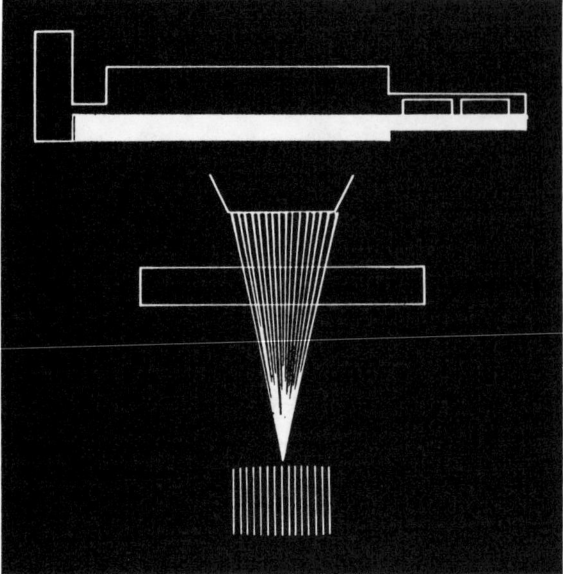

Figure 3.6 Manufacture of glass fibre. The story starts with molten glass, heated in a modern furnace. starts with molten glass, heated in a modern furnace. As it is pulled through tiny holes, that liquid mass is transformed into fibres smaller in diameter than a human hair. These fine fibres are then assembled into textile yarn, or into reinforcement products, including glass fibre roving, mat, chopped strands, glass web and milled fibres material (*Owens-Corning Fiberglas*)

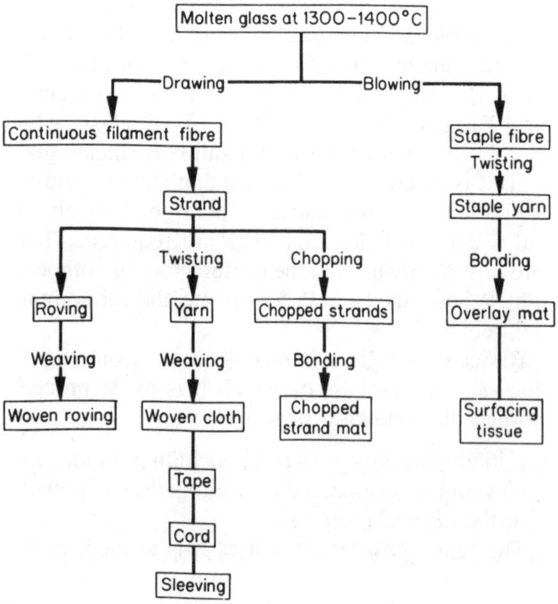

Molten glass at 1300–1400°C
Drawing
Blowing
Continuous filament fibre
Staple fibre
Twisting
Strand
Staple yarn
Twisting
Chopping
Bonding
Roving
Yarn
Chopped strands
Overlay mat
Weaving
Weaving
Bonding
Woven roving
Woven cloth
Chopped strand mat
Surfacing tissue
Tape
Cord
Sleeving

Figure 3.7 Derivation of glass fibre reinforcement

referred to as bushings, each containing several hundreds of fine holes. As the glass flows out of the bushings under gravity it is attenuated at high speed.

After fiberizing, the filaments are coated with a chemical treatment usually referred to as a forming size. The filaments are then drawn together to form a strand which is wound on a removable sleeve on a high-speed winding head (Figure 3.6).

The basic packages are usually referred to as cakes and form the basic glass fibre which, after drying, is processed into the various reinforcement products (Figure 3.7). Most reinforcement materials are manufactured from continuous filaments ranging in fibre diameter from 5 to 13 micrometres. The fibres are made into strands by the use of size. In the case of strands which are subsequently twisted into weaving yarns, the size lubricates the filaments as well as acting as an adhesive. These textile sizes are generally removed by heat or solvents and replaced by a chemical finish before being used with polyester resins. For strands which are not processed into yarns it is usual to apply sizes which are compatible with moulding resins.

Glass reinforcements are supplied in a number of forms, including chopped strand mats, needled mats, bidirectional materials such as woven rovings and glass fabrics, and rovings which are used for chopping into random lengths or as high-strength directional reinforcements (Figures 3.8–3.11).

3.19 TYPES OF REINFORCING MATERIAL

3.19.1 Woven fabrics

Glass fibre fabrics are available in a wide range of weaves and weights. Lightweight fabrics produce laminates with higher tensile strength and modulus than heavy fabrics of a similar weave. The type of weave will also influence the strength (due, in part, to the amount of crimp in the fabric), and usually satin weave fabrics, which have little crimp, give stronger laminates than plain weaves which have a higher crimp. Satin weaves also drape more easily and are quicker to impregnate. Besides fabrics made from twisted yarns, it is now the practice to use woven fabrics manufactured from rovings. These fabrics are cheaper to produce and can be much heavier in weight (Figure 3.12).

3.19.2 Chopped strand mat

Chopped strand glass mat (CSM) is the most widely used form of reinforcement. It is suitable for moulding the most complex forms. The strength of laminates

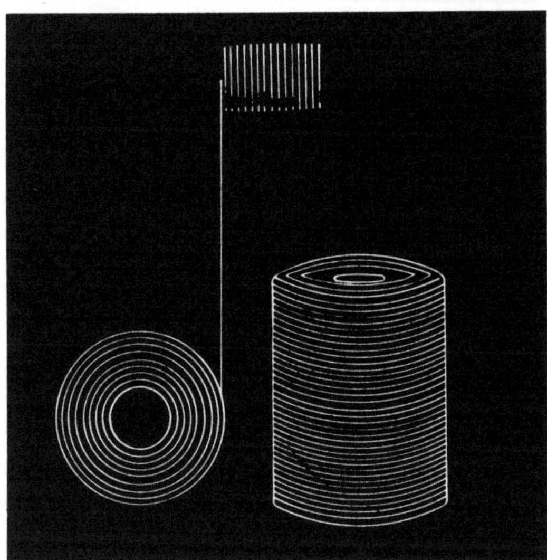

Figure 3.8 Manufacture of rovings. Rovings can be supplied to suit a variety of processes, including projection moulding, continuous laminating, filament winding, pultrusion and as reinforcement for sheet moulding compound. These rovings consist of continuous glass strands, gathered together without any mechanical twist and would to form a tubular, cylindrical package (*Owens-Corning Fiberglas*)

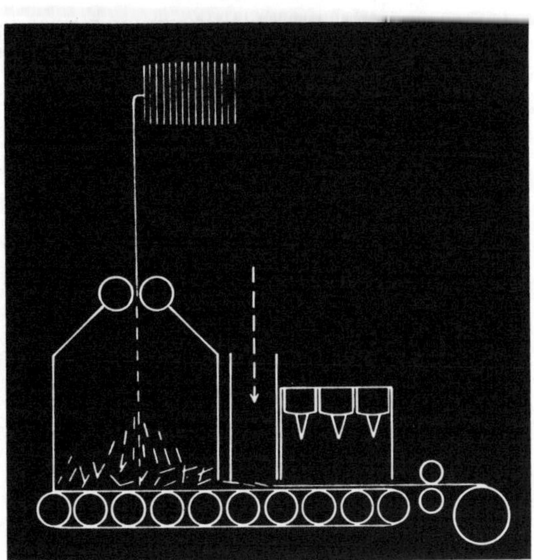

Figure 3.10 Manufacture of chopped strands. Chopped strands are widely used to reinforce thermoplastic compound (GRTP), polyester bulk moulding compounds (BMC) and in the manufacture of wet-laid glass webs. Chopped strands consist of continuous glass strands chopped to a desired length and are available with a wide variety of surface treatments to ensure compatibility with most resin systems. They are generally solvent and heat resistant and offer excellent flow properties (*Owens-Corning Fiberglas*)

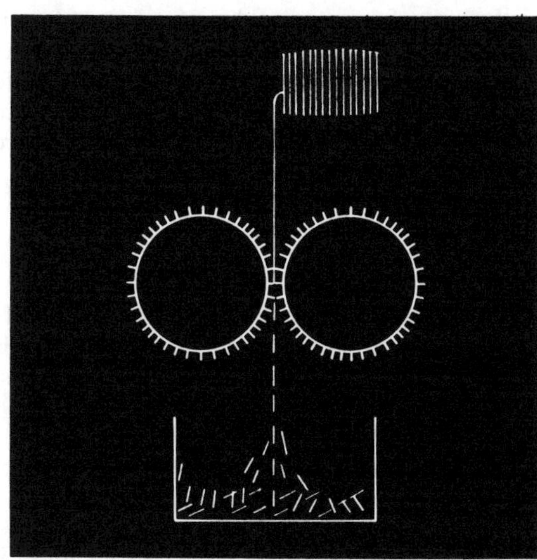

Figure 3.9 Manufacture of mats. Glass fibre mats are used as resin reinforcement in contact and compression moulded application. Chopped strand mats, made from fine chopped glass strands bonded with a powder of emulsion binder, are used in both areas (*Owens-Corning Fiberglas*)

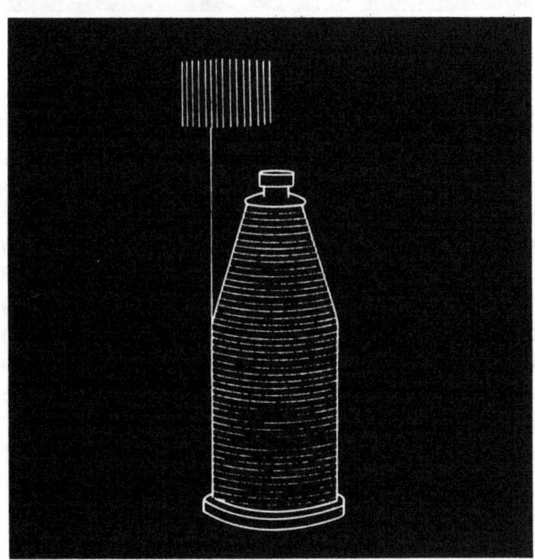

Figure 3.11 Manufacture of yarn (*Owens-Corning Fiberglas*)

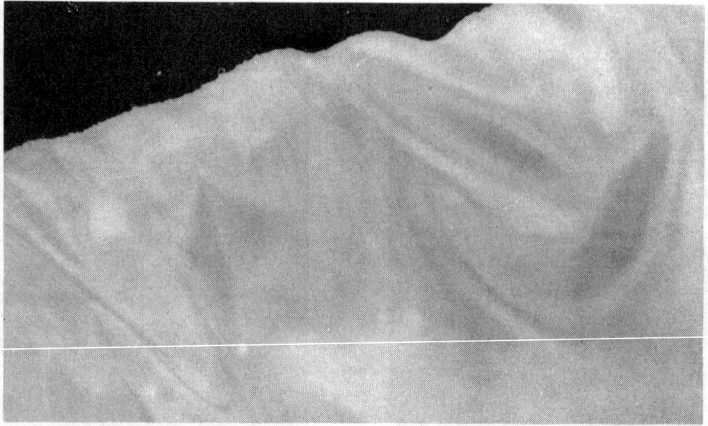

Figure 3.12 Woven fabric (*Scott Bader Co. Ltd*)

Figure 3.13 Rovings (*Owens-Corning Fiberglas*)

made from chopped strand mat is less than that with woven fabrics, since the glass content which can be achieved is considerably lower. The laminates have similar strengths in all directions because the fibres are random in orientation. Chopped strand mat consists of randomly distributed strands of glass about 50 mm long which are bonded together with a variety of adhesives. The type of binder or adhesive will produce differing moulding characteristics and will tend to make one mat more suitable than another for specific applications.

3.19.3 Needle mat

This is mechanically bound together and the need for an adhesive binder is eliminated. This mat has a high resin pick-up owing to its bulk, and cannot be used satisfactorily in moulding methods where no pressure is applied. It is used for press moulding and various low-pressure techniques such as pressure injection, vacuum and pressure bag.

3.19.4 Rovings

These are formed by grouping untwisted strands together and winding them on a 'cheese'. They are used for chopping applications to replace mats either in contact moulding (spray-up), or translucent sheet manufacture of press moulding (preform). Special grades of roving are available for each of these different chopping applications. Rovings are also used for weaving, for filament winding and for pultrusion processes. Special forms are available to suit these processes (Figure 3.13).

3.19.5 Chopped strands

These consist of rovings prechopped into strands of 6 mm, 13 mm, 25 mm or 50 mm lengths. This material is used for dough moulding compounds, and in casting resins to prevent cracking (Figure 3.14).

3.19.6 Staple fibres

These are occasionally used to improve the finish of mouldings. Two types are normally available, a compact form for contact moulding and a soft bulky form for press moulding. These materials are frequently used to reinforce gel coats. The weathering properties of translucent sheeting are considerably improved by the use of surfacing tissue (Figure 3.15).

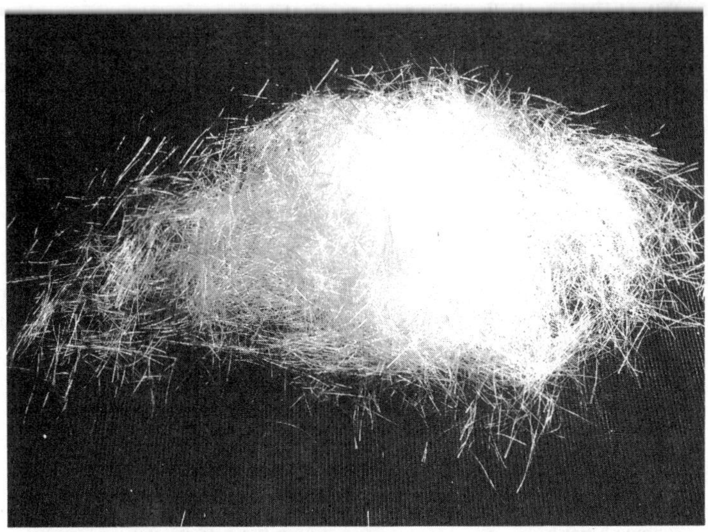

Figure 3.14 Chopped strands (*Owens-Corning Fiberglas*)

Figure 3.15 Staple fibres (surface mat) (*Owens-Corning Fiberglas*)

3.19.7 Application of these materials

Probably chopped strand mat is most commonly used for the average moulding. It is available in several different thicknesses and specified by weight: 300, 450 and 600 g/m^2. The 450 g/m^2 is the most frequently used, and is often supplemented with the 300 g/m^2. The 600 g/m^2 density is rather too bulky for many purposes, and may not drape as easily, although all forms become very pliable when wetted with the resin.

The woven glass fibre cloths are generally of two kinds, made from continuous filaments or from staple fibres. Obviously, most fabricators use the woven variety of glass fibre for those structures that are going to be the most highly stressed. For example, a moulded glass fibre seat pan and squab unit would be made with woven material as reinforcement, but a detachable hard top for a sports car body would more probably be made with chopped strand mat as a basis. However, it is quite customary to combine cloth and mat, not only to obtain adequate thickness, but because if the sandwich of resin, mat and cloth is arranged so that the mat is nearest to the surface of the final product, the appearance will be better.

The top layer of resin is comparatively thin, and the weave of cloth can show up underneath it, especially if some areas have to be buffed subsequently. Chopped fibres do not show up so prominently, but some fabricators compromise by using the thinnest possible cloth (surfacing tissue as it is known) nearest the surface, on top of the chopped strand mat. When moulding, these orders are of course reversed, the tissue going on to the gel coat on the inside of the mould, followed by the mat and resin lay-up.

It is important to note that if glass cloths or woven mat are used, it is possible to lay up the materials so that the reinforcement is in the direction of the greatest stresses, thus giving extra strength to the entire fabrication. In plain weave cloths, each warp and weft thread passes over one yarn and under the next. In twill weaves, the weft yarns pass over one warp and under more than one warp yarn; in 2 × 1 twill, the weft yarns pass over one warp yarn and under two warp yarns. Satin weaves may be of multishaft types, when each warp and weft yarn goes under one and

Table 3.20 Advantages and disadvantages of glass fibre reinforcement

	Advantages	Uses	Disadvantages
ROVINGS	Unidirectional strength	Spray-up Local longitudinal strength Mechanical bond for bulkheads Making tubes	Limited use in hand lay-up
WOVEN ROVINGS	Easy to handle Bidirectional strength High glass content High impact resistance	To increase longitudinal and transverse strength	Poor interlaminar adhesion Traps air, causes voids High cost
CHOPPED STRAND MAT	Multidirectional strength Low cost Good interlaminar bond Can be moulded into complex shape	Various General-purpose reinforcement	
WOVEN CLOTH	High strength Smooth finish	Sheathing As a fire barrier	Very high cost difficult to 'wet out'
SURFACE TISSUE	Fine texture	Reinforcing gelcoat	Low strength

over several yarns. Unidirectional cloth is one in which the strength is higher in one direction than the other, and balanced cloth is a type with the warp and weft strength about equal. Although relatively expensive the woven forms have many excellent qualities, including high dimensional stability, high tensile and impact strength, good heat, weather and chemical resistance, low moisture absorption, resistance to fire and good thermoelectrical properties. A number of different weaves and weights are available, and thicknesses may range from 0.05 mm to 9.14 mm, with weights from 30 g/m^2 to 1 kg/m^2, although the grades mostly used in the automotive field probably have weights of about 60 g/m^2 and will be of plain, twill or satin weave.

The advantages and disadvantages of glass fibre reinforcement materials are indicated in Table 3.20.

3.19.8 Carbon fibre

This is another reinforcing material. Carbon fibres possess a very high modulus of elasticity, and have been used successfully in conjunction with epoxy resin to produce low-density composites possessing high strength.

3.20 RESINS USED IN REINFORCED COMPOSITE MATERIALS

The first manmade plastics were produced in this country in 1862 by Alexander Parkes and were the forerunner of celluloid. Since then a large variety of plastics have been developed commercially, particularly in the last 25 years. They extend over a wide range of properties. Phenol formaldehyde is a hard thermoset material; polystyrene is a hard, brittle thermoplastic; polythene and plasticized polyvinyl chloride (PVC) are soft, tough thermoplastic materials; and so on. Plastics also exist in various physical forms. They can be bulk solid materials, rigid or flexible foams, or in the form of sheet or film. All plastics have one important common property. They are composed of macromolecules, which are large chain-like molecules consisting of many simple repeating units. The chemist calls these molecular chains polymers. Not all polymers are used for making plastic mouldings. Manmade polymers are called synthetic resins until they have been moulded in some way, when they are called plastics (Figure 3.16).

Most synthetic resins are made from oil. The resin is an essential component of glass fibre reinforced plastic. The most widely used is unsaturated polyester resin, which can be cured to a solid state either by catalyst and heat or by catalyst and accelerators at room temperature. The ability of polyester resin to cure at room temperature into a hard material is one of the main reasons for the growth of the reinforced plastics industry. It was this which led to development of the room temperature contact moulding methods which permit production of extremely large integral units.

Polyester resins are formulated by the reaction of

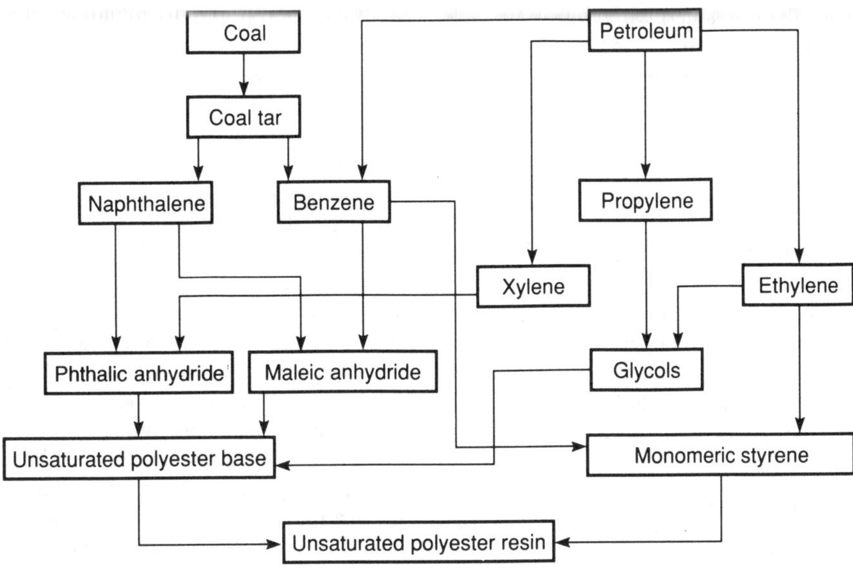

Figure 3.16 Production of unsaturated polyester resin (*Scott Bader Co. Ltd*)

organic acids and alcohols which produces a class of material called esters. When the acids are polybasic and the alcohols are polyhydric they can react to form a very complex ester which is generally known as polyester. These are usually called alkyds, and have long been important in surface coating formulations because of their toughness, chemical resistance, and endurance. If the acid or alcohol used contains an unsaturated carbon bond, the polyester formed can react further with other unsaturated materials such as styrene or diallyl phthalate. The result of this reaction is to interconnect the different polyester units to form the three-dimensional cross-linked structure that is characteristic of thermosetting resins. The available polyesters are solutions of these alkyds in the cross-linking monomers. The curing of the resin is the reaction of the monomer and the alkyd to form the cross-linked structure. An unsaturated polyester resin is one which is capable of being cured from a liquid to a solid state when subjected to the right conditions. It is usually referred to as polyester.

3.20.1 Catalysts and accelerators

In order to mould or laminate a polyester resin, the resin must be cured. This is the name given to the overall process of gelation and hardening, which is achieved either by the use of a catalyst and heating, or at normal room temperature by using a catalyst and an accelerator. Catalysts for polyester resins are usually organic peroxides. Pure catalysts are chemically unstable and liable to decompose with explosive violence. They are supplied, therefore, as a paste or liquid dispersion in a plasticizer, or as a powder in an inert filler. Many chemical compounds act as accelerators, making it possible for the resin-containing catalyst to be cured without the use of heat. Some accelerators have only limited or specific uses, such as quaternary ammonium compounds, vanadium, tin or zirconium salts. By far the most important of all accelerators are those based on a cobalt soap or those based on a tertiary amine. It is essential to choose the correct type of catalyst and accelerator, as well as to use the correct amount, if the optimum properties of the final cured resin or laminate are to be obtained.

3.20.2 Pre-accelerated resins

Many resins are supplied with an in-built accelerator system controlled to give the most suitable gelling and hardening characteristics for the fabricator. Pre-accelerated resins need only the addition of a catalyst to start the curing reaction at room temperature. Resins of this type are ideal for production runs under controlled workshop conditions.

The cure of a polyester resin will begin as soon as a suitable catalyst is added. The speed of the reactions will depend on the resin and the activity of the catalyst. Without the addition of an accelerator, heat or ultraviolet radiation, the resin will take a considerable time to cure. In order to speed up this reaction at room temperature it is usual to add an accelerator.

The quantity of accelerator added will control the time of gelation and the rate of hardening.

There are three distinct phases in the curing reaction:

Gel time This is the period from the addition of the accelerator to the setting of the resin to a soft gel.

Hardening time This is the time from the setting of the resin to the point when the resin is hard enough to allow the moulding or laminate to be withdrawn from the mould.

Maturing time This may be hours, several days or even weeks depending on the resin and curing system, and is the time taken for the moulding or laminate to acquire its full hardness and chemical resistance. The maturing process can be accelerated by post-curing.

3.20.3 Fillers and pigments

Fillers are used in polyester resins to impart particular properties. They will give opacity to castings and laminates, produce dense gel coats, and impart specific mechanical, electrical and fire resisting properties. A particular property may often be improved by the selection of a suitable filler. Powdered mineral fillers usually increase compressive strength; fibrous fillers improve tensile and impact strength. Moulding properties can also be modified by the use of fillers; for example, shrinkage of the moulding during cure can be considerably reduced. There is no doubt, also, that the wet lay-up process on vertical surfaces would be virtually impossible if thixotropic fillers were not available.

Polyester resins can be coloured to any shade by the addition of selected pigments and pigment pastes, the main requirement being to ensure thorough dispersion of colouring matter throughout the resin to avoid patchy mouldings.

Both pigments and fillers can increase the cure time of the resin by dilution effect, and the adjusted catalyst and promoter are added to compensate.

3.20.4 Releasing agents

Releasing agents used in the normal moulding processes may be either water-soluble film-forming compounds, or some type of wax compound. The choice of releasing agent depends on the size and complexity of the moulding and on the surface finish of the mould. Small mouldings of simple shape, taken from a suitable GRP mould, should require only a film of polyvinyl alcohol (PVAL) to be applied as a solution by cloth, sponge or spray. Some mouldings are likely to stick if only PVAL is used. PVAL is available as a solution in water or solvent, or as a concentrate which has to be diluted, and it may be in either coloured or colourless form.

Suitable wax emulsions are also available as a releasing agent. They are supplied as surface finishing pastes, liquid wax or wax polishes. The recommended method of application can vary depending on the material to be finished. Hand apply with a pad of damp, good quality mutton cloth or equivalent, in straight even strokes. Buff lightly to a shine with a clean, dry, good quality mutton cloth. Machine at 1800 rpm using a G-mop foam finishing head. Soak this head in clean water before use and keep damp during compounding. Apply the wax to the surface. After compounding, remove residue and buff lightly to a shine with a clean, dry, good quality mutton cloth.

Wax polishes should be applied in small quantities since they contain a high percentage of wax solids. Application with a pad of clean, soft cloth should be limited to an area of approximately 1 square metre. Polishing should be carried out immediately, before the wax is allowed to dry. This can be done either by hand or by machine with the aid of a wool mop polishing bonnet.

Frekote is a semi-permanent, multirelease gloss finish, non-wax polymeric mould release system specially designed for high-gloss polyester mouldings. It will give a semi-permanent release interface when correctly applied to moulds from ambient up to 135°C. This multirelease interface chemically bonds to the mould's surface and forms on it a microthin layer of a chemically resistant coating. It does not build up on the mould and will give a high-gloss finish to all polyester resins, cultured marble and onyx. It can be used on moulds made from polyester, epoxy, metal or composite moulds. Care should be taken to avoid contact with the skin, and the wearing of suitable clothing, especially gloves, is highly recommended. These products must be used in a well-ventilated area.

3.20.5 Adhesives used with GRP

Since polyester resin is highly adhesive, it is the logical choice for bonding most materials to GRP surfaces.

Suitable alternatives include the Sika Technique, which is a heavy-duty, polyurethane-based joining compound. It cures to a flexible rubber which bonds firmly to wood, metal, glass and GRP. It is ideal for such jobs as bonding glass to GRP or bonding GRP and metal, as is often required on vehicles with GRP bodywork. It is not affected by vibration and is totally waterproof. The Araldite range includes a number of industrial adhesives which are highly recommended for use with GRP. Most high-strength impact adhesives (superglues) can be used on GRP laminates.

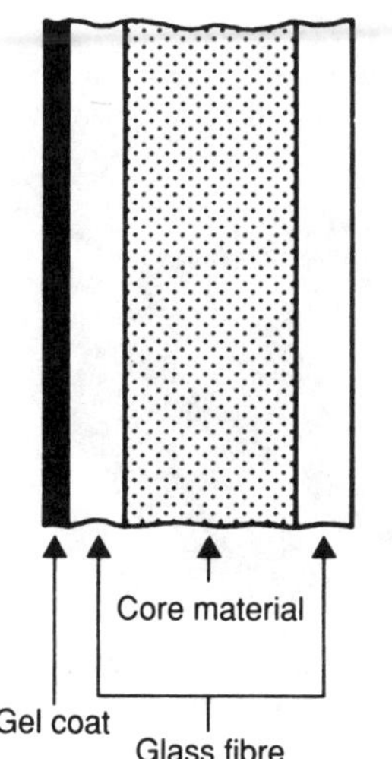

Figure 3.17 Typical sandwich construction (*Scott Bader Co. Ltd*)

Most other adhesives will be incapable of bonding strongly to GRP and should not be used when maximum adhesion is essential.

3.20.6 Core materials

Core materials, usually polyurethane, are used in sandwich construction, that is basically a laminate consisting of a foam sheet between two or more glass fibre layers (Figure 3.17). This gives the laminate considerable added rigidity without greatly increasing weight. Foam materials are available which can be bent and folded to follow curved surfaces such as vehicle bodies. Foam sheet can be glued or stapled together, then laminated over to produce a strong box structure, without requiring a mould. Typical formers and core materials are paper rope, polyurethane rigid foam sheet, scoreboard contoured foam sheet, Termanto PVC rigid foam sheet, Term PVC contoured foam sheet, and Termino PVC contoured foam sheet.

3.20.7 Formers

A former is anything which provides shape or form to a GRP laminate. They are often used as a basis for stiffening ribs or box sections. A popular material for formers is a paper rope, made of paper wound on flexible wire cord. This is laid on the GRP surface and is laminated over to produce reinforcing ribs, which give added stiffness with little extra weight. The former itself provides none of the extra stiffness; this results entirely from the box section of the laminate rib. Wood, metal, or plastic tubing and folded cardboard can all be used successfully as formers. Another popular material is polyurethane foam sheet, which can be cut and shaped to any required form (Figure 3.18).

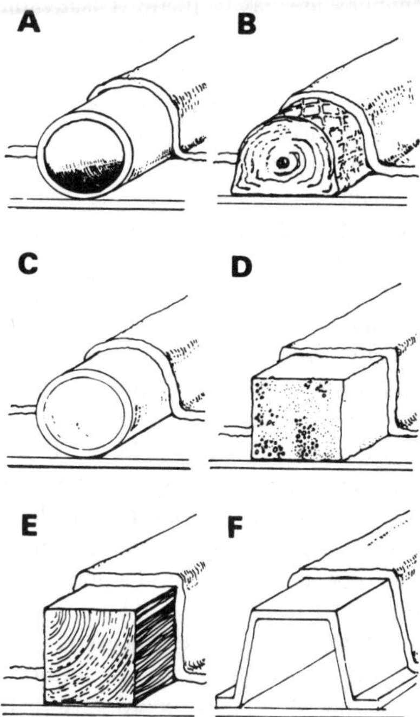

Figure 3.18 Typical formers: (a) metal tube (b) paper rope (c) cardboard tube (d) foam strip (e) wood (f) folded cardboard (*Scott Bader Co. Ltd*)

3.21 MOULDING TECHNIQUES FOR REINFORCED COMPOSITE LAMINATES

3.21.1 Contact moulding

This is the oldest, simplest and most popular fabrication technique for the automotive, reinforced plastic body industry. It is normally used for relatively short runs, but it has also been adapted successfully for series production. It is the only production method which takes full advantage of the two most important characteristics of polyester resin, namely that it can be set without heat and without pressure. A considerable industry has been built around contact moulding, which has facilitated the cost effective production of large one-piece mouldings, particularly for low production runs. Contact moulding advantages are that a minimum of equipment is required, tooling is inexpensive, there are practically no size restrictions, and design changes are easily made. Disadvantages are that the labour context is high and the quality of the moulding depends on the skill of the operators. The lay-up and curing times are comparatively slow, and only one good surface finish is achieved.

The contact moulding process is carried out in the following manner. A master pattern or model is made, representing in all its dimensions the finished product. This could be, for example, a full-size wood model of the type of body shell or cab shell required, or it might equally well be a steel or aluminium panel-beaten structure of composite type, or even a plaster model reinforced with wire mesh. From this is made a master mould, which must be female or concave for the most part, and this would in all probability be made in reinforced plastics similar to those used for the final product. It is important to differentiate, as a matter of terminology, between 'mould' and 'moulding', one being the production tool and the other the product itself.

An important aspect of the process is that the surface of the mould will inevitably be faithfully reproduced in the moulding, and accordingly if the mould is bumpy or rough so will be the final article. It does not usually matter if the unseen or partly hidden side of the moulding is rough (indeed it usually is), but for the displayed surface to be unsightly is not normally tolerable. This is why a female mould is usually used. If the original pattern was very smooth, so will be the inside of the mould and, therefore, the outside of the moulding. This is why so much trouble is taken over the pattern. If in wood it is tooled and sanded to perfection, and if in metal it is panel beaten with the greatest possible skill, then ground and polished if necessary. It is of the greatest importance that separation of the moulding from the mould be easy. A special compound or a polish such as carnauba wax and/or silicone lubricant can be used; plastics film is occasionally used as a separating membrane, as it will not adhere strongly to either the mould or the moulding. For contact moulding, the equipment is relatively simple and inexpensive. Contact moulding can be further subdivided into hand lay-up and spray lay-up.

The application of the release agent to the mould is followed by brush or spray application of a gel coat of resin. There has been a constant improvement and development in polyester resins, and among other things this has led to the introduction of successful semi-flexible gel coats, which are of particular interest to the motor industry. The gel coat is a continuous skin on the working face of a moulding. It is almost pure resin and its object is to give a good finish as well as to protect the working surface from corrosion and other damage. It also hides the fibre pattern of the reinforcement. The gel coat can be colour impregnated or otherwise specially formulated, for example for extra abrasion or impact resistance. It should be as even in thickness as possible, as thicker areas are very susceptible to accidental damage, while thin patches can lower the resistance of the structure to moisture and to chemical attack.

Figure 3.19 Application of release agent to mould surface (*Scott Bader Co. Ltd*)

Even in hand lay-up the spray method may be used for this stage and for the application of the separating agent, so that there is a small element of mechanization. A fine surfacing tissue may be applied to the gel coat while wet, or it may simply be allowed to gel as it is. Further resin is sprayed on or brushed on, and mat or woven cloth, which has been carefully cut to patterns, is laid in position. Consolidation and air removal are then effected by manual means. It is customary to use rollers made up of split washers for this operation, which is an extremely important one if consistency and strength of the moulding are to be obtained. More mat or cloth is added in order to build up the requisite thickness of reinforced plastics, and the moulding is allowed to set or cure. Curing normally takes place at room temperature, but sometimes under a certain degree of heat if the process is to be accelerated. It should be remembered that curing is itself a heat-producing process. Contact moulding can also be carried out by simple mechanical means, but the general principle is always that of bringing the materials into contact with the mould, without the use of any dies.

The following is a summary of the contact moulding process:

1 The master mould must be spotlessly clean.
2 A release agent is applied to the entire surface of the mould face (Figure 3.19).

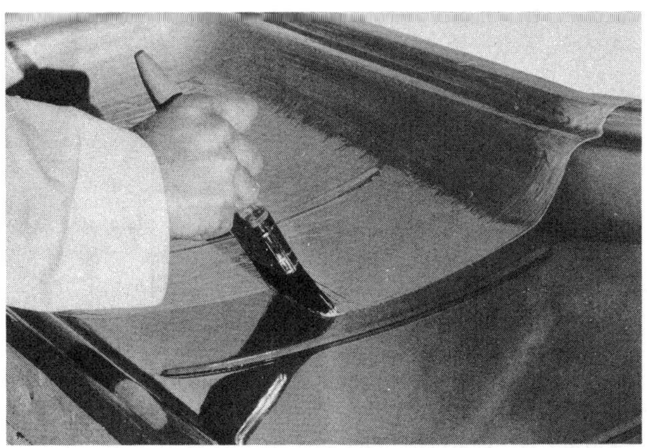

Figure 3.20 Application of gel coat covering (*Scott Bader Co. Ltd*)

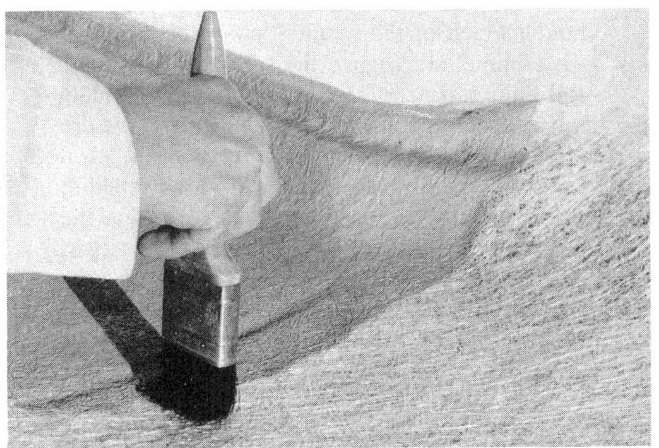

Figure 3.21 Impregnation of glass mat with resin (*Scott Bader Co. Ltd*)

Figure 3.22 Rolling impregnated mat (*Scott Bader Co. Ltd*)

Figure 3.23 Final mould and moulding (*Scott Bader Co. Ltd*)

3 Gel coat covering is applied by brush or spray (Figure 3.20).
4 Catalyst is added to the resin and the catalysed resin is smoothed over the gel-coated mould.
5 Glass fibre mat, precut to the exact size, is laid on the mould and a further small quantity of resin is poured on to the mat. With brushes and hand roller, the resin is drawn through the mat (Figure 3.21). A second layer of mat is applied and the drawing-through process is repeated until the required thickness is achieved. It is critical that all air bubbles be removed by brushing and rolling (Figure 3.22).
6 The mould is allowed to cure naturally or heat is used to speed up the curing process.
7 After curing, the moulds are broken and the completed sections are removed (Figure 3.23).
8 They are then trimmed and ready for use.

3.21.2 Spray-up technique

A development from the basic manual contact process which is employed with increasing frequency in the automotive body industry is known as spray-up (Figure 3.24). In this method, rovings are automatically fed through a chopping unit and the resultant chopped strands are blown or carried by the sprayed resin stream on to the mould. The glass and resin mix applied in this way is consolidated, and the air pockets or bubbles are removed by manual rolling, as in simple hand lay-up.

There are several commercial spraying systems available, where the glass fibre and resin are depos-

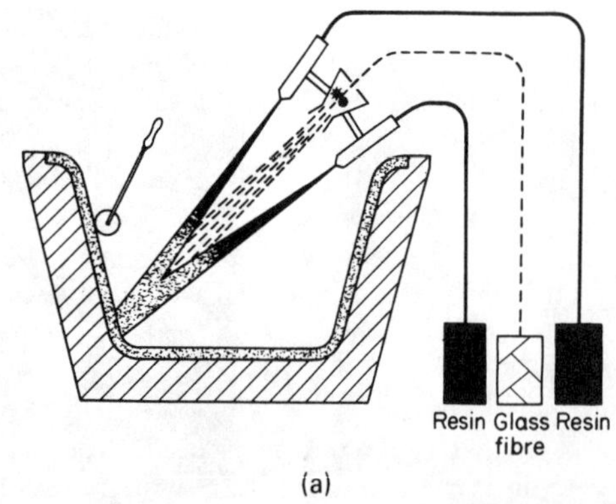

Figure 3.24 The spray-up technique (*BP Chemicals (UK) Ltd*)

ited simultaneously on the mould face. They consist of two principal types. In the *twin-pot system* a twin-nozzle spray gun is used, and in order to prevent gelation in the gun the resin is divided into two parts, one of which is catalysed and the other accelerated. The two streams of resin spray converge near the surface of the mould simultaneously with a stream of glass fibre ejected by a glass rovings chopper. In the other type, the catalyst injector system, accelerated

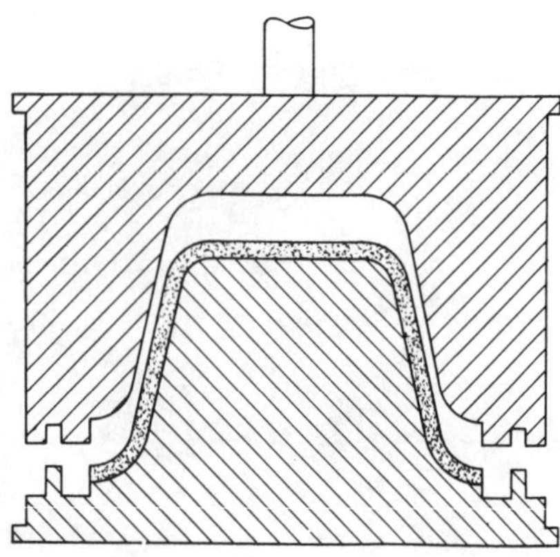

Figure 3.25 Hot press moulding

resin is sprayed from a single-nozzle gun, but liquid catalyst is metered into the resin in the gun itself. A glass rovings chopper delivers the reinforcement to the mould surface as in the former system.

Although much of the manual labour of the hand lay-up is eliminated by using the spray process, thorough rolling is still necessary, not only to consolidate the deposited glass resin mixture, but also to ensure that the accelerated and catalysed portions of the resin are adequately mixed. Considerable skill is needed to control the thickness of the laminate when using the resin glass spray gun. Spraying reduces labour costs, especially when the volume of production is large enough to keep the equipment in constant use.

3.21.3 Hot press moulding

This process involves the use of chopped strand glass mat, pre-impregnated with polyester resin, which is then in general principle formed in presses in a similar manner to that used for forming steel sheet (Figure 3.25). In this case, however, the dies, which are preheated, have to remain closed for the curing cycle of the pre-impregnated mat, which may involve a period from 15 to 30 seconds. Hot press moulding using matched dies have good finish on both surfaces. Further, this method enables high glass content and uniform dimensional properties and appearances to be achieved at lower cost than by other methods for runs above 1000 units. Such a process reduces the labour content of producing panels, but much increases the initial tooling charge.

3.21.4 Production composite moulding processes

There are many fibreglass reinforced plastic moulding processes available to the designer.

Each process has its own characteristics, as well as its own limitations as to part size, shape, production rate, compatible reinforcements, and suitable resin systems. The most common moulding processes are as follows (Figure 3.26): hand lay-up, spray-up, resin transfer moulding, compression moulding, injection moulding thermoplastics, injection moulding thermosets, pultrusion, and reinforced reaction injection moulding.

The three techniques used in the production of body panels are as follows:

Cold press mouldings This is used in the manufacture of the boot lid. The boot lid is formed by cold pressing a mineral reinforced resin-coated glass fibre mat in a gel-coated mould, forming a component which is very stiff in relation to its weight (see Figure 3.27).

Reinforced reaction injection moulding The RRIM technique is used for all the vertical body panels such as the front and rear wings, front grille and bumper assembly, and rear panel and bumper assembly. RRIM polyurethane has the properties of good recovery from deformation, outstanding resistance to wear, impact and abrasion, and a fast cycle time in manufacture. The use of this material for all exposed corners of the car helps to reduce minor body damage repair (see Figure 3.26h).

Pressure assisted resin injection moulding The bonnet is a pressure-assisted moulding of sandwich construction with polyester resin exterior on either side of a rigid urethane core. The underside of the moulding is an intumescent fire barrier which is a major safety factor for an engine compartment cover (see Figure 3.26e).

Lotus Cars Ltd

Lotus have been producing reinforced composite motor cars since 1957. In 1973 the company introduced the vacuum-assisted resin injection (VARI) system for vehicle body manufacture. The first VARI moulded production car was the Lotus Elite introduced in 1974; since then developments have continued in the processes, tooling and techniques of producing composite vehicle bodies.

The Lotus VARI process provides a method of moulding fibre reinforced composite panels from matched tooling. The process can be used to manufacture large body panels with integrated foam structures and captive metal fixings, using relatively low-cost tooling. As there is no dependency upon platen size and press tonnage – an obvious limitation of other processes – there are no panel size limitations. This means that the vehicle body structures can be moulded in much fewer sections, minimizing panel assembly times and costs. Lotus mould their complete body shell in a few pieces, each of which has an integral moulded structure. This structure uses isophthalic polyester resin in conjunction with continuous filament mat. In addition, woven and unidirectional glass reinforcements are used in areas where more specific loadings are required. Kevlar aramid materials are also included where particularly high strengths are needed.

The VARI tooling, which can be either metal faced or constructed entirely from composite materials, is designed so that each tool becomes its own press. The pressure is created by vacuum which draws the tools together and holds the male and female throughout the moulding cycle. Therefore no mechanical clamping mechanisms are involved. After closing the tools the resin is injected using a machine which dispenses precise quantities of catalysed materials. After the curing cycle the vacuum is reversed and the tools open, releasing the moulded panel or body section.

3.22 DESIGNING REINFORCED COMPOSITE MATERIALS FOR STRENGTH

The ability to design and fabricate large structures as a whole, rather than as an assembly of components, is one of the chief advantages that glass resin laminates bring to the designer. This is supported by the ease of modifying the material thickness at specific locations and, taking advantage of the properties of the various types of reinforcement, by building in extra strength at any point and in any direction (Figure 3.28). The skill of the operator is an important factor.

Strength of glass fibre reinforced plastic laminates in any direction is dependent on the orientation of fibre reinforcement to that direction (Figure 3.29). When chopped strand mat is used, its random fibre arrangement can be expected to give roughly equal mechanical properties in all planes; however, maximum strength will in practice be parallel to the plane of the laminate. Plain woven roving gives optimum mechanical properties at right angles, while unidirectional roving mat shows highest strength along the roving; as the roving is continuous and uncrimped, this last type will be stronger than other types of reinforcement.

There is still an unfortunate tendency on the part of designers to use a traditional design, known to be satisfactory for wood or metal, for reinforced plastics which have, of course, completely different properties and processing characteristics (Tables 3.21, 3.22). This may give glass fibre reinforced plastic mouldings of incorrect shape, since although conventional

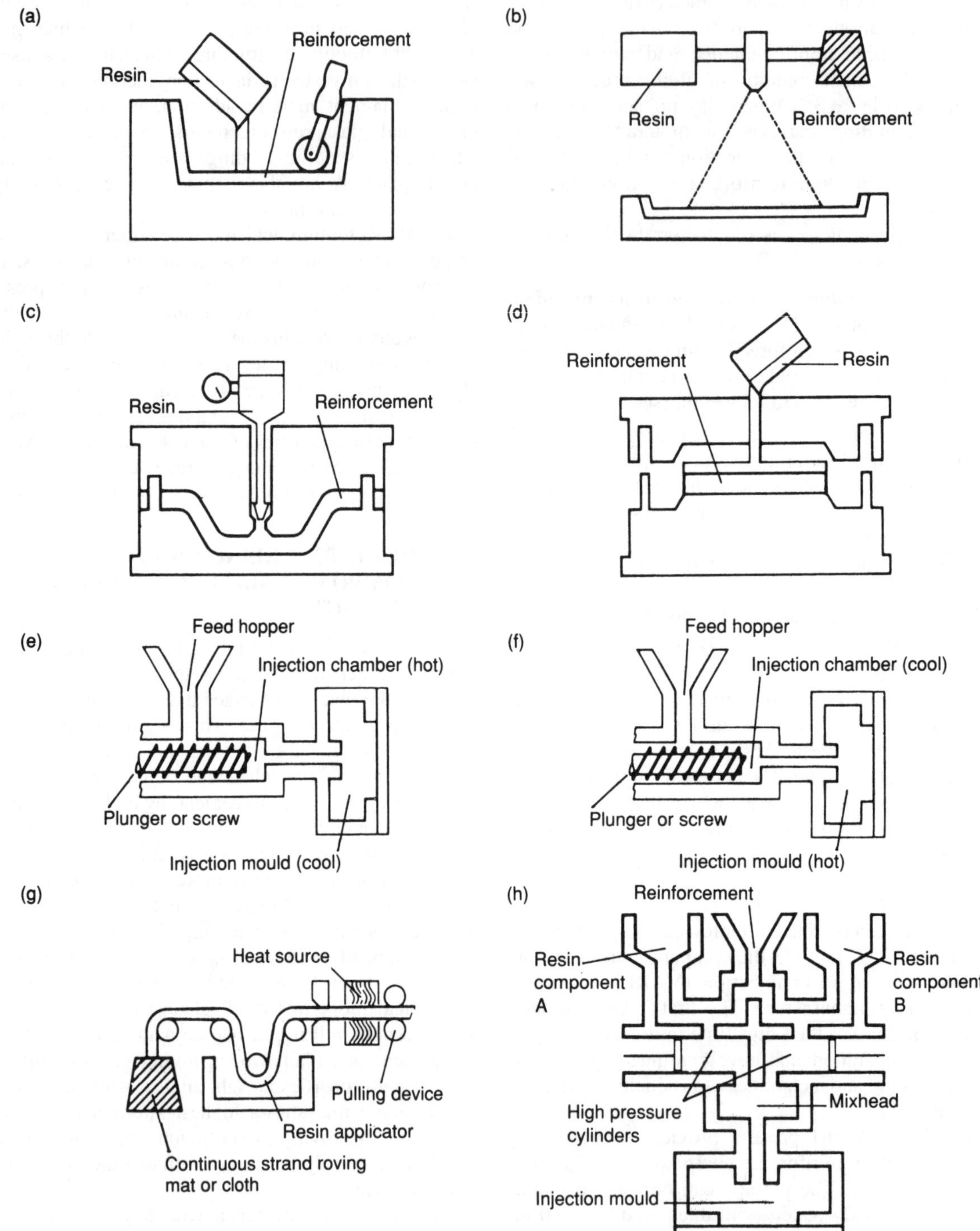

Figure 3.26 Composite moulding processes (*Owens-Corning Fibreglass*). (a) Hand lay-up; (b) spray-up; (c) resin transfer moulding; (d) compression moulding; (e) injection moulding thermoplastics; (f) injection moulding thermosets; (g) pultrusion; (h) reinforced reaction injection moulding

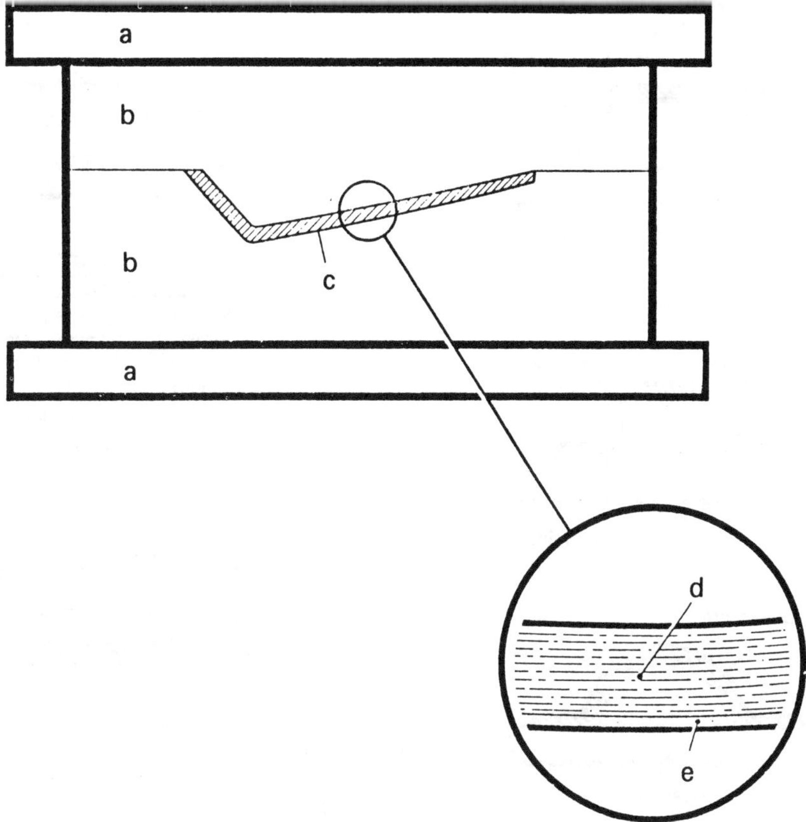

Figure 3.27 Cold press process: (a) 50 tonne hydraulic press (b) Kirksite metal mould (c) moulding (d) continuous filament glass (e) polyester surface finish (*Reliant Motor PLC*)

materials are well suited to straight lines and flat surfaces, the properties of glass fibre reinforced plastic components are improved by the introduction of curvature, and if possible double curvature, to the design. The specific strength (tensile strength weight ratio) of glass-polyester laminates is high, but rigidity tends to be on the low side. It may therefore be necessary to design for additional rigidity rather than for optimum tensile strength. This can be effected by various means, of which increased overall moulding thickness is perhaps the least desirable as it is wasteful and may well add unnecessary weight. The use of simple or compound curves in the design may be the answer, or perhaps local corrugations can be introduced as in metal designs; these can often be incorporated into the overall styling, especially in vehicle bodies. Localized thickening, particularly towards the edges of a panel, will contribute usefully to the stiffness of the moulding. A common practice to achieve extra rigidity is the integral moulding of ribs into the reverse face of the laminate. These ribs, which are often used in large boat hulls, can be solid or hollow; for solid ribs a permanent core of glass fibre, wood or plastic foam can be laminated in, while hollow ribs are achieved by use of removable tube or simple former of cardboard or similar material (see Figure 3.18).

Large panels can be made considerably more rigid by the employment of a sandwich form of construction, whereby two layers of glass fibre reinforced plastic are separated by a thick but relatively weak lightweight material. The benefit here derives from the fact that stiffness is a direct function of thickness. Other advantages of this method are increased heat and sound insulation. Stress analysis is usually based on the tensile strength at which crazing of the resin matrix occurs. This corresponds to the yield point of conventional materials. It is usual to allow a safety factor of between 1:3 and 3:5, depending on the conditions of service. Frequently the stressing of a moulding is so complex that it defeats analysis. In such cases it may be necessary to base the design on the

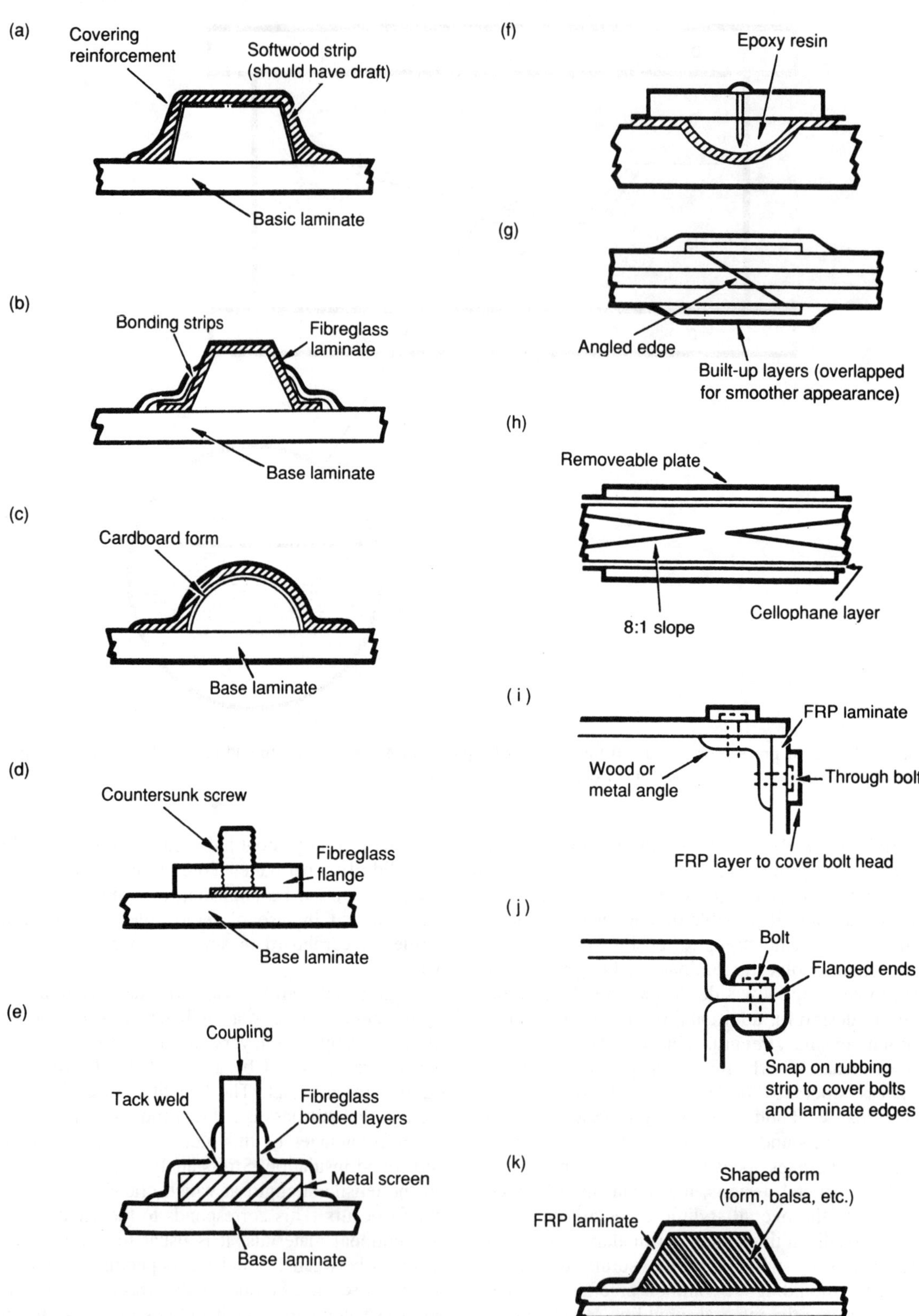
(a)
Covering reinforcement
Softwood strip (should have draft)
Basic laminate
(b)
Bonding strips
Fibreglass laminate
Base laminate
(c)
Cardboard form
Base laminate
(d)
Countersunk screw
Fibreglass flange
Base laminate
(e)
Coupling
Tack weld
Fibreglass bonded layers
Metal screen
Base laminate
(f)
Epoxy resin
(g)
Angled edge
Built-up layers (overlapped for smoother appearance)
(h)
Removeable plate
8:1 slope
Cellophane layer
(i)
FRP laminate
Wood or metal angle
Through bolt
FRP layer to cover bolt head
(j)
Bolt
Flanged ends
Snap on rubbing strip to cover bolts and laminate edges
(k)
Shaped form (form, balsa, etc.)
FRP laminate

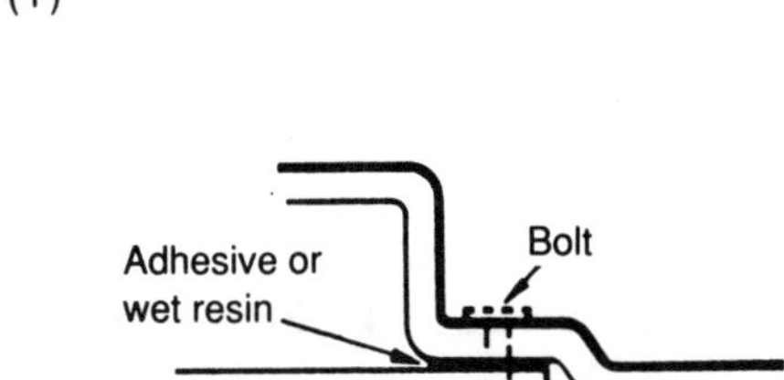

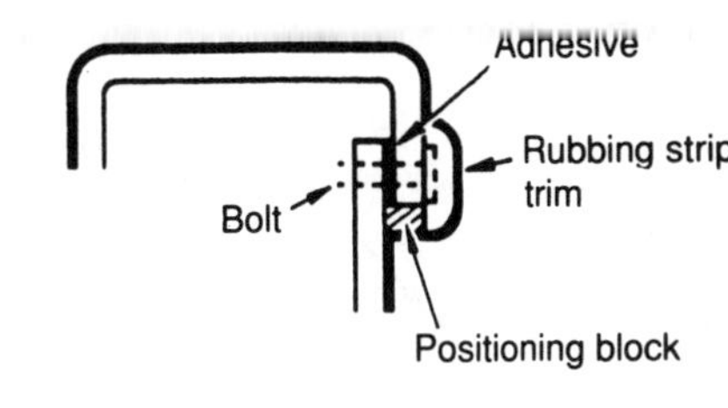

Figure 3.28 Design consideration for reinforced composite mouldings (*Owens-Corning Fiberglas*). Designs may incorporate large areas of sandwich construction with cores for vibration dampening, stiffening, thermal inculation or flotation. The type of core material will depend upon the specific property required. Care must be exercised to ensure adequate adhesion between the laminate and the core material used. Unusual and acute contours, expensive to cut in metal dies, are both practical and economical in fibreglass spray-up.

(a) Encased. Wood strips, plywood or metal stiffeners can be pressed into the part while wet and covered with resin and reinforcement to anchor and protect from deterioration. Softwoods are normally used since hardwoods are difficult to bond. Note: if stiffeners are added after cure, bonded surfaces should be sanded to assure good adhesion.
(b) Separately fabricated fibreglass stiffeners can be pressed into layers to be reinforced while it is still wet. While fibreglass adheres well to itself, additional spray-up is needed to tie into part.
(c) Integral. Light, low-cost forms (cardboard mailing tubes, folded cardboard, balsa wood) pressed into the mould during spray-up give stiffening shape to fibreglass. Thickness of spray-up over the form should be at least 3 mm. Exact thickness will depend upon strength and stiffness requirements of the part.
(d) Bolting flange. This can be made by cutting a section of a 13 mm laminate. This laminate can then be tapped for studs or countersunk for cap screws. The assembly is then placed into a wet layer of spray-up material. The flange can be strengthened by adding layers.
(e) Threaded coupling. A collar of expanded metal mesh (for coarse metal screen) is tack-welded to a threaded pipe coupling. Then, this assembly is pressed into a wet spray-up layer and cover layers are added for increased torque resistance. Large beads on the tack welds will act to 'key' the laminate to prevent twisting.
(f) Nailing or screw insert. Into a depression in the spray-up part, a shrink-resistant epoxy resin is cast to hold nails or other mechanical fasteners. Mould surface in depression should be purposely left rough to give maximum gripping to insert. Additional layers can then be sprayed over insert.
(g) Simple butt. When premoulded sheets are joined, edges are angled to give more bonding area. Then laminate layers are built up in thickness on both sides to desired strength.
(h) Optimum appearance. For smooth surface a double-tapered joint cured between removable plates is required. Slope of angle should be 8:1. Cellophane placed between plate and laminate will facilitate plate removal.
(i) The simple corner uses a premoulded sheet of *Fibreglas* reinforcement and resin. For strength, a wood or metal angle may be incorporated or the butt itself carried around corner.
(j) Flanged edges bolted together require trim to hide joint. Though bolts are shown. Self-tapping screws or rivets may be used to hold parts together.
(k) Built-up stiffeners incorporated in laminate may be cardboard, balsa, various foams, plastic, or sheet metal. They should be designed to spread loads over wide area.
(l) Integral stiffeners are most easily incorporated as flanges at a joint to provide dual function. Pads or gussets are desirable where stiffeners intersect a flat, flexible area.
(m) Lapped sections are held in place mechanically with adhesive or wet resin between sections as seal. Trim hides edges. Fastening device may be bolts, screws or rivets.

The best procedure is a corner integral with the structure, with no joints. This provides optimum performance opportunity.

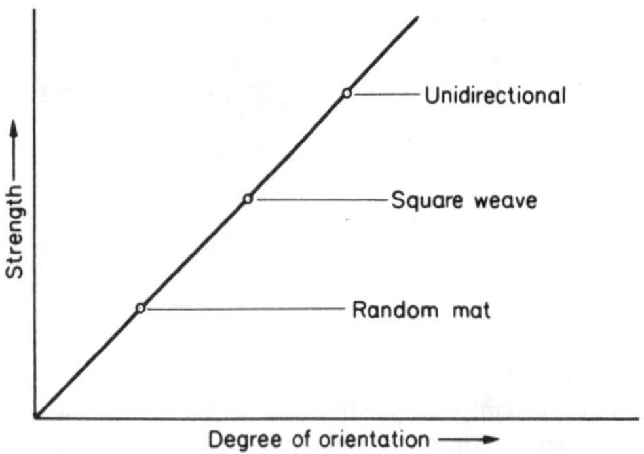

Figure 3.29 Orientation of fibre reinforcement

known performance of a similar structure and to produce a prototype moulding for testing under service conditions. Where metal inserts are to be incorporated in the moulding, allowance should be made in the design of the joint for the greatly differing yields of the two materials under similar loading. A wider insert will result in lower stress per centimetre at the load transfer point. The stress in the resin laminate can be taken up to a point where strain in both materials is similar by thickening the laminate where it approaches and surrounds the metal insert. Ideally the base of the insert should be about four times as wide as it is long.

It is better to achieve load transfer by adhesion rather than by mechanical interlocking; mechanical methods are satisfactory, however, if only small loads are involved. Joining of two glass fibre reinforced plastic components to each other can be effected by adhesive or mechanical means, or by a combination of both. In adhesive joint design the bonding area should be as large as possible. For ordinary butt or scarf joints, extra reinforcement should be provided by lamination of extra layers of resin and glass over the joint on both sides; in general it is better to use overlapping joints (Figure 3.30). Exposure of glass fibre by roughening in each case will enhance bond strength. Cured laminates are commonly bonded by means of epoxy-type adhesives. A polyester resin adhesive may be used, but here it is necessary to ensure that the adhesive films are thick enough to avoid problems of undercure. This can be done by including a single sheet of glass cloth or mat with open texture in the joint. Most kinds of mechanical fastener (nuts and bolts, self-tapping screws, rivets and others) can usually be employed without any trouble, provided the load is spread by means of large heads and/or large washers. To prevent laminate crushing it is a good plan to use spacers for bolted connections. Provided the extent of the bonded area is taken into account, and thorough cleaning and roughening are carried out, there is no reason why drilled and tapped metal plates, or special tapped inserts, should not be laminated into the moulding or on the reverse side.

3.23 COMMON FAULTS IN MOULDED LAMINATES

Many complaints concerning the appearance and performance of these mouldings stem from the basic cause that the resin is undercured. There are, however, several problems in the form of visible flaws or other defects, the remedies of which will become apparent from the analysis of the causes.

3.23.1 Wrinkling

This is caused by solvent attack on the gel coat by the monomer in the laminating resin due to the fact that the gel coat is undercured. Wrinkling can be avoided by ensuring that the resin formulation is correct and the gel coat is not too thin, and by controlling temperature and humidity and keeping the work away from moving air, especially warm air. If the workshop is equipped with hot air blowers, these should be directed away from the moulds (Figure 3.31).

3.23.2 Pin-holing

Surface pin-holing is caused by small air bubbles which are trapped in the gel coat before gelation. It occurs when the resin is too viscous, or has a high filler content, or when the gel coat resin wets the release agent imperfectly (Figure 3.32).

3.23.3 Poor adhesion of the gel coat resin

Unless the adhesion of the gel coat to the backing laminate is very poor, this defect will be noticed only when the structure is being handled and pieces of gel coat flake off. Areas of poor adhesion can be detected sometimes by the presence of a blister, or by local undulations in the surface when it is viewed obliquely. Poor gel coat adhesion can be caused by inadequate consolidation of the laminate; by contamination of the gel coat before the glass fibre is laid up; or, more generally, by the gel coat being left to cure for too long (Figure 3.33).

3.23.4 Fibre pattern

The pattern of the composite reinforcement is sometimes visible through the gel coat or prominently

Table 3.21 Properties of fibreglass composites and alternative materials

Material	ASTM test method	Glass fibre by weight	Specific gravity	Density	Tensile strength	Tensile modulus	Elongation	Flexural strength	Flexural modulus	Compressive strength	Impact strength Izod	Hardness	Flammability
					10^3 psi	10^6 psi	%	10^3 psi	10^6 psi	10^3 psi	ft lb/in notched at 73°F	Rockwell (except where noted)	
		%	D792	lb/in^3	D638	D638	D638	D790	D790	D695	D256	D785	UL-94
Glass fibre reinforced thermosets	Polyester SMC (compression)	30.0	1.85	0.066	12.00	1.70	1.0	26.00	1.60	24.00	16.00	Barcol 68	5V
	Polyester SMC (compression)	20.0	1.78	0.064	5.30	1.70	.4	16.00	1.40	23.00	8.20	Barcol 68	5V
	Polyester SMC (compression)	50.0	2.00	0.072	23.00	2.27	1.7	45.00	2.00	32.00	19.40	Barcol 68	5V
	Polyester BMC (compression)	22.0	1.82	0.065	6.00	1.75	.5	12.80	1.58	20.00	4.26	Barcol 68	5V
	Polyester BMC (injection)	22.0	1.82	0.065	4.86	1.53	.5	12.65	1.44	–	2.89	Barcol 68	VO
	Epoxy (filament wound)	80.0	2.08	0.061	80.00	4.00	1.6	100.00	5.00	45.00	45.00	M98	VO
	Polyester (pultruded)	55.0	1.69	0.060	30.00	2.50	–	30.00	1.60	30.00	25.00	Barcol 50	VO
	Polyurethane, milled fibres (RRIM)	13.0	1.07	0.038	2.80	–	140.0	–	0.037–0.053	–	–	SD* 65–75	VO
	Polyurethane flaked glass (RRIM)	23.0	1.17	0.042	4.41	–	38.9	–	0.15	–	2.10	–	VO
	Polyester (spray-up/lay-up)	30.0	1.37	0.049	12.50	1.00	1.3	28.00	0.75	22.00	13.00–15.00	Barcol 50	VO
	Polyester, woven roving (lay-up)	50.0	1.64	0.059	37.00	2.25	1.6	46.00	2.25	27.00	33.00	Barcol 50	VO
Glass fibre reinforced thermoplastics	Acetal	25.0	1.61	0.058	18.50	1.25	3.0	28.00	1.10	17.00	1.80	M79	HB
	Nylon 6	30.0	1.37	0.049	24.00	1.05	3.0	29.00	1.11	24.00	2.20	R121	HB
	Nylon 6/6	30.0	1.48	0.053	23.00	1.20	1.9	35.00	0.80	26.50	2.20	M95	HB
	Polycarbonate	10.0	1.26	0.045	12.00	.75	9.0	16.00	0.60	14.00	2.00	M80	V-1
	Polypropylene	20.0	1.04	0.037	6.50	.54	3.0	8.30	0.52	25.00	1.10	R103	HB
	Polyphenylene sulphide	40.0	1.64	0.059	22.00	2.05	3.0	37.00	1.90	21.00	1.50	R123	V-O/5V
	Acrylonitrile butadiene styrene (ABS)	20.0	1.22	0.044	11.00	.90	2.0	15.50	0.87	14.00	1.20	R107	HB
	Polyphenylene oxide (PPO)	20.0	1.21	0.043	14.50	.92	5.0	18.50	0.75	17.60	1.80	R107	HB
	Polystyrene acrylonitrile (SAN)	20.0	1.22	0.044	14.50	1.25	1.8	19.00	1.10	17.50	1.10	R122	HB
	Polyester (PBT)	30.0	1.52	0.054	19.00	1.20	4.0	28.00	1.17	18.00	1.80	R118	HB
	Polyester (PET)	30.0	1.56	0.056	21.00	1.30	6.6	32.00	1.25	25.00	1.80	R120	HB
Unreinforced thermoplastics	Acetal	–	1.41	0.051	8.80	.41	40.0	13.00	0.38	16.00	1.00	M78-M80	HB
	Nylon 6	–	1.12	0.040	11.80	.38	30.0	15.70	0.39	13.00	0.60	R119	HB
	Nylon 6/6	–	1.13	0.041	11.50	.40	60.0	17.00	0.42	15.00	0.80	R120, M83	V-2
	Polycarbonate	–	1.20	0.043	9.50	.34	110.0	13.50	0.34	12.50	16.00	M70	V-2
	Polypropylene	–	.89	0.032	5.00	.10	200.0	5.00	0.13–0.20	3.50	1–20	R50–96	HB
	Polyphenylene sulphide	–	1.30	0.045	9.50	.48	1.0	14.00	0.55	16.00	0.50	R123	V-O
	Acrylonirtrile butadiene styrene (ABS)	–	1.03	0.037	6.00	.30	5.0	11.00	0.35–0.40	10.00	3–6	R107–115	HB
	Polyphenylene oxide (PPO)	–	1.10	0.039	7.80	.38	50.0	12.80	0.33–0.40	12.00	5.00	R115	V-1
	Polystyrene acrylonitrile (SAN)	–	1.05	0.038	9.50	.40	.5	14.00	0.55	14.00	0.30–0.45	M80–85	HB
	Polyester (PBT)	–	1.31	0.047	8.20	.28	50.0	12.00	0.33–0.40	8.60	.80	M68–78	HB
	Polyester (PET)	–	1.34	0.048	8.50	.40	50.0	14.00	0.35–0.45	11.00	0.25–0.65	M94–101	HB
Metals	ASTM A-606 HSLA steel (cold rolled)	–	7.75	0.280	65.00	30.00	22.0	–	–	65.00	–	B80	–
	SAE 1008 low-carbon steel (cold rolled)	–	7.86	0.280	48.00	30.00	37.0	–	–	48.00	–	B34–52	–
	AISI 304 stainless stell	–	8.03	0.290	80.00	28.00	40.0	–	–	80.00	–	B88	–
	TA 2036 aluminum (wrought)	–	2.74	0.099	49.00	10.20	23.0	–	–	49.00	–	R80	–
	ASTM B85 aluminum (die cast)	–	2.82	0.102	48.00	10.30	2.5	–	–	48.00	–	Brinell 85	–
	ASTM AZ91B magnesium (die cast)	–	1.83	0.066	33.00	65.00	3.0	–	–	33.00	–	Brinell 85	–
	ASTM AG40A zinc (die cast)	–	6.59	0.238	41.00	10.90	10.0	–	–	41.00	–	Brinell 82	–

Table 3.22 Compatibility of materials and processes for fibreglass composites

	Thermosets					Thermoplastics									
	Polyester	**Polyester SMC**	**Polyester BMC**	**Epoxy**	**Polyurethane**	**Acetal**	**Nylon 6**	**Nylon 6/6**	**Polycarbonate**	**Polypropylene**	**Polyphenylene sulphide**	**ABS**	**Polyphenylene oxide**	**Polystyrene**	**Polyester PBT**
Injection moulding	•		•	•	•	•	•	•	•	•	•	•	•	•	•
Hand lay-up	•			•											
Spray-up	•			•											
Compression moulding	•	•	•	•							•				
Preform moulding	•			•											
Filament winding	•			•											
Pultrusion	•			•											
Resin transfer moulding	•													•	•
Reinforced reaction injection moulding	•			•	•		•								

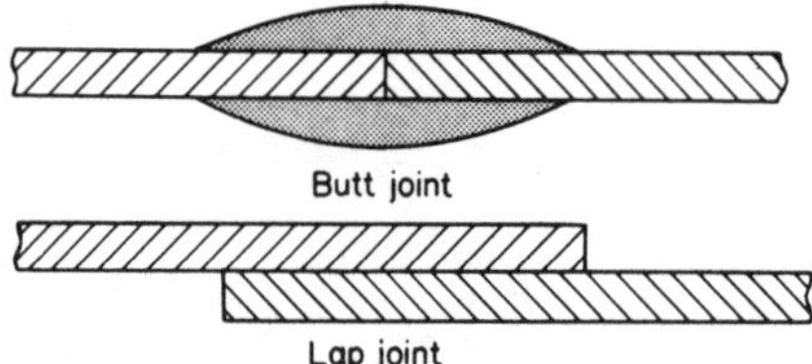

Figure 3.30 Bonded joints

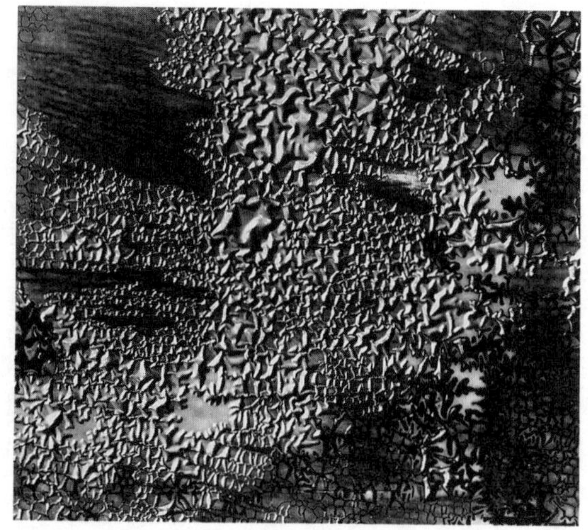

Figure 3.31 Wrinkling (*Scott Bader Co. Ltd*)

Figure 3.32 Pinholing (*Scott Bader Co. Ltd*)

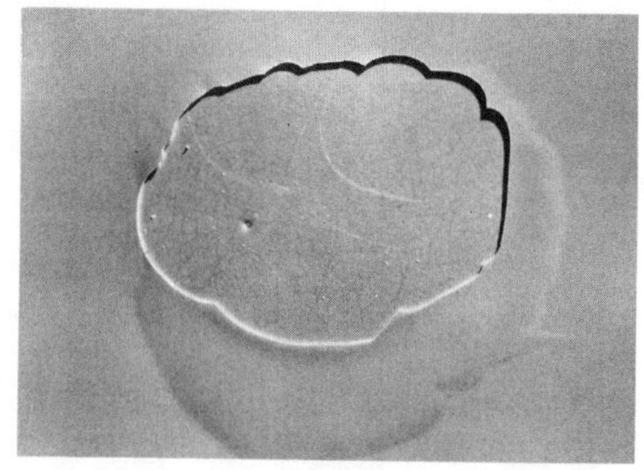

Figure 3.33 Poor adhesion of the gel coat (*Scott Bader Co. Ltd*)

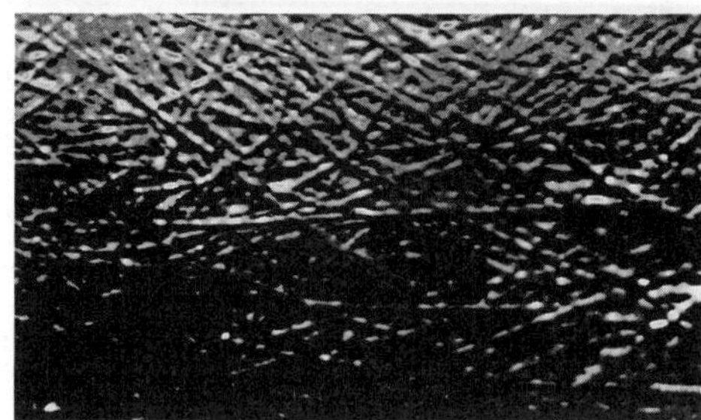

Figure 3.34 Fibre pattern (*Scott Bader Co. Ltd*)

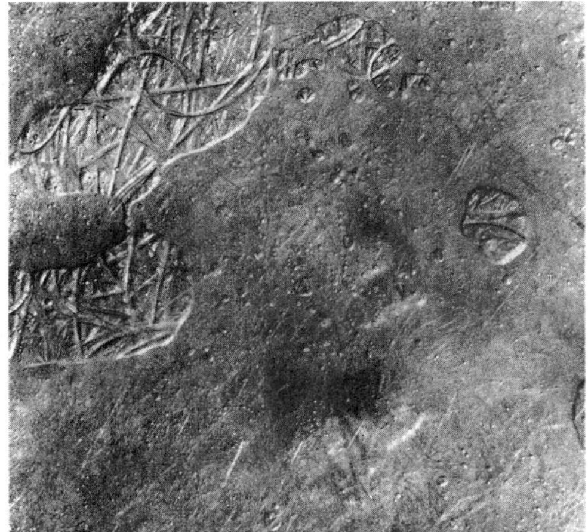

Figure 3.35 Severe blisters (*Scott Bader Co. Ltd*)

Figure 3.36 Crazing (*Scott Bader Co. Ltd*)

noticeable on its surface. This usually occurs when the gel coat is too thin or when the reinforcement has been laid up and rolled before the gel coat has hardened sufficiently, or when the moulding is removed too early from the mould (Figure 3.34).

3.23.5 Fish eyes

On a very highly polished mould, particularly when silicone modified waxes are used, the gel coat is almost non-existent. This shows up as patches of pale colour usually up to 6 mm in diameter. It can also occur in long straight lines following the strokes of the brush during application. This fault is rarely experienced when a PVAL film is correctly applied.

3.23.6 Blisters

The presence of blisters indicates that there is delamination within the moulding and that the air or solvent has been entrapped. Blisters which extend over a considerable area may also indicate that the resin is undercured, and this type of blister may not form until some months after moulding. Blisters can also occur if the moulding is subjected to an excessive amount of radiant heat during cure. A possible cause of this defect is the use of MEKP rather than cyclohexanone peroxide paste. If, on the other hand, the blister is below the surface, the cause is likely to be imperfect wetting of the glass fibre by the resin during impregnation. This would be due to the fact that insufficient time had been allowed for the mat to absorb the resin before rolling. Blisters of this kind can usually be detected by inspection as soon as the moulding has been removed from the mould (Figure 3.35).

3.23.7 Crazing

Crazing can occur immediately after manufacture or it may take some months to develop. It appears as fine hair cracks in the surface of the resin. Often the only initial evidence of crazing is that the resin has lost its surface gloss. Crazing is generally associated with resin-rich areas and is caused by the use of an unsuitable resin or resin formulation in the gel coat. The addition of extra styrene to the gel coat resin is a common cause.

Alternatively the gel coat resin may be too hard with respect to its thickness. In other words, the thicker the gel coat the more resilient the resin needs to be. Crazing which appears after some months of exposure to the weather or chemical attack is caused by either undercure, the use of too much filler, or the use of a resin which has been made too flexible (Figure 3.36).

3.23.8 Star cracking

This is the result of having an over-thick gel coat, and occurs when the laminate has received a reverse

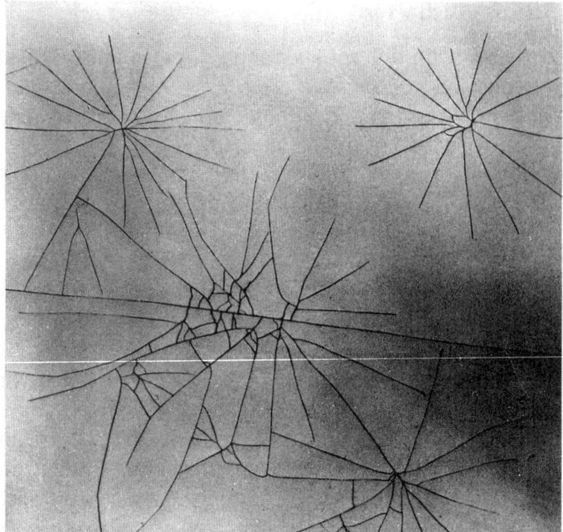

Figure 3.37 Star cracks (*Scott Bader Co. Ltd*)

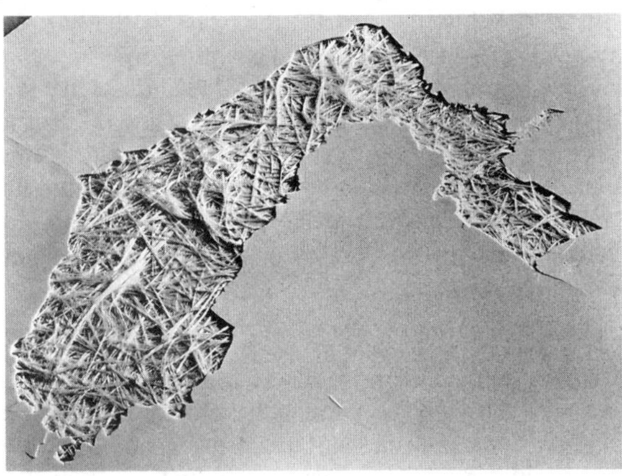

Figure 3.38 Internal dry patch (*Scott Bader Co. Ltd*)

impact. Gel coats should never be more than 0.5 mm thick (Figure 3.37).

3.23.9 Internal dry patches

These can be caused by attempting to impregnate more than one layer of mat at a time. The presence of internal dry patches can be readily confirmed by tapping the surface with a coin (Figure 3.38).

3.23.10 Leaching

This is a serious fault. Leaching occurs after exposure of the laminate to the weather, and is characterized by a loss of resin from the laminate, leaving the glass fibres exposed to attack by moisture. Leaching indicates either that the resin used has not been adequately cured, or that it is an unsatisfactory resin for that particular application.

3.23.11 Cracking, shrinking and discolouring

The identification of this fault is visual. Deep cracks appear in the resin and the colour changes from a green to a mauve purple and are hot to the touch. The cause of this fault can be a large build-up of resin due to drainage or to excess application; both lead to extreme exotherm. Alternatively, incorrect amounts of catalyst or accelerator may be used, usually an excess. The effect of the build-up of exotherm is to cause excessive shrinking of the moulding and internal stress. The extreme exotherm can also damage and distort the mould itself. This can be prevented by: avoiding large amounts of resin (but where this is necessary the build-up should be gradual; using the correct amount of ingredients; being aware of any increase in workshop temperatures; and noting any variations in the correct percentages of catalyst or accelerator used.

3.23.12 Low-rigidity laminate

The first identification is by touch; the laminate will feel spongy and more flexible than usual. Alternatively, check by applying a Barcol hardness tester. The common causes of this fault are low resin content or undercure of the resin. To prevent this fault, ensure the correct ratio of resin to glass and eliminate any draughts. Large areas should be made to gel quicker to cut down styrene loss, and correct proportions of catalyst and accelerator must be used.

3.24 SAFETY PRECAUTIONS

The handling of polyester resin, glass fibre, and ancillary materials such as catalysts, presents several hazards which can be reduced to a minimum if the correct precautions are taken. Most glass fibre materials and resins are perfectly safe to use provided the potential hazards are recognized and reasonable precautions are adopted. Normally you will have no problems if you follow these rules:

1 Do not let any materials come into contact with the skin, eyes or mouth.
2 Do not inhale mist or vapours, and always work in a well-ventilated workshop.
3 Do not smoke or use naked flames in the workshop.

3.24.1 Storage precautions

Liquid polyester resins are flammable but not highly flammable, most of them having a flashpoint of 31°C. Resins and accelerators should preferably be kept in a brick-built store conforming to the normal fire regulations for a paint store. The storage life of polyester resin is about six to 12 months provided the resin is kept below 20°C in the dark (in metal drums). Storage at higher temperatures, even for only a few days, will considerably reduce the shelf life.

Catalysts are organic peroxides and present a special fire hazard. They should be stored in a separate area, preferably in a well-ventilated fire-resisting compartment. If kept reasonably cool they will not burn or explode. In case of fire in the vicinity, they can be kept safe by drenching the containers with water.

3.24.2 Operating precautions

Most polyester resins contain monomeric styrene, which is a good grease solvent and may cause irritation to the skin. The most effective method of protecting hands is the use of a barrier cream or rubber gloves, and this is strongly recommended. Resin can be removed from the hands with proprietary resin removing creams, or with acetone followed immediately by a wash in warm soapy water. In sufficient concentration styrene vapour is irritating to the eyes and respiratory passages, and therefore workshops should be well ventilated. When resin is sprayed a gauze mask should be worn to protect the mouth and nose. This also applies to trimming operations when resin and glass dust can cause irritation.

Catalysts are extremely irritating to the skin and can cause burns if not washed off immediately with plenty of warm water. Particular care must be taken with liquid catalysts to avoid splashing, spilling or contact with the eyes. Protective goggles should be worn as a necessary precaution.

3.24.3 Workshop conditions

The building should not be damp and it should be adequately heated and ventilated. Good head room is desirable and sufficient space should be allowed for all operations. The floor area should be divided into sections as follows: preparation of reinforcement, mixing of resins, moulding, trimming and finishing. Resin and curing agents should be stored away from the working area in a cool place, observing the necessary precautions for flammable liquids and keeping in mind the special hazards associated with organic peroxides. Glass fibre should be stored and tailored under dry conditions and separately from the moulding area. The temperature of the building should be controlled between 15°C and 25°C. Ventilation should be good by normal standards, but draughts and fluctuations in temperature must be avoided, so doors and windows should not be used for ventilation control. Dust extraction in the trimming section should be of the down-draught type. Cleanliness is important both for health of the operators and for preventing contamination of resin and reinforcement.

As far as possible, health and comfort depend in the first place on planned extraction, and in the second place on workshop education in the nature of the materials used. Almost all the offence comes from the styrene vapour and glass filaments, both of which advertise their presence before the concentration reaches a danger level. As far as is known the only real source of physical harm is the dust produced in grinding, but all the materials and byproducts contribute to discomfort, and sensible evasive action is essential.

3.24.4 Spillage and disposal

Most of the following products are covered by the terms of the Deposit of Poisonous Wastes Act:

Polyester resin Absorb spillages in dry sand and dispose by landfill or controlled incineration.

Furane resin Extinguish all naked lights, open doors and windows. Absorb spillage into sand or chalk. Pack into drums, seal and store prior to collection by specialized chemical disposal company.

Catalyst Absorb into vermiculite, remove to landfill or controlled incineration. Wash down remaining traces with copious water.

Accelerator Absorb into dry sand and dispose by landfill or controlled incineration.

Release agents Wash down with water.

Mould cleaner Absorb into sand or earth, remove to landfill or controlled incineration. Flush contaminated area with water.

3.24.5 Fire hazards

Many resins and associated products are either flammable or contain flammable additives. Styrene, catalyst and acetone are particularly dangerous. Do not smoke or use naked lights, oil burners or similar heating devices in the working area. If a fire does start, do not attempt to put it out with water unless it is a catalyst. Dry powder extinguishers can be used on accelerator, mould cleaner, acetone, resins and release agents.

Fires can be started if catalysed but uncured resins are thrown away. The waste resin will continue to cure and the heat generated by the curing process can ignite other waste materials. Therefore unwanted resin should be left in a safe place until it is fully cured; it can then be discarded without risk of fire.

QUESTIONS

1 What would the following alloy steels be used for: (a) high-tensile steel (b) manganese steel (c) chrome-vanadium steel?
2 List the properties of commercially pure aluminium.
3 Explain why, in the construction of a motor vehicle, commercially pure aluminium has a very limited application.
4 Identify the grades of hardness in aluminium sheet and state how the hardness is achieved.
5 Explain how you would identify the following: (a) low-carbon steel (b) aluminium alloy (c) stainless steel.
6 Describe the difference between laminated safety glass and toughened safety glass.
7 Give three requirements of a body sealing compound, and describe one type of sealer used in vehicle repair.
8 Suggest reasons why stainless steel is sometimes used for trim and mouldings.
9 Explain the difference between hide and PVC materials.
10 Explain what is meant by micro-alloyed steel or HSS.
11 Give reasons why car manufacturers are using zinc-coated steels.
12 Name the three main groups of stainless steel.
13 Explain the following terms in relation to plastic: (a) monomer (b) polymer (c) copolymer.
14 Explain the difference between thermoplastics and thermosetting plastics.
15 Which safety glass, used for vehicle windscreens, shatters into small segments on impact?
16 Describe the basic properties required of a body joint sealing compound.
17 Identify the group of plastics that can be softened or remoulded by the application of heat.
18 Steel panels can be strengthened without adding weight. Name and explain the process.
19 Describe three different ways in which the surface of steel can be protected.
20 State the reasons why certain metals need to be protected from the effects of the atmosphere.
21 Describe how some metals can resist attack by the atmosphere.
22 What is the alloying effect when zinc and copper are added to aluminium?
23 Explain the different properties of heat-treatable and non-heat-treatable aluminium alloys.
24 State the reasons, other than weldability, why low-carbon steel is chosen in preference to aluminium as a vehicle body shell material.
25 Define the term 'HSLA steel'.
26 Define what is meant by the term 'non-ferrous metal'.
27 Explain the difference in properties between low-carbon steel and alloy steel.
28 Describe the two processes which can be used to join plastic.
29 Explain where plastic can be used on a vehicle.
30 State the applications where natural rubber has been replaced by synthetic materials in the automobile industry.
31 Explain the following abbreviations: RRIM, VARI, PVAL.
32 State three advantages of reinforced composite materials when used in vehicle body construction.
33 List the main physical properties of glass fibre composite materials.
34 Explain the advantages and disadvantages of reinforced composite materials when used as an alternative to low-carbon steel.
35 Explain briefly the function of (a) the releasing agent (b) the gel coat (c) the catalyst.
36 Describe the process of contact moulding.
37 Explain the purpose of pre-accelerated resins.
38 With the aid of a sketch, show the lay-up of a laminate in the mould.
39 Describe the type of tools that would be used to trim the edges of reinforced laminates.
40 Describe the sequence of repair to a damaged glass fibre composite body panel.
41 Describe how a patch mould can be used during the repair to damage of a GRP laminate.
42 Name the types of materials that could be used to reinforce polyester resin.
43 Name three reinforcing materials that could be added to a GRP moulding to give strength to the laminate.
44 In the automobile industry, why is GRP limited in use to a small specialist sector?
45 State the reasons why certain manufacturers of sports cars prefer to make their vehicle bodies from GRP.

4

Automotive finishing and refinishing

4.1 HISTORY OF AUTOMOTIVE FINISHING

No repair to a vehicle is complete until it has been painted to match the rest of the vehicle and is rendered undetectable. This part of the operation is carried out by the spray painter, who must have a knowledge of the type of materials used in the repair shop in order to help him select the best process for refinishing the vehicle concerned. Nowadays the spray painter has the help of the paint manufacturers, who can supply him with literature to cover every painting process, but his predecessor, the coach painter, had to have a very solid understanding of the materials at his disposal.

As the term 'coach painting' implies, this is a craft which dates back long before the days of the motor car. Reference is made to coach painting in the diary of Samuel Pepys, in which he makes mention of the buying and repairing of a second-hand horse-drawn carriage. The amount charged to him for the repainting was sixty pounds which, four hundred years ago, would be a fairly large sum of money. The time involved for the repainting was one month which, bearing in mind the type of paint used, was not unreasonable.

In those days the painter not only mixed his own colours but actually manufactured his own paints, using a pestle and mortar to grind the pigment and oil together. The choice of materials available was rather limited, but it is to the credit of the craftsmen of those days that the finished appearance was of a very high standard and extremely durable. Perhaps the best protective pigment available to the craftsman was white lead, and he made full use of it. The lead paste was mixed with linseed oil and applied by brush, one coat every second day, being too slow in drying to allow for more frequent coatings. Several coats were applied, each being rubbed down smooth prior to the application of the next coat. When the work was judged to be ready for the colour coats, these were also applied in several layers, being too transparent to cover in one solid coat.

When the painting was completed, the sign writer took over and embellished the coach with line work and heraldic emblems. One of the most widely used materials was silver leaf; gold leaf was not then available. Following this part of the work, as many as seven coats of varnish were applied. The varnish, being rather yellow in colour, tended to enhance the silver leaf by tinting it amber and giving it the appearance of gold. This slow, laborious and costly painting process continued almost without alteration right up to the end of the nineteenth century and the birth of the motor car. Paint manufacturers had, however, come into being, with a consequent improvement in the range and quality of materials at the disposal of the painter.

The early motor car, like its predecessor the coach or carriage, was of coach-built construction, and the existing methods of painting were suitable for this type of vehicle. However, as the motor car increased in popularity and demand for it grew, a new and faster method of production had to be found. This was achieved by the advent of pressed steel construction, but the paint process caused a bottleneck to production and so research was carried out to solve this problem.

The answer came with the development of cellulose lacquer which, though not a complete answer, was nevertheless an extremely fast drying material which allowed for several coats to be applied in one day (speed, of course, being the main criterion). Being so rapid in drying, cellulose was not suitable for brushing purposes and so the application of paint using a spray gun came into its own. By 1930 all new motor cars were being finished by this method. The material, however, was lacking in solid content, and consequently several coats had to be applied to achieve a coating of worthwhile thickness. Another time consuming factor was that the finished vehicle had to be burnished to obtain a high gloss.

Around 1935 cellulose-lacquer-based paints were combined with other synthetic materials to produce a paint which dried in 30 minutes, had better 'build' qualities and thus required fewer coats. It also reduced the burnishing time and so eased the bottleneck which still existed in the paint section of the production line. During the Second World War a great deal of research was carried out, and success achieved, with

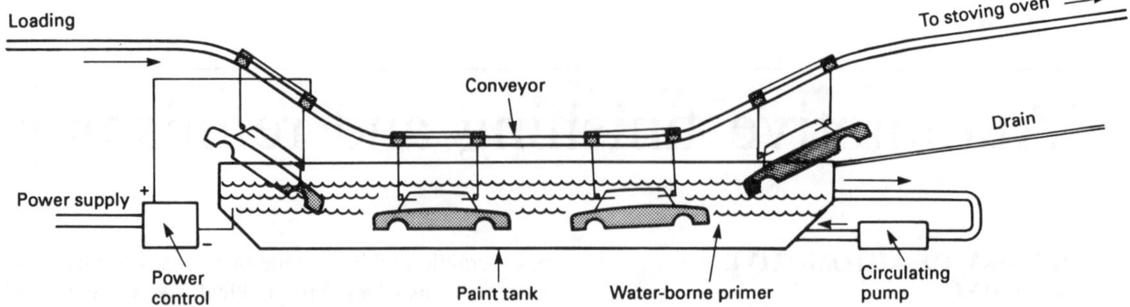

Figure 4.1 Electrodeposition of priming coat

thermosetting paints which could be force dried at elevated temperatures. These paints provided a hard glossy finish, required fewer coats than the cellulose materials and were more chemically resistant. A further advantage was that the finish required no burnishing or polishing. In all, this was a paint ideally suited to the expanding motor vehicle industry, and by the early 1950s all new motor cars were being finished in these stoving synthetics.

As well as improvements in finishing materials, changes in painting techniques were being evolved. Perhaps the most revolutionary changes were introduced in the application of the priming paints, mainly in the field of dip application. In this method the entire body shell is completely immersed in a tank of priming paint (which is specially formulated for this purpose), is withdrawn, allowed to drain, and is then passed on to a stoving oven for baking.

Stoved synthetic finishes became the accepted finish on new motor vehicles, but difficulties were experienced in refinishing damaged areas as a result of colour fading. Though the colours did not fade drastically, they did, however, fade sufficiently to give the refinisher a difficult job to obtain a perfect match.

In 1963 Vauxhall introduced a finish on their new Viva model which the paint manufacturers claimed had better colour stability. This was the acrylic resin stoving finish which was produced with the cooperation of the plastics industry (being of the thermoplastic type). By 1965 Ford had changed all of their colours to a high-bake acrylic finish, which was a product of the paint industry only, being thermosetting. BLMC, Rootes, Standard, Triumph and Rover followed suit by changing most of their colours to the acrylic range. Acrylic paints, as well as possessing good colour stability, are durable, have good gloss and are easily polished.

The method by which the priming coat on modern vehicle body shells is applied is known as electrodeposition (Figure 4.1). A large dip tank containing 2500 litres of a water-borne paint is included in the production line. An overhead conveyor carries the body shells from the precleaning area to the dip tank. The paint is charged with electricity and the shell is earthed through the conveyor. The thinner of the paint, being water, acts as an electrolyte; the paint solids, i.e. pigment and binder, are ionized and are attracted to the earthed car body. An even coating of paint is thus applied, even on thin metal edges. The thickness of the coating can be varied according to the electrical potential introduced. When the car body moves out of the tank, surplus paint drains out of it and the shell is then rinsed off under sprinklers, which does not affect the electrodeposited coating. The car body is then dried off and baked.

As to the future of motor vehicle finishing, it seems reasonable to expect water-borne paints to be developed to such an extent that they will become the accepted finish on new motor vehicles. Looking even further ahead, it could be that the car body will be formed entirely of a moulded plastic which could be self-coloured. Should this come about, damaged areas could be removed and replaced with a new section which is already coloured to match the rest of the car. However, there will still be a place for the refinisher, as car owners will, in all probability, desire the occasional colour change on their car. In all, there have been many developments since the days of the coach painter and his home-made paints.

Legislation is now forcing all users of paint, from the car factory to the bodyshop, to move away from solvent-based products. Under the Environmental Protection Act water-based base coats are not required by law until 1998. However, the laws affecting car manufacturers on solvent emissions are already extremely strict, leading to the use of water-borne paint materials much sooner than they will be in bodyshops.

The introduction of COSHH, the Environmental Protection Act and the new Health and Safety Regulations to protect the environment will bring about many changes to the accident repair industry, especially in the refinishing of vehicles.

As the industry comes to terms with these Regulations and the mounting pressure from the authorities to conform and comply with the solvent emission laws, the accident repair industry also has an obligation to protect the environment.

4.2 TERMS USED IN SPRAY PAINTING

In order to be able to appreciate more fully the descriptions of processes and practices in the paint shop, the reader should make himself acquainted with the following trade terms and items of equipment.

Air delivery The actual volume of compressed air delivered by the compressor after making allowances for losses due to friction. It is measured in litres per second.

Air duster A tool which, when fitted to an air line, is useful for blowing water from recesses and for drying a surface quickly prior to painting.

Air pressure The pressure of air which has been mechanically compressed. It is measured in bars or pounds per square inch (psi).

Air receiver A reservoir or storage tank to contain compressed air.

Atomization The breaking up of paint or other materials into very fine particles. Good atomization is essential in spray painting.

Double-header coating This results from the practice of spraying one coat immediately after another without allowing a flash-off period.

Dry coating Several thin coats of paint can be applied fairly rapidly if they are sprayed 'dry'. This is done by increasing the ratio of atomizing air to paint at the spray head of the gun. Dry coating is particularly useful when carrying out local repairs to paintwork.

Feather edging The rubbing down of a damaged area of paintwork until there is no perceptible edge between the paint and the substrate.

Feathering the gun To ease the pressure on the gun trigger while spraying, thereby reducing the volume of paint passing through the fluid tip. This is done mainly when spraying local areas to achieve a feather edge.

Flash off To allow the greater part of the more volatile solvents in a sprayed coat of lacquer or enamel to evaporate before proceeding with the application of another coat or with stoving.

Fluid cup A container for the paint attached to the spray gun in conventional spray painting, or a separate item in pressure feed systems connected to the spray gun by means of a fluid hose.

Fluid nozzle The orifice in the fluid tip.

Ground coats The paint coats between the primer and finishing coats. Ground coats are usually of a similar colour to the enamel.

Guide coat A thin coating applied as evenly as possible over a surface to be rubbed down. Following rubbing down, no trace of the guide coat should remain, so that complete flattening of the surface is achieved. The guide coat should obviously be of a contrasting colour to that over which it is applied.

Hold-out The degree of imperviousness of a dried paint film. Some filler coatings in particular are porous and so they tend to absorb the binder or medium of finishing coats, thus reducing their effectiveness as a glossy finish.

Matt finish A surface finish which has no glossy effect.

Shrinkage This refers to the manner in which some paints decrease in size not only vertically but also horizontally. As with sinkage, this is caused by solvent evaporation. Nitrocellulose materials are particularly prone to this phenomenon, which can affect the adhesion to the substrate.

Sinkage This trade term can have two interpretations. When paint is applied over a particularly porous surface it will sink into it, and if this paint is a finishing material the gloss will be impaired. The other explanation of the term brings in solvent evaporation. The solvents or thinner added to paints to reduce the viscosity evaporate during the drying process and consequently some of the liquid content of the paint vanishes. When this happens the paint film becomes thinner and projections on the substrate come through it.

Tacking off To wipe over a surface with a specially treated cloth, which is slightly sticky, to remove dust. Tacking off is essential before applying finishing coats.

Viscosity The degree of resistance to flow of a liquid. More simply, it refers to the thickness or otherwise of a fluid such as paint.

4.3 BASIC COMPOSITION OF PAINT

Pigments Fine solid particles which do not dissolve in the binder. They give colour and/or body to the paint. Some pigments possess good anticorrosive properties and are used in paints designed to give protection

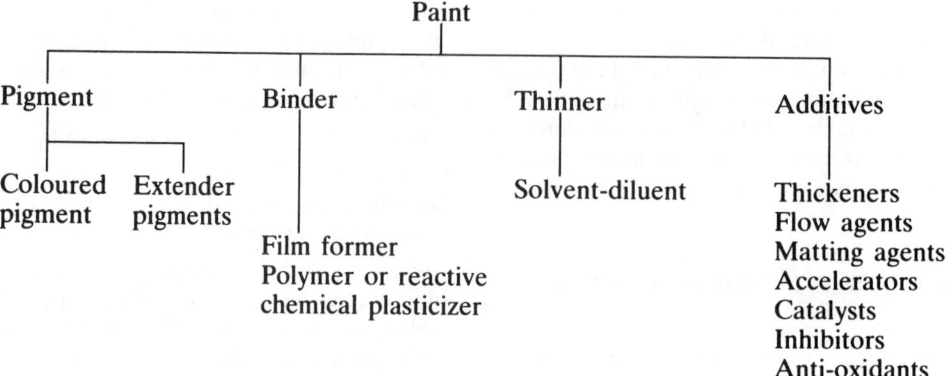

Figure 4.2 Composition of paint

to the substrate. Extenders are cheaper than pigment, but when used in the correct proportions they carry out many useful functions such as improvement of adhesion and ease of sanding.

Binder Reacts to form a film, and binds the pigments together and to the surface. The binder is often referred to as the medium of the paint.

Thinner Some of the liquid of the paint is often withheld from the paint container and supplied separately as a thinner. The user adds thinners to adjust the viscosity to suit his requirements.

Additives Small quantities of substances which are added to carry out special jobs. Wax in varnish creates a matt finish, and silicones in metallic paint give a hammer finish. Figure 4.2 illustrates the composition of paint.

4.4 TYPES OF PAINT

Cellulose synthetic

This dries by the evaporation of the solvent. The main advantage of this material is rapid air drying. However, there are a number of disadvantages. The coating dries rapidly only when thin films are applied, otherwise drying is delayed by solvent retention. The high proportion of solvent used (60 per cent in most cases) results in shrinkage which causes the film to adhere poorly to the substrate. The absence of chemical change means that the dried film does not increase in chemical resistance, and is readily softened by the original solvent.

Oil paints

The drying of an oil paint depends on the ability of certain drying oils to dry by a reaction that involves atmospheric oxygen, a process which is confined to relatively thin films.

Synthetic paints

These are mixtures of drying oils and synthetic resins. The most obvious limitation of a paint based solely on a drying oil is slow drying. To improve this property and to give tougher films and improve the gloss, a resin is added to the oil and they are cooked together for a period so that they chemically combine. The varnishes produced can be divided into two main classes based on their oil-to-resin content: long oil and short oil.

Stoving paints

These are also mixtures of oils and resins that require exposure to an elevated temperature to produce a cure (dried film). The time of exposure is mainly dependent on the temperature: 60 minutes at a temperature of not less than 138°C; and 10 minutes at a temperature of not less than 205°C.

Blacking paints

Chassis black is a cheap black paint generally based on bitumen. It has good adhesive qualities on bare metal and is a good rust inhibitor.

Tyre black is also a cheap black paint, being of low viscosity. Several proprietary brands are available.

Two-pack paints

These are probably the most widely used paints in the vehicle refinishing trade, with more than 80 per cent of refinishers preferring them. They present special health hazards, and the user should be equipped with an air-fed mask and face visor to prevent inhalation of the vapours when spraying (Figure 4.3). A canister mask of the CC type can be used as an alternative,

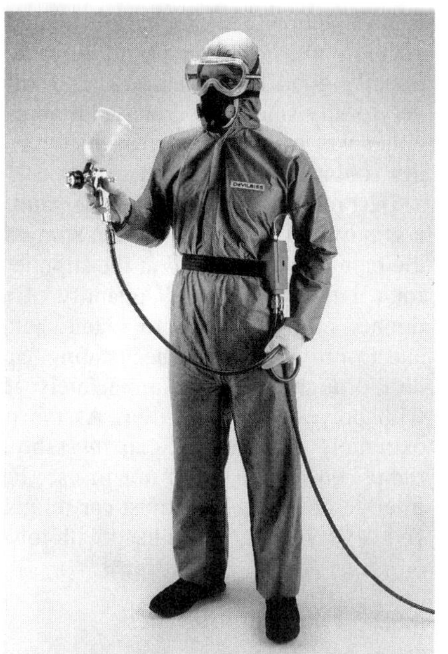

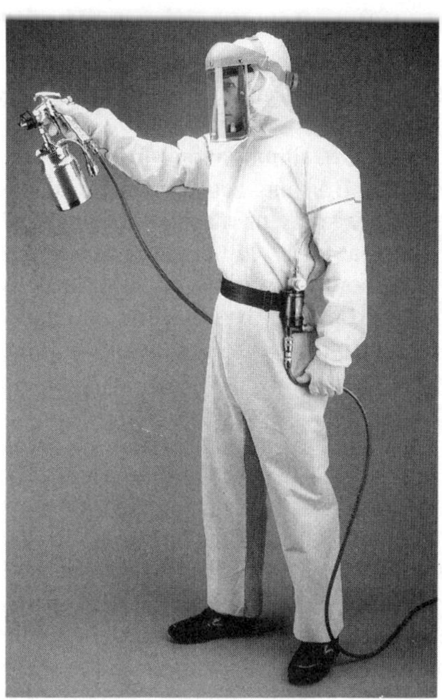

Figure 4.3 DeVilbiss Pulsafe breathing air kit showing half-mask and visor outfits (*DeVilbiss Automotive Refinishing Products*)

but these can prove to be expensive as the canister is only useful for 15 minutes' continuous use and should then be discarded. Precautions should also be taken to prevent the mixed material or spray vapours making contact with the skin, as this can cause dermatitis.

These paints consist of a base material and a catalyst or activator. When they are mixed together, a chemical reaction takes place which results in complete polymerization. Two-pack (or 2K) paints have a limited pot life after mixing, but when curing is complete they can equal stoving paints in hardness and durability. They are characterized by high solids content and low solvent content, which results in high build and good scratch filling with the minimum number of coats, thus resulting in savings on labour time and overspray wastage.

The gloss from the gun is good and no burnishing or polishing is needed unless dirt is present in the finish. Should this be the case, the finish can be wet flatted with P1200 paper, using soap as a lubricant, and burnishing can be carried out using fine rubbing compound and a polishing machine of 6000 rev/min. A clean, dry lambswool mop is recommended for best results.

Materials included in this group are: acrylic and polyurethane primer undercoats and finishes, including base-coat-and-clear finishes; epoxy resin primers and finishes; and polyester spraying fillers. It is common practice in most refinishing paint shops to force dry these materials using low-bake ovens on large areas for 30 minutes of 60°C, or using infrared lamps on small areas.

Low-bake finishes

These are modified stoving paints which can be completely cured at a temperature between 66 and 93°C. The material was formulated for the refinisher to enable him to match more closely the original finish of the car manufacturers. These are now losing favour to two-pack materials.

4.5 MATERIALS USED IN REFINISHING

4.5.1 Primers and dual-purpose primers

Primer

The priming coat is the first coat of paint on any surface. Its functions are to gain maximum adhesion to the substrate, to provide a sound base for subsequent coatings, and on metals to act as a corrosion inhibitor. The priming paint should be selected to suit the type of surface on to which it is to be applied.

Etch primer

The best types of etch primer are two-pack materials, the base and activator being supplied separately. They are mixed 10 minutes before use, but have a limited pot life (about six hours), although long-life etch primers are now available. Brushing activators are available for those shops where spray painting is not practical. Etch primers have a fairly high water absorption characteristic and should be coated with surfacer or filler after the appropriate drying time, to avoid moisture absorption from the surrounding atmosphere. Special thinners are provided for etch primers and should be kept for this purpose only.

The pigment for this type of paint is zinc chromate which makes it an ideal primer for aluminium, although it can be used also with good effect on most other metals. The activator contains phosphoric acid which etches the surface, thereby ensuring good adhesion. An extremely thin coating gives best results. Once the base and activator have been mixed together they should on no account be returned to the tin of base.

Primer surfacer

A primer surfacer does the work of both a primer and a surfacer.

Self-etching primer surfacers are available which eliminate the need for using an etching primer and then coating over with surfacer or filler. Self-etching primer surfacers have gained in popularity, the main advantage being improved adhesion to metal substrates when compared with standard primer surfacers.

Primer filler

A primer filler is similar in function to a primer and a filler.

Polyester primer filler is a two-pack paint, first used extensively on the European continent. When introduced into Britain it was viewed with a certain amount of distrust, as the claims made for it (primer, stopper, surfacer rolled into one) appeared too good to be true. There was good cause for this distrust, as one of the difficulties encountered was that blistering occurred when it was used with nitrocellulose materials. Another factor against it was that it required four hours to cure. However, modifications have been made to it which have resolved the problem of blistering and reduced the curing time to two hours at an ambient temperature of 25°C, which can be still further reduced to 15 minutes when force dried.

This paint is particularly suitable for use on rough surfaces where heavy coatings are required to level up the surface with the minimum of effort. Several coatings can be applied wet-on-wet as it dries by catalyst action. The amount of sinkage after drying is virtually nil. Polyester primer/filler possesses exceptionally good build qualities, and when dry can be very easily smoothed down with abrasive paper used either wet or dry. Face masks should be worn when dry rubbing is carried out.

Best results are obtained if the paint is applied with a gravity-fed spray gun. When spraying local repairs the paint can be applied at the supplied viscosity, but for a large area a small quantity of the appropriate thinner may be added to obtain better atomization and to provide a smoother coating. Spray guns must be thoroughly cleaned immediately after being used with polyester primer filler. As the paint solvent is extremely volatile, the cap must be screwed down tightly on the tin when not in use. Polyester primer filler can be used with most car refinishing processes and is very useful when used with low-bake enamels.

4.5.2 Fillers and levellers

Surfacer

A surfacer is applied over the primer. Its function is to build up the coating thickness, while filling up minor defects such as scratches.

Filler

Filler is a heavy-bodied material used for levelling defects which are too deep to be filled economically with surfacer. Fillers are manufactured to suit the method of application: there are spraying, brushing and knifing types. Fillers are available as both single-pack or two-pack, cellulose and synthetic types.

Stopper

Stopper is a putty-like substance used for filling up defects too deep for satisfactory levelling with either filler or surfacer.

Though deep indentations are normally filled up by the body repair worker, the painter is sometimes required to carry out levelling work. A two-pack stopper, usually based on polyester resin and a catalyst, is used for this purpose. It dries rapidly in heavy layers, unlike cellulose or oil-based stoppers which must be applied in thin layers with a drying period between applications. Polyester stopper is intended for use on bare metal or over high-baked primers. It cannot be used as an intercoat stopper over etching primers, or between enamel coats. It can, however, be coated over with most standard paint systems used in refinishing, including cellulose synthetic, coach finish and low-bake synthetics. The normal curing time of a polyester stopper is about 30 minutes at 20°C.

4.5.3 Sealers

There are three types of sealers used by the refinisher: standard, isolators and bleed inhibitors.

Standard sealers have a low pigment and high binder content. They are supplied ready for use and are applied over the final coat of surfacer or spray filler to provide hold-out of the finishing material and promote higher gloss. They also reduce the risk of crazing when applying acrylic lacquer-type finish over cellulose-based undercoats.

Isolating sealers are more heavily pigmented and are used to avoid a reaction between different types of paint systems, for example when applying a lacquer-type paint over a synthetic enamel.

Bleed-inhibiting sealers contain carbon black pigment which is able to absorb floating colour. These sealers are recommended when carrying out a colour change over a colour which is suspected of being a bleeder, that is some of the pigment or dyestuff will float into the new coating and discolour it. Several reds, particularly those containing organic pigments, are prone to this behaviour.

Whichever sealer is used, it can only do its job if it is a continuous film. For this reason it must not be flatted down, though a light denib, carefully done, is permissible.

4.5.4 Finish

Finish is the term used to describe the finishing colour coats. They have a comparatively low pigment content as opposed to surfacers and fillers. The high percentage of the vehicle or binder provides the glossy effect.

4.5.5 Abrasive papers

The type of abrasive paper mainly used by the spray painter is known as wet-or-dry paper. The abrasives used are silicon carbide and aluminium oxide, and these are attached to a treated paper backing by means of waterproof resin glue.

Wet-or-dry paper is available in various grades ranging from 80D (coarse) to P1200 (very fine): low numbers denote coarse grades and high numbers identify the finer grades. This type of abrasive paper is normally used in conjunction with water to avoid a build-up of paint particles which would affect the abrading effectiveness of the paper.

In addition to wet-or-dry paper, the spray painter should have in his stock emery paper and production papers. These are much coarser than wet-or-dry papers and are used for rougher work such as removing rust or mill scale. These two papers are most effective when used with power tools in the form of circular discs, being attached to the pad with specially formulated disc adhesives. As they are normally used dry, the particles are attached to the paper or cloth backing with a good quality hide glue.

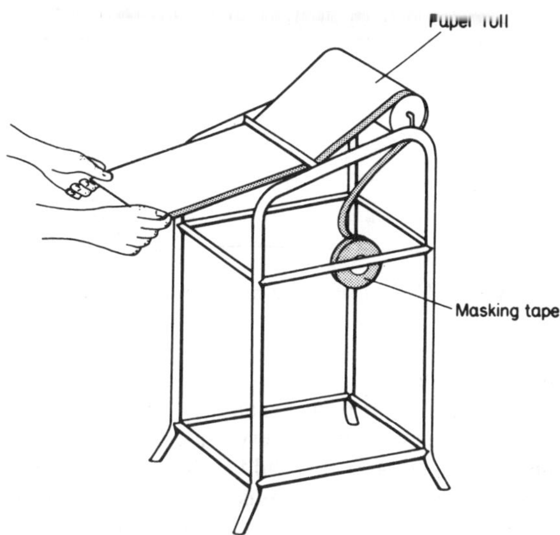

Figure 4.4 Masking machine

Abrasive papers fall into two categories, based on the amount of space between the particles. If these are widely spaced the paper is referred to as open coated and is used to remove paint or rust which tend to fill up the spaces on finer grades. When the particles are tightly packed the paper is known as close coated and is used to rub down smoother surfaces.

4.5.6 Masking

Masking tape
This is a paper tape, one side of which is coated with an adhesive of a non-drying composition. It is supplied in rolls in a variety of widths, but the most widely used are those measuring 20 mm and 25 mm. Wider tapes are considered to be uneconomical, though narrower tapes such as 13 mm do have limited uses. Where two-tone work is carried out, a finer edge can be achieved by the use of gummed paper than is possible with masking tape, but the time spent on removing it makes it less popular.

Masking machines are also available in which a roll of masking paper and various widths of masking tape are mounted. As the paper is pulled out of the machine, a strip of the tape is automatically attached to one of the edges (Figure 4.4).

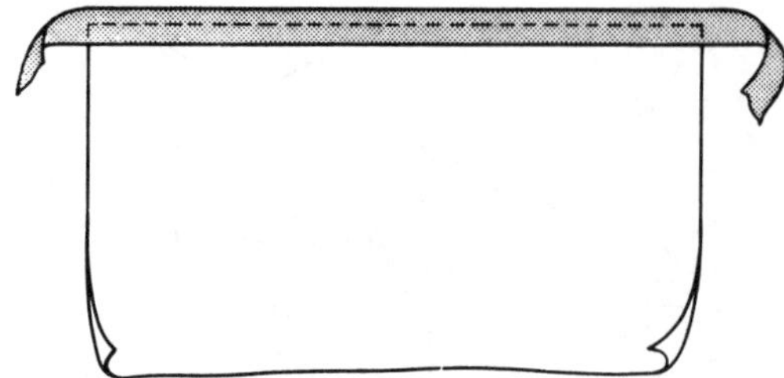

Figure 4.5 Method of fixing masking tape to paper

Masking paper
Brown paper (kraft paper) is an ideal masking material, though newspaper is very widely used for this purpose (Figure 4.5). It should be borne in mind that newspaper is extremely porous and subject to solvent penetration, and it is therefore recommended that at least two layers be used over existing paintwork. It has been common practice in some shops to oil or grease the newspaper, but this cannot be recommended because of the risk of grease contamination to those areas to be sprayed.

4.5.7 Burnishing

Burnishing compound
There are both burnishing pastes and polishing liquids. These are generally emulsions of mineral oils and water with the addition of an emulsifying agent. They also contain mild abrasives to 'cut down' the final coat of enamel and promote a good lustre. Burnishing compounds are also used during the carrying out of local paint repairs to remove overspray.

Standard compounds and creams or liquids contain ammonia to keep them fresh in the tin, but ammonia can cause staining of clear coatings. Special ammonia-free compounds and liquids are available for the burnishing of clear-over-base finishes.

Mutton cloth
In days gone by, butchers and slaughterers wrapped joints of meat in this material – hence its name. Mutton cloth is available in different grades: coarse, medium and fine. The coarser grades are used with rubbing compound to remove overspray when carrying out localized paint repairs. The fine grades are used with fine compounds and creams for final burnishing to promote a deep gloss.

4.5.8 Solvents

The paint shop should be well stocked with the appropriate solvents for the types of paint to be used. The importance of using the correct thinner for a particular type or make of paint cannot be overstressed, as many painting defects can be traced to the incorrect use of solvents. Solvents can prove to be an expensive item, and in consequence a cheaper form of cleaning solvent should be stocked for the purposes of cleaning spray guns and equipment.

Solvents fall into two categories, high boilers and low boilers. Those which require a high temperature to bring them up to their boiling point tend to be slow in evaporating and consequently slow in drying. Low boilers, on the other hand, evaporate and dry quickly. Low boilers are best used with primers, surfacers and fillers where it is required to build up the coating thickness fairly rapidly. High boilers can be used in finishing coats to promote better flow, helping to eliminate an orange-peel defect. A good quality solvent, however, will contain both high and low boilers in well-balanced proportions.

4.6 SPRAY PAINTING EQUIPMENT

The items of equipment in a spray painting shop are basically as follows: air compressing unit, air line, air filter, pressure regulator; air hose and finally spray gun.

4.6.1 Air compressor

When selecting an air compressor for a particular workshop, one should calculate the volume of compressed air that will be required to operate the various tools throughout the workshop, such as rubbing-down tools and spray guns. The size of the compressor chosen should be capable of giving a higher free air delivery than is required.

In a workshop where more than one activity is carried out, such as panel beating and spray painting, it is advisable to install a stationary two-stage compressor. This should be bolted to the floor at least 300 mm from any wall, and sited where it can receive an ample supply of clean dry air. A two-stage compressor consists of an electric or fuel oil motor, the compression unit itself, and a storage tank. In addition to these basic components there can also be various refinements in the way of safety devices and operating switches.

Taking a piston-type compressor as an example, the operational sequence is as follows. The motor drives the compressing pistons, which are situated within two cylinders set above the storage tank. Air at normal atmospheric pressure is drawn into the first cylinder via the air intake, to which is attached a

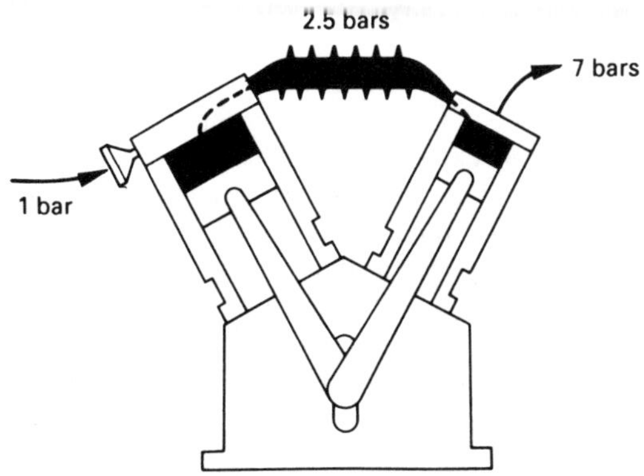

Figure 4.6 Single-acting two-stage compressor

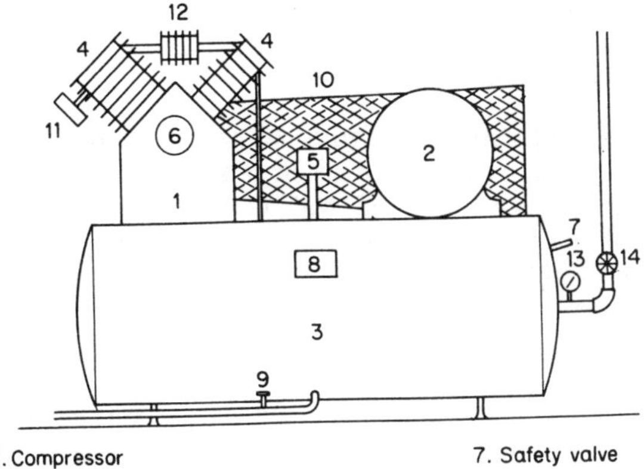

Figure 4.7 Principal parts of a stationary two-stage compressor

filter. The air is then compressed to an intermediate pressure by the action of the piston moving upwards and reducing the volume of the cylinder (Figure 4.6). When a sufficient pressure has been attained, the air is pumped through a chamber (or intercooler) into the second smaller cylinder, where it is compressed even further by piston action and finally discharged into the storage tank for subsequent use (Figure 4.7).

The compressed air is stored in the receiver (or storage tank) for a cooling-off period, during which moisture and vapours within the air will condense and collect on the floor of the tank. The very act of compressing air will generate heat, and so the cooling period is essential. In a further effort to keep the air cool, the compression cylinders have external fins to provide a greater surface area for the heat generated within them to dissipate into the atmosphere. In addition to storing the compressed air and providing a cooling-off period, the air receiver acts as a buffer between the compressor and the spray gun, blanketing out pulsations so the compressed air can be drawn from it at a steady even pressure.

It is important that the size of the air receiver be in direct proportion to the size of the compressor. Contrary to popular belief, air volume is more important to the spray painter than is air pressure. A typical suction-fed spray gun will require 3–4.5 litres of free air delivered (FAD) per second to allow it to operate satisfactorily. The gun must have this volume of air from the air receiver regardless of the pressure. In a typical two-stage piston-type compressor, the free air delivered will only be 70 to 75 per cent that of the air displaced by the piston in the first cylinder, and therefore it has only 70 to 75 per cent volumetric efficiency. Free air delivery is in consequence the prime factor to consider when purchasing a compressor.

4.6.2 Air lines

The compressed air is drawn from the air receiver and led to the air transformer through a galvanized tube known as an air line. The three main requirements of an air line installation are:

1 Low pressure drop between the compressor plant and the points of air consumption.
2 Minimum of air leakage.
3 High degree of contamination filtering throughout the system.

Pressure drop in an air line is caused by the frictional action of the air molecules on the inside surface of the pipe, as well as pressure build-up at angles. The greater the distance between the compressor and consumption points, the greater will be the pressure drop. For this reason the air line must be of sufficient internal diameter to carry a satisfactory volume of compressed air into the spray shop. For the average refinishing shop an internal diameter of 25 mm will suffice. Pressure drops can be increased where angles are introduced into the system, and in consequence the number of angles should be kept to a minimum. Also the more couplings there are in the installation, the greater is the risk of air leakage with a consequential drop in pressure.

It is inevitable that some moisture vapour will be drawn into the air line from the air receiver, and for this reason the air line should slope down towards the

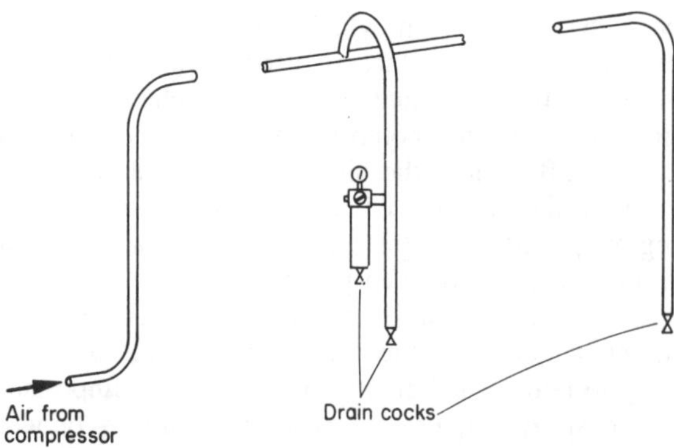

Figure 4.8 Air line installation

Figure 4.9 DeVilbiss DVFR-1 filter regulator assembly and DVF2-2 filter regulator coalescer (*DeVilbiss Automotive Refinishing Products*)

Figure 4.10 Single-braid air hose

air receiver to enable condensed moisture to run into it rather than to the air transformer. The service line leading to the transformer should be tapped from the top of the main air line, and should preferably be in the form of a U bend rather than elbow joints with right-angled bends on them (Figure 4.8).

4.6.3 Air transformer

This consists of two units: a condenser or filter, and a pressure regulator. The condenser allows the compressed air to expand into a chamber, thus assisting cooling, and then removes moisture from it by means of a removable filter. A drain cock is situated at the base of the chamber to drain off accumulated impurities periodically. The regulator is a reducing valve with which to reduce the air pressure from the compressor to that required for spraying. The air transformer should be fitted with a pressure gauge giving an accurate reading of the pressure of air passing through the regulating valve (Figure 4.9). A small lightweight filter can be fitted to the handle of the spray gun as an additional safeguard in conditions where exceptional humidity exists.

4.6.4 Air hose

The compressed air is led from the transformer to the spray gun by means of an air hose. This consists of a rubber tube covered with cotton braid enclosed within a rubber covering; the three layers are vulcanized into one (Figure 4.10). Multibraid hoses are available for high-pressure work. At each end of the hose are couplings for attachment to the transformer and the spray gun.

Though the interior wall of the air hose is smooth, it will still create a certain amount of resistance to the flow of air, particularly when long lengths are used. Hoses with an interior diameter of 6 mm should never be used in lengths exceeding 4 m because of the high pressure drop encountered. For example, if the air transformer is delivering air at a pressure of 3.5 bar (50 psi) the pressure drop over 4 m would be 0.7 bar (10 psi). However, using an air hose of diameter 8 mm over the same length and at the same supplied pressure, the drop would only be 0.2 bar (3.5 psi). From this it can be seen that a hose of 8 mm diameter will give best results on lengths exceeding 4 m. When working in a spray booth, the spray painter rarely, if ever, requires an air hose greater than 6 m in length.

4.7 TYPES OF SPRAY GUN

Of all the tools and techniques used in paint shops, the spray gun and spray painting have provided the most satisfactory method of applying paint. Unless there is a complete change in the design and construction of the motor vehicle, the spray gun will be used for many years to come. The spray gun, like all tools, is only effective in the hands of a skilled operator, and therefore a painter should know as much as possible about what has become the main tool of the trade.

A spray gun is a precision instrument which uses compressed air to atomize the fluid paint and break it up into small particles. The air and paint enter the gun through separate passages, mix, and are then ejected at the front of the gun.

Spray guns can be divided into groups:

1 By methods of paint supply, such as suction feed, gravity feed or pressure feed.
2 Those with detached or attached paint containers.
3 Internal mix or external mix types.
4 Bleeder or non-bleeder types.

The most widely used spray gun is the suction-feed, external-mix, non-bleeder model.

4.7.1 Paint supply methods

Suction-feed gun
With this type, a stream of compressed air creates a vacuum at the fluid tip which allows atmospheric pressure within the fluid cup to force the paint up the fluid tube to the fluid tip and air cap. The paint container (fluid cup) is limited to one litre (1000 cm^3) capacity to enable the gun to be handled without fatigue. The suction-feed spray gun is easily identified, as the fluid tip protrudes slightly beyond the air cap (Figure 4.11).

Gravity-feed gun
The fluid cup is mounted above the spray head and paint is fed to the gun by the force of gravity.

The fluid cup is usually limited to 0.5 litre capacity, which makes the gravity-feed gun unsuitable for the spraying of large areas. However, it is very useful for painting local repairs where heavy-bodied fillers are applied, and where rapid colour changes are necessary (Figure 4.12).

Pressure-feed gun
This type of gun sprays paint that has been forced from the paint container by compressed air. The air cap of this gun is not designed to create a vacuum, and the fluid tip is flush with the front of the air cap (Figure 4.13). Pressure-feed guns are used where a large quantity of a particular paint is to be sprayed, or where the material is too viscose or heavy to be siphoned from the fluid container as in suction-feed guns.

The fluid container or pressure vessel is connected to the gun by means of a reinforced fluid hose, and normally ranges in size from 2 to 25 litres. The smaller pressure vessels can be carried in the operator's free

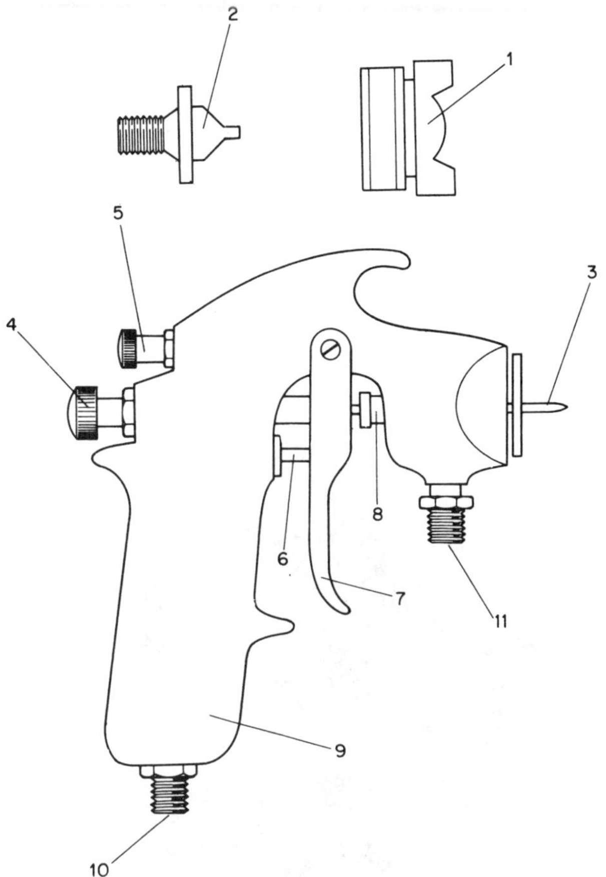

Figure 4.11 Typical suction-feed spray gun

1 Air cap
2 Fluid tip
3 Fluid needle
4 Fluid control screw
5 Spreader control
6 Air valve
7 Trigger
8 Fluid packing nut
9 Gun body
10 Air inlet
11 Fluid inlet

hand (Figure 4.14), but ones from 10 litres capacity upward can be mounted on wheels for easy portability (Figure 4.15).

4.7.2 Mix methods

Internal-mix gun
This gun mixes air and paint inside the air cap, and is used with low air pressure to apply slow-drying materials (Figure 4.16).

External-mix gun
This is the most widely used type of gun, and can be used to spray most types of paint. It is the best type of gun for spraying quick-drying materials.

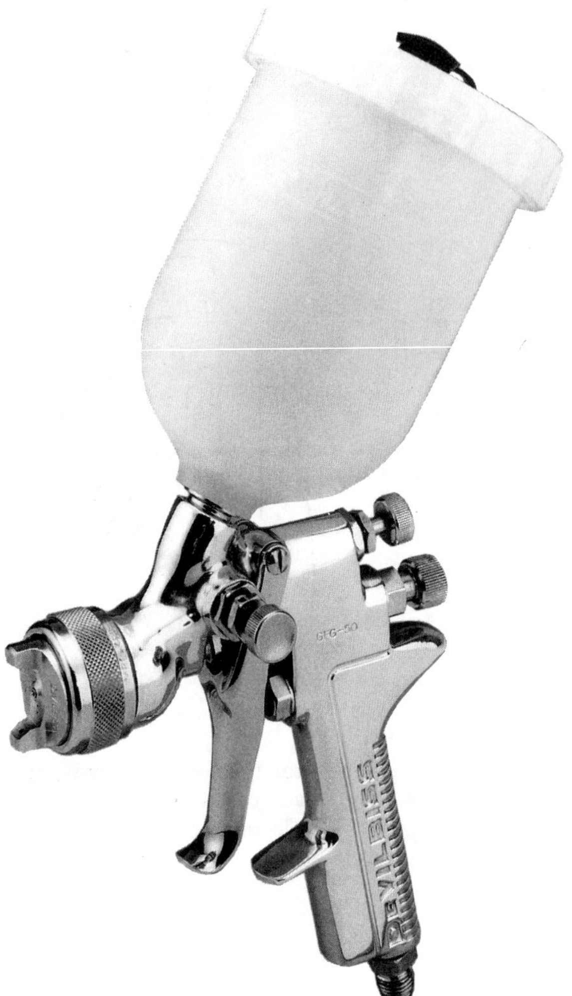

Figure 4.12 Gravity-feed spray gun (*DeVilbiss Automotive Refinishing Products*)

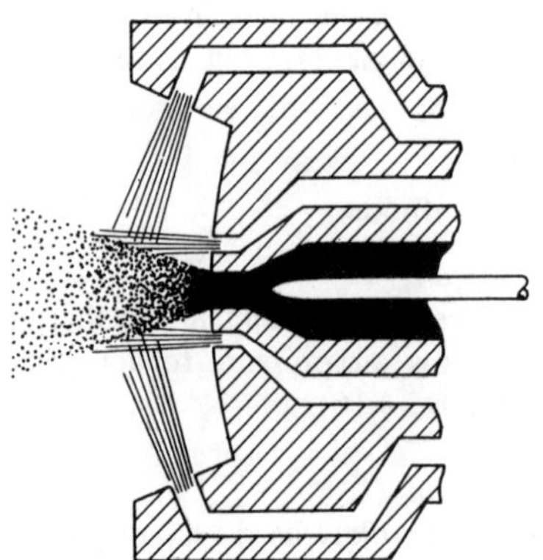

Figure 4.13 Pressure-feed spray head

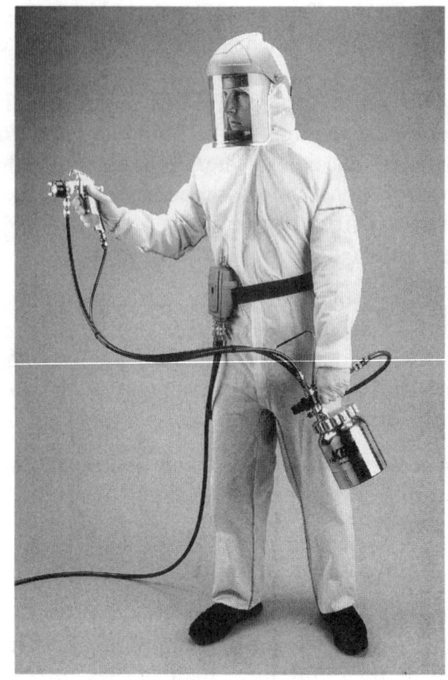

Figure 4.14 Remote cup with gun (2.3 litre capacity) (*DeVilbiss Automotive Refinishing Products*)

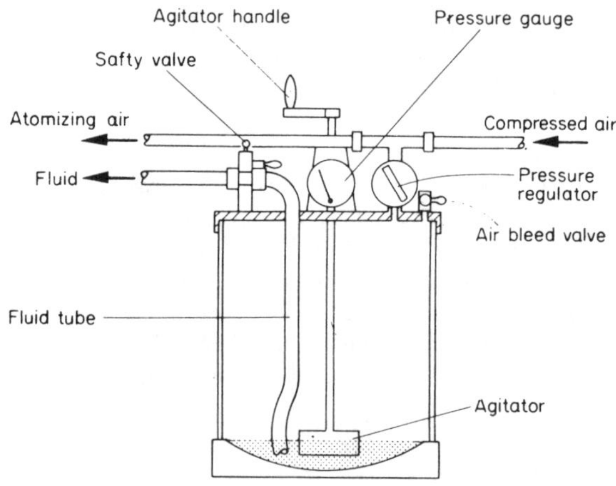

Figure 4.15 Typical pressure-feed tank

The air and paint mix beyond the air cap, and perfect atomization can be achieved (Figure 4.17).

4.7.3 Bleed methods

Bleeder gun

This is designed without an air valve. Air continually passes through the gun, thus preventing a build-up of

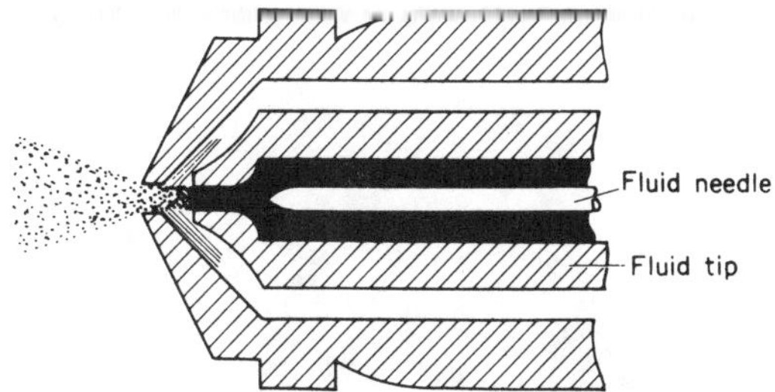

Figure 4.16 Internal-mix spray head

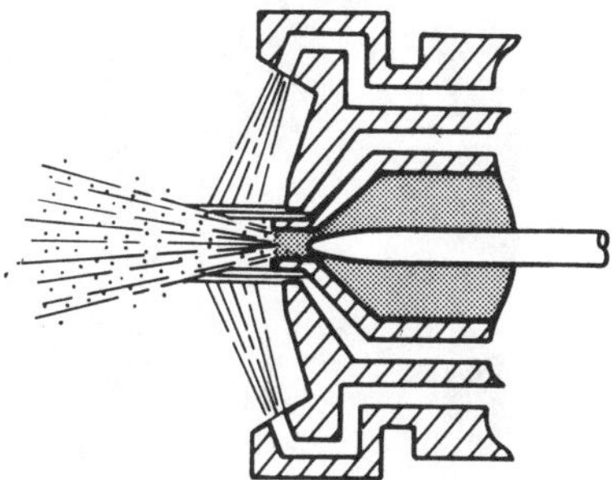

Figure 4.17 External-mix suction-feed spray head

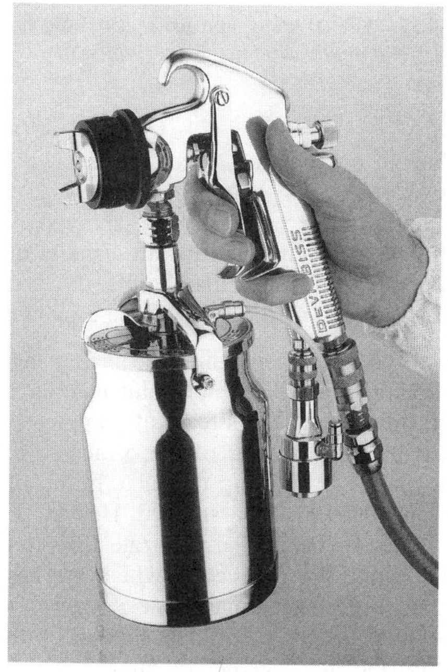

Figure 4.18 High-volume low-pressure (HVLP) cup gun (*DeVilbiss Automotive Refinishing Products*)

pressure in the air hose. The gun is usually used with small compressors having a limited output and having no unload or pressure switch. The trigger on the gun controls the flow of paint only.

Non-bleeder gun

These are equipped with an air valve. The trigger controls both the flow of compressed air passing through the gun and the flow of paint.

4.7.4 High-volume low-pressure (HVLP) spray guns

Legislation which is part of the Environmental Protection Act requires that spray painters must reduce the amount of paint vapours being released into the atmosphere as a result of their working activities (see Chapter 1).

In conventional air-atomized spray painting, about 35 per cent of the mixed paint actually reaches the job surface. The remaining 65 per cent is either extracted to the atmosphere or collects on the workshop floor and walls. Of this waste material, 30 per cent is solid while 70 per cent is classified as volatile organic compounds (VOCs). It is with these VOCs that the legislation is concerned.

In an effort to conform to the EPA, spray gun manufacturers have developed spray guns which atomize the paint at low air pressure, that is 0.6 bar (10 psi) as opposed to the usual 4–5 bar (60–75 psi). This reduced air pressure results in greatly reduced overspray, and as a further bonus 65 per cent of the mixed paint reaches the surface. Savings on paint wastage are obvious.

Though atomization is achieved with much lower pressure, these spray guns still require large volumes of compressed air to operate them. A typical air pressure at the air transformer may be 4 bar (60 psi), but this is reduced in the gun body by means of an air restrictor which reduces the air velocity at the gun outlet in the ratio of about 6:1. Other manufacturers like DeVilbiss do not have restrictors in the guns and only need 25 psi at the inlet to give 10 psi at the cap (Figures 4.18, 4.19).

A special air cap is provided with the gun. This has a pressure gauge attached to it which enables the spray painter to adjust the outlet pressure. When the pressure has been adjusted to 10 psi, the air cap is removed and a more conventional type fitted (Figure 4.20). The fluid cup is pressurized and can be either attached to the gun (Figure 4.18) or remote from it (Figure 4.21). Gravity types are also available (see Figure 4.22). The basic parts of these types of spray guns are seen in Figure 4.23.

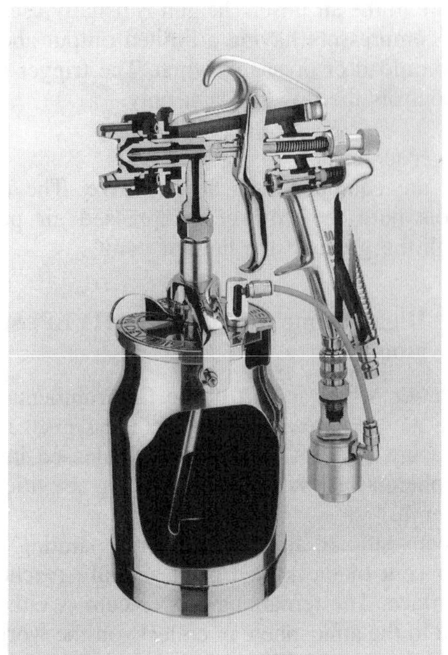

Figure 4.19 Sectional view of an HVLP cup gun (*DeVilbiss Automotive Refinishing Products*)

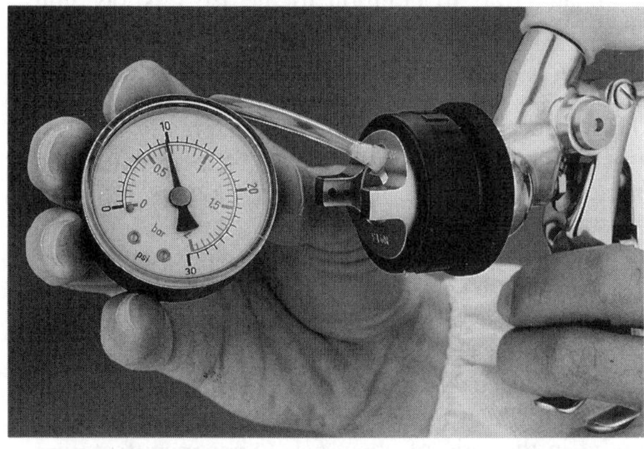

Figure 4.20 Air cap with pressure gauge attached (*DeVilbiss Automotive Refinishing Products*)

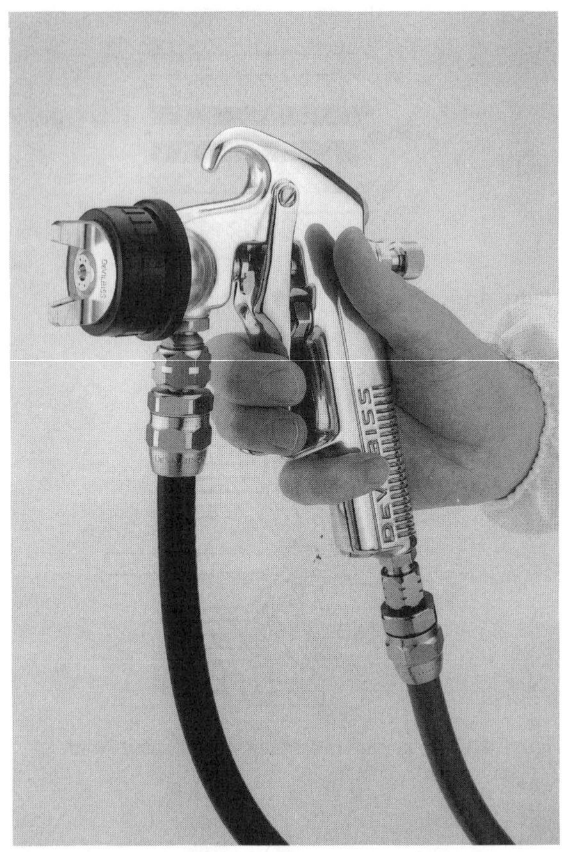

Figure 4.21 HVLP spray gun using a remote fluid cup (*DeVilbiss Automotive Refinishing Products*)

4.8 BASIC PARTS OF A STANDARD SPRAY GUN

Air cap This is the nozzle at the front of the gun that directs compressed air into the stream of paint, thus atomizing it and forming the spray pattern. Air caps are designed to give a wide variety of spray patterns and sizes, and in addition ensure perfect atomization over a wide range of paint viscosities. The choice of an air cap (Figure 4.24) depends on the following:

1 Volume of compressed air available.
2 Type of paint feed system being used.
3 Type and volume of paint to be sprayed.
4 Size of the fluid tip (air caps are usually designed to operate with a particular fluid tip).
5 Nature and size of the surface to be painted.

Fluid tip This is situated behind the air cap and meters out the paint. The volume of paint passing through the fluid tip and into the stream of compressed air is governed by the diameter of the orifice in the fluid tip. The choice of fluid tip depends, in the main, on the type of material to be sprayed. Heavy, coarse or fibrous materials require large nozzle sizes to prevent clogging, while thin materials, which are applied at low pressures, require small nozzle sizes to prevent an excessive flow of paint. The fluid tip provides a seating for the fluid needle (Figure 4.24).

Fluid needle This seats in the fluid tip, its function being to start and stop the flow of paint. For the gun

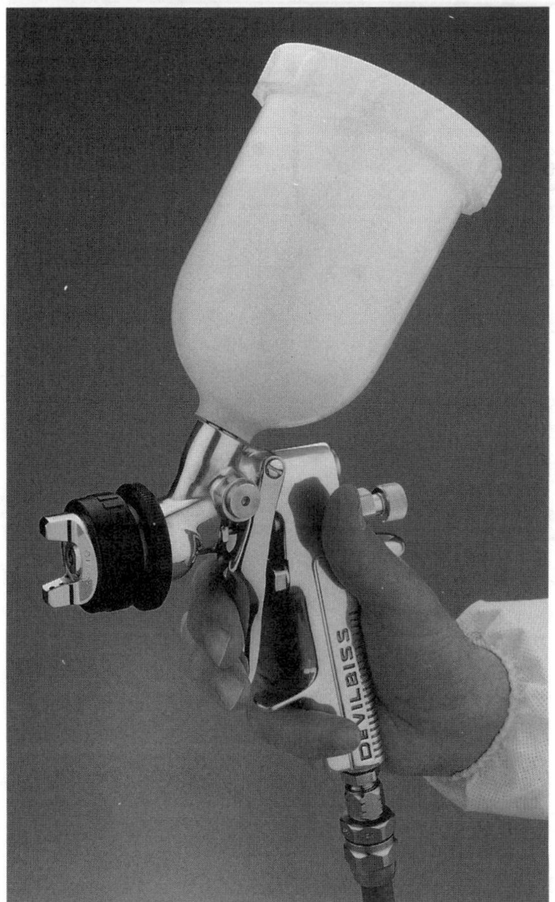

Figure 4.22 HVLP gravity-fed spray gun (*DeVilbiss Automotive Refinishing Products*)

to operate efficiently, the fluid tip and fluid needle should be selected as a pair, and should be of the same size (Figure 4.24).

Fluid control screw This is an adjustment control which limits the length of travel of the fluid needle, governing the flow of paint from the fluid tip.

Trigger The function of the trigger is to operate the air valve and also the fluid needle.

Air valve This is situated in the handle of the spray gun (or gun body) directly behind the trigger by which it is operated. Its function is to control the passage of air through the gun.

Gun body This can be regarded as a basic frame on to which the spray painter will mount a suitable set-up to suit his requirements.

Spreader control This is of great importance in controlling the volume of air passing to the horn holes of the air cap. If air is cut off from the horn holes a narrow jet of paint giving a spot pattern is ejected, but when air is allowed to pass through the horn holes a fan spread is obtained, the width of fan varying according to the volume of air (Figure 4.25).

4.9 SPRAY GUN MAINTENANCE AND CLEANING

4.9.1 Cleaning a suction-feed gun

First, loosen the gun from the paint container, allowing the fluid tube to remain in the container, and unscrew the air cap a few turns. Holding a piece of cloth over the air cap, pull the trigger to divert air down the fluid tube and drive the paint back into the container. Next, empty paint from the container and rinse it out with the appropriate solvent and an old paint brush. Pour a small quantity of the solvent into the container and spray it through the gun to flush out the fluid passages. Remove the air cap and immerse it in clean solvent, then dry it out by blowing with compressed air.

If the holes in the air cap become blocked with dried paint, a stiff brush moistened with solvent will usually remove the obstruction. If not, a toothpick or sharpened matchstick can be used to clean out the hole. On no account must wire or a nail be used, for this can distort the holes and permanently damage the air cap, resulting in a distorted spray pattern.

4.9.2 Cleaning a gravity-feed gun

Remove the cup lid, empty out surplus paint and replace it with a small quantity of solvent. Replace the lid, unscrew the air cap a few turns and, holding a piece of cloth over the air cap, pull the trigger. Air will be diverted into the fluid passage, causing a boiling action and flushing it clean. Spray solvent through the gun and clean the air cap. Finally, with a solvent-soaked rag wipe the outside of the paint container clean.

4.9.3 Cleaning a pressure-feed set-up

Shut off the air supply to the pressure tank, release pressure in it and loosen the lid. Unscrew the air cap a few turns, hold a piece of cloth over and pull the trigger. The pressure will force the paint from the gun and fluid hose back into the tank. Empty surplus paint from the tank and pour a small quantity of solvent, replace the lid firmly and pressurize the tank at about 0.3 bar (5 psi). The pressure will force the solvent to

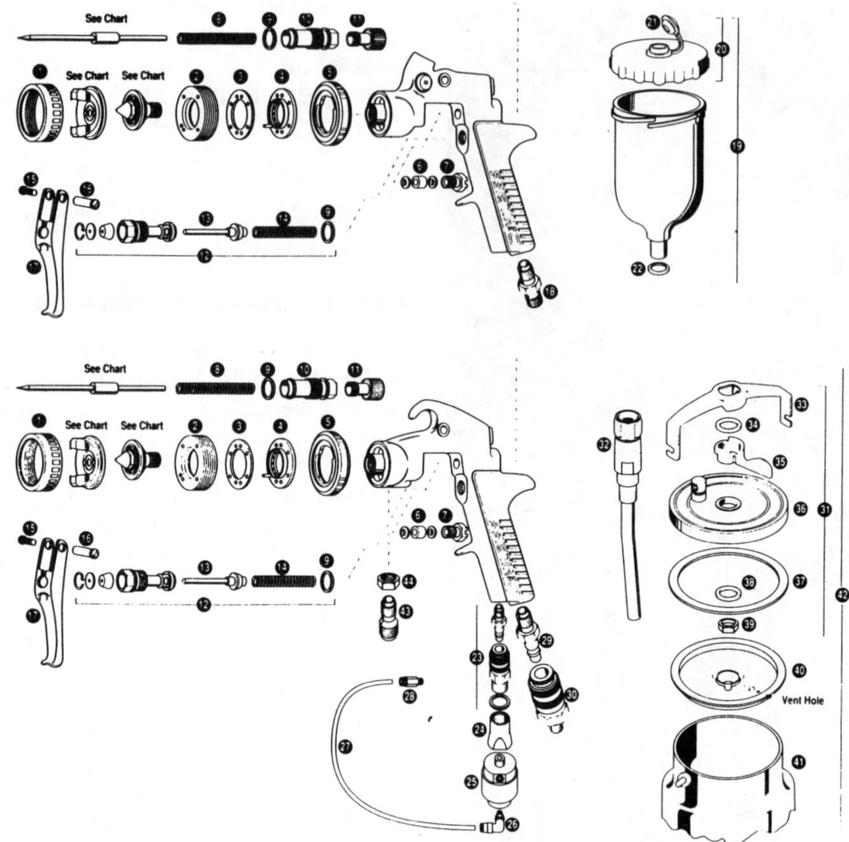

Figure 4.23 Basic parts of the HVLP spray guns (*DeVilbiss Automotive Refinishing Products*)

1 Nylon air cap retaining ring
2 Front air cap baffle (nylon)
3 Kit of 2 baffle seals
4 Rear baffle assembly
5 Spreader/fan adjustment ring
6 Fluid needle packing set
7 Fluid needle packing gland
8 Kit of 5 fluid needle springs
9 Kit of 5 gun body bushing gaskets
10 Fluid needle body bushing
11 Fluid needle adjusting screw
12 Complete air valve assembly
13 Air valve stem
14 Kit of 3 air valve (trigger) springs
15 Kit of 5 trigger pivot screws
16 Kit of 5 female trigger pivot studs
17 Chrome plated trigger
18 Male/male air connector $\frac{1}{4}$ in BSP
19 1 pint nylon gravity cup assembly
20 Gravity cup lid assembly
21 Kit of 5 drip check lids
22 Kit of 5 O-rings for gravity cup
23 Mini QD female valve and male stem
24 Adaptor to fit regulator to QD valve
25 Cup pressure regulator
26 Kit of 2 regulator elbow connectors
27 Kit of 5 regulator to cup tubes
28 Cup pressure tube connector
29 QD stem
30 QD valve
31 Complete cup lid assembly $\frac{3}{8}$ in BSP
32 Fluid tube $\frac{3}{8}$ in BSP female fitting
33 1 quart KR cup yoke
34 Kit of 5 washers
35 Cam lever
36 KR 1 quart pressure cup lid
37 Kit of 3 KR cup lid gaskets
38 Washer
39 Fluid tube retaining nut
40 Kit of 5 drip free diaphragms
41 1 quart (1.14 litre) PTFE lined pressure cup
42 KR cup and lid assembly (PTFE lined)
43 Fluid inlet connector $\frac{3}{8}$ in BSP
44 Fluid inlet locknut

Figure 4.24 Air cap, fluid tip and fluid needle (*DeVilbiss Automotive Refinishing Products*)

the gun. Hold a piece of cloth over the air cap and pull the trigger, and the higher atomizing pressure will force the solvent back into the pressure tank. If this is repeated about a dozen times, the purging action will clean out the fluid passages. Disconnect the fluid hose and blow it out with compressed air. Steep and clean the air cap and finally dry out the tank with a piece of cloth.

Whichever type of gun is used it must not be immersed in solvent, as this causes the lubrication oil to be washed away and will cause paint leakage from the gun. In addition to this, the air passage could become blocked with pigment sludge.

4.9.4 Lubrication

After the gun has been cleaned, a drop of oil should be applied to the fluid needle packing, air valve packing, and trigger fulcrum screw.

A spray gun will function efficiently provided that it is clean and well maintained, but neglect will eventually cause the gun to malfunction.

4.10 SPRAY GUN MOTION STUDY

The spray gun is not a difficult tool to master, but a study of the following text and the accompanying diagrams will be invaluable to the inexperienced sprayer. Any person who is using a spray gun for the first time should obviously spray a few practice panels such as disused car doors, wings, bonnets and so on to get the feel of the gun before undertaking actual work.

4.10.1 Spraying flat surfaces

The gun should be held at right angles to the work and at a distance of 150–200 mm. Should the gun be held too close to the surface this will result in too much paint being applied, causing runs and sags. Holding the gun too far from the surface creates excessive overspray and a sandy finish (Figure 4.26). The relationship of gun distance and stroke speed is easily understood, and with a little practice the sprayer is able to adjust his speed of movement to suit the distance between the gun and the surface. The gun distance should be kept as constant as possible, and arcing of the gun must be avoided to obtain an even coating thickness. The correct gun action is acquired by keeping the wrist flexible (Figure 4.27). Do not tilt the gun; hold it perpendicular to the surface. Tilting will give an uneven spray pattern resulting in lines across the work (Figure 4.28).

When spraying a panel, the technique of 'triggering' the gun must be mastered. The stroke is started off the panel and the trigger is pulled when the gun reaches the edge of the panel. The trigger is released at the other edge of the panel but the stroke is carried on for a short distance before reversing for the second stroke. This triggering action must be practised and perfected to avoid a build-up of paint at the panel edges and to reduce paint wastage due to overspray.

The method of spraying a panel is shown in Figure 4.29. Note that the gun is aimed at the top edge of the panel, and from then on the aiming point is the bottom of each previous stroke. This gives the 50 per cent overlap necessary to obtain a wet coating. An alternative method is shown in Figure 4.30. In this method the ends of the panel are first sprayed with single vertical strokes, the panel then being completed with horizontal strokes. This technique reduces overspray and ensures complete coverage of the surface.

Long panels such as those encountered on furniture vans require a different approach. A certain amount of arcing is permitted to avoid a build-up of paint where the strokes overlap, and the triggering of the gun is very important. The length of each horizontal stroke is 450–900 mm approximately, or whatever the sprayer can manage comfortably. Figure 4.31 shows the method of overlapping with the panel being sprayed in separate sections, each section overlapping the previous one by about 100 mm.

When spraying level surfaces such as car roofs and bonnets, always start on the near side and work to the

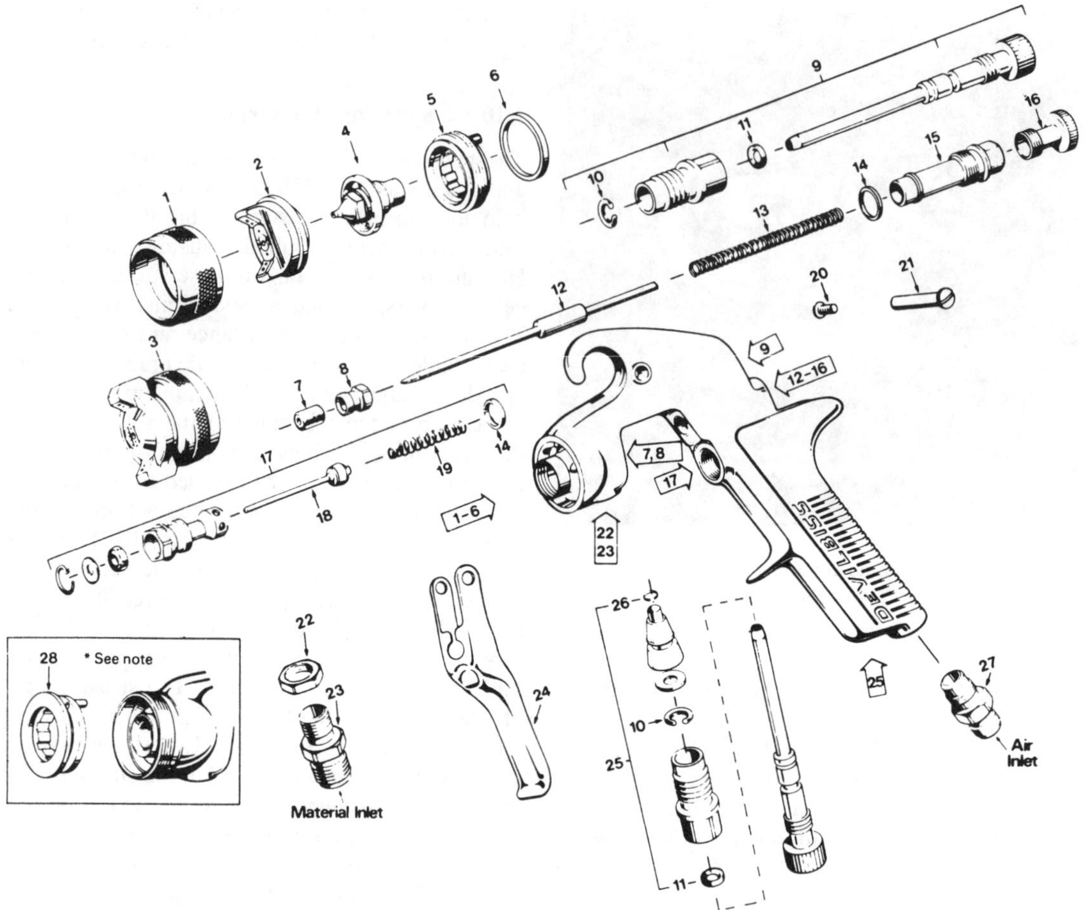

Figure 4.25 Basic parts of a standard spray gun (*DeVilbiss Automotive Refinishing Products*)

1 Retaining ring for air cap
2 Air cap
3 Air cap and retaining ring
4 Corrosion resistant fluid tip and gasket
5 Baffle
6 Kit of five seals
7 Kit of five JGA-7 fluid needle packings
8 Fluid needle packing nut
9 Valve assembly
10 Kit of five circlips
11 Kit of five O-rings
12 Fluid needle
13 Spring
14 Kit of five gaskets
15 Gun body bushing
16 Fluid needle adjusting screw
17 Air valve assembly
18 Air valve
19 Kit of three springs
20 Kit of five screws
21 Kit of five trigger bearing studs
22 Locknut
23 Connector $\frac{3}{8}$ in BSP
24 Trigger
25 Air flow valve
26 Retaining ring
27 Connector $\frac{1}{4}$ in BSP
28 Baffle

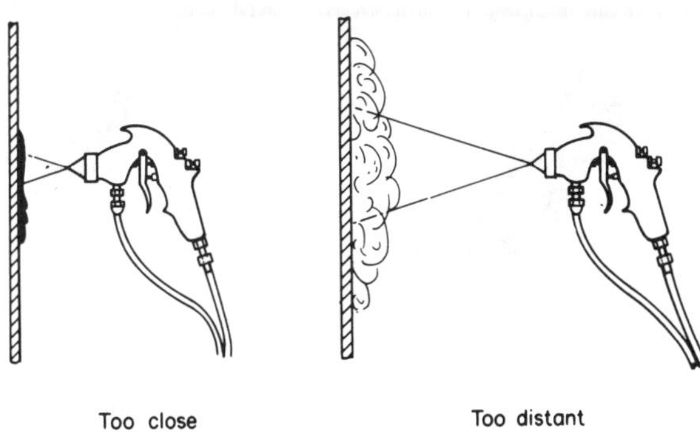

Figure 4.26 Spray distance

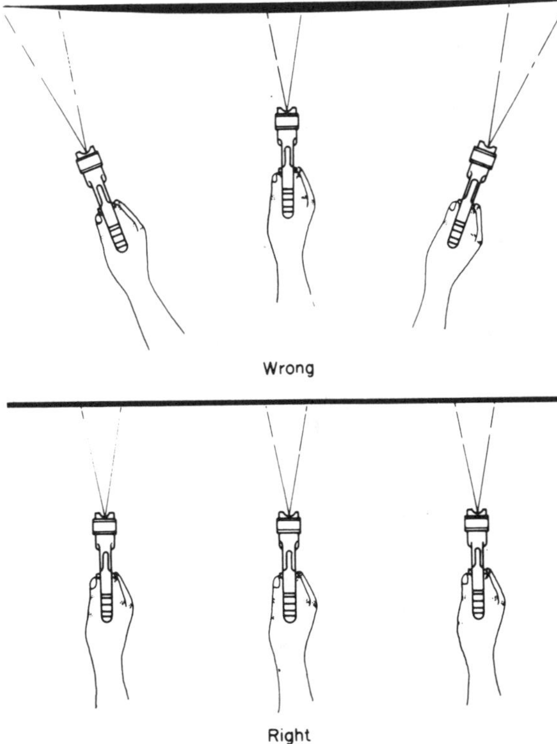

Figure 4.27 Maintaining spray distance

far side to redissolve any overspray. A certain amount of gun tilting is usually unavoidable when reaching across a car roof and overspray is thus created.

4.10.2 Spraying curved surfaces

As previously stated, the gun should be kept at right angles to the surface and as near a constant distance from it as possible (Figure 4.32).

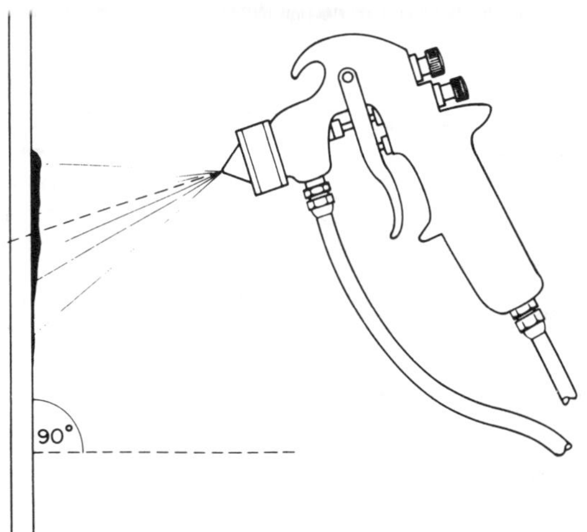

Figure 4.28 Tilting spray gun

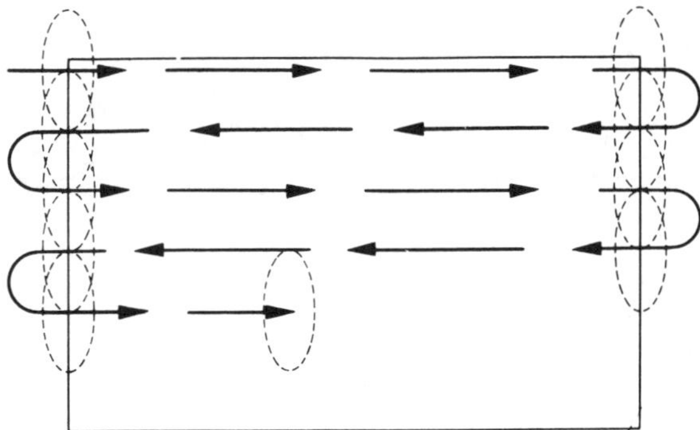

Figure 4.29 Panel spraying method

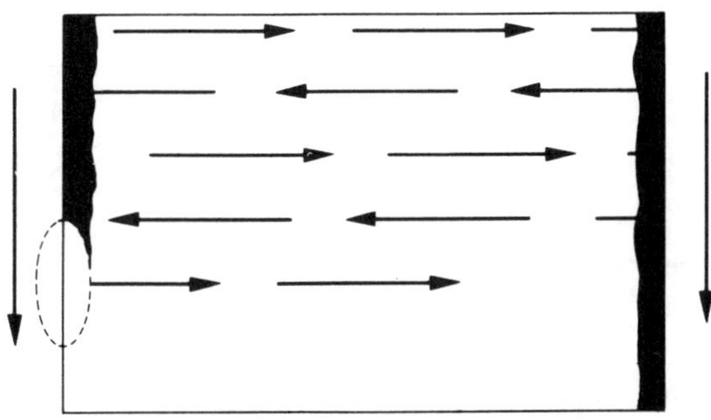

Figure 4.30 Alternative panel spraying method

4.10.3 Spraying external corners

Figure 4.33 shows the method of spraying the edges and corners of a panel, the centre being sprayed like a plain panel. Figure 4.34 shows the technique used to paint vertical corners, the point to watch being to half trigger the gun to avoid applying too much paint.

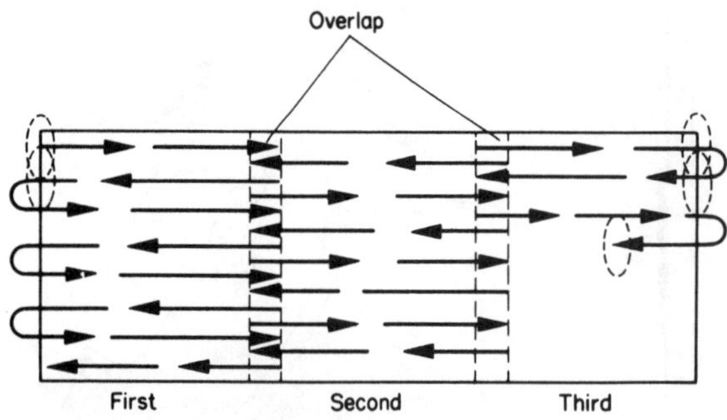

Figure 4.31 Spraying large panels

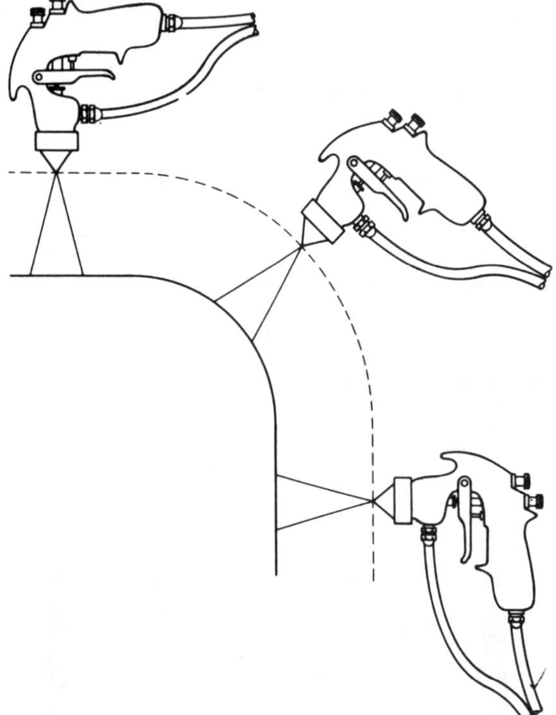

Figure 4.32 Spraying curved surface

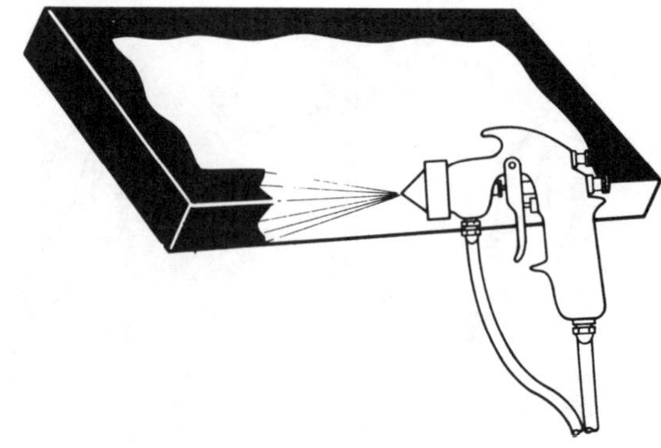

Figure 4.33 Spraying external corners

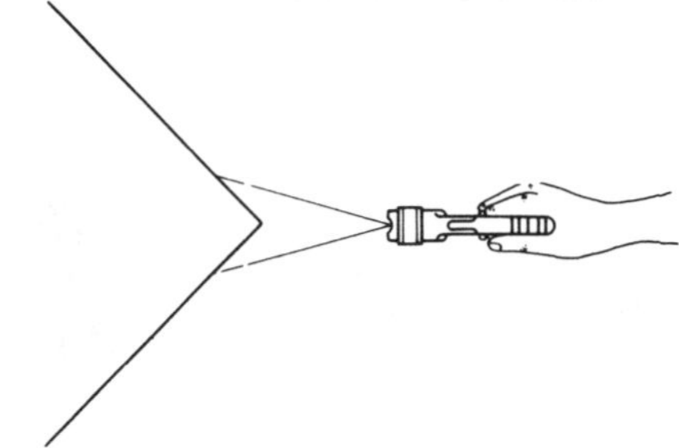

Figure 4.34 Spraying vertical corners

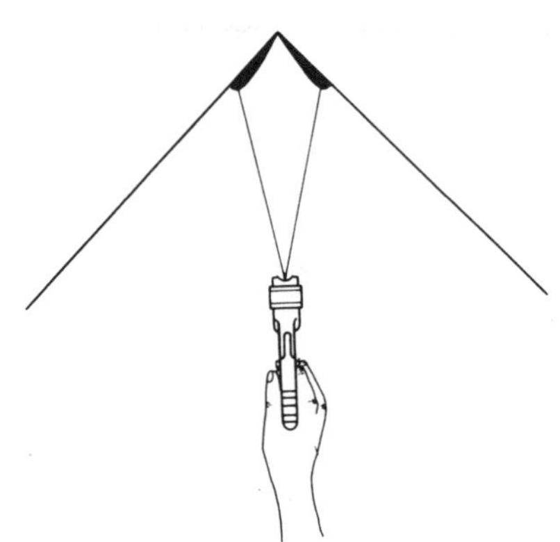

Figure 4.35 Spraying internal corners

4.10.4 Spraying internal corners

Spraying directly into a corner (Figure 4.35) gives an uneven coating but is satisfactory for most work. When an even coating is necessary, such as with metallic finishes, it is better to spray each face separately, starting with a vertical stroke at the edge of the panel (Figure 4.36). The vertical stroke should be followed with short horizontal strokes in order to avoid overspraying or double coating the adjoining surface.

4.10.5 Spraying sequence

An automobile should be sprayed in sections spraying one section at a time before moving on to the next one. Figure 4.37 shows a typical method, but this may vary depending on the size and shape of the vehicle concerned (Figures 4.38, 4.39). The painter must decide on his approach before commencing to spray and then work methodically round the car,

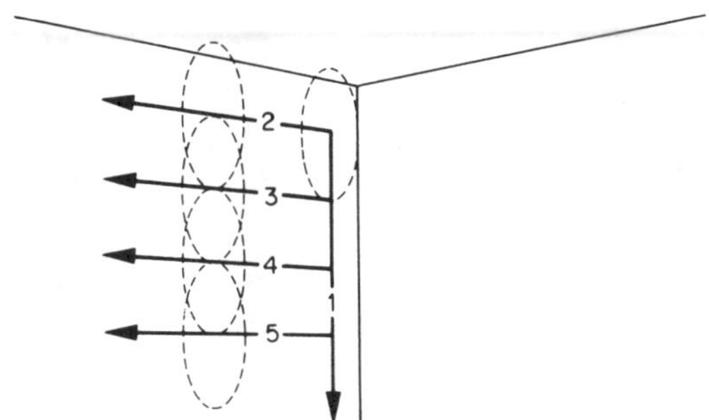

Figure 4.36 Spraying internal corners: method for better finish

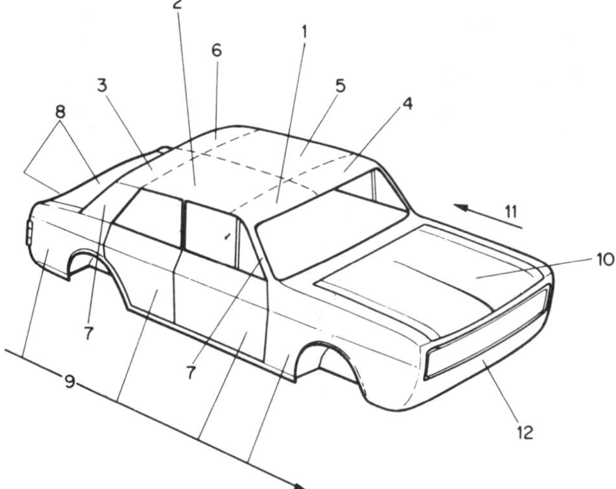

Figure 4.37 Suggested sequence of spraying a car

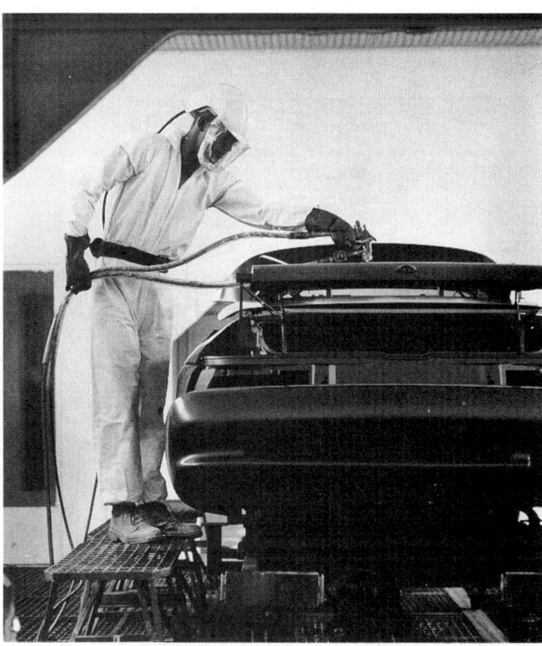

Figure 4.38 Spraying a vehicle roof (*Racal Safety Ltd*)

Figure 4.39 Spraying a vehicle wing (*DeVilbiss Automotive Refinishing Products*)

finishing at a point where an overlap is least noticeable. As the bonnet of the car is the panel which attracts most attention by the customer, most spray painters prefer to spray this last to avoid the risk of overspray falling on it.

4.11 SPRAYING DEFECTS

No matter how excellent a spraying equipment is, sooner or later some small trouble shows itself which, if it were allowed to develop, would mar the work done. However, this trouble can usually be very quickly rectified if the operator knows where to look for its source. The following sections contain the causes and remedies of all the troubles most commonly encountered in spraying.

4.11.1 Fluttering spray

Sometimes the gun will give a fluttering or jerky spray (Figure 4.40), caused by an air leakage into the paint supply line. This may be due to the following (numbers correspond to those on the figure):

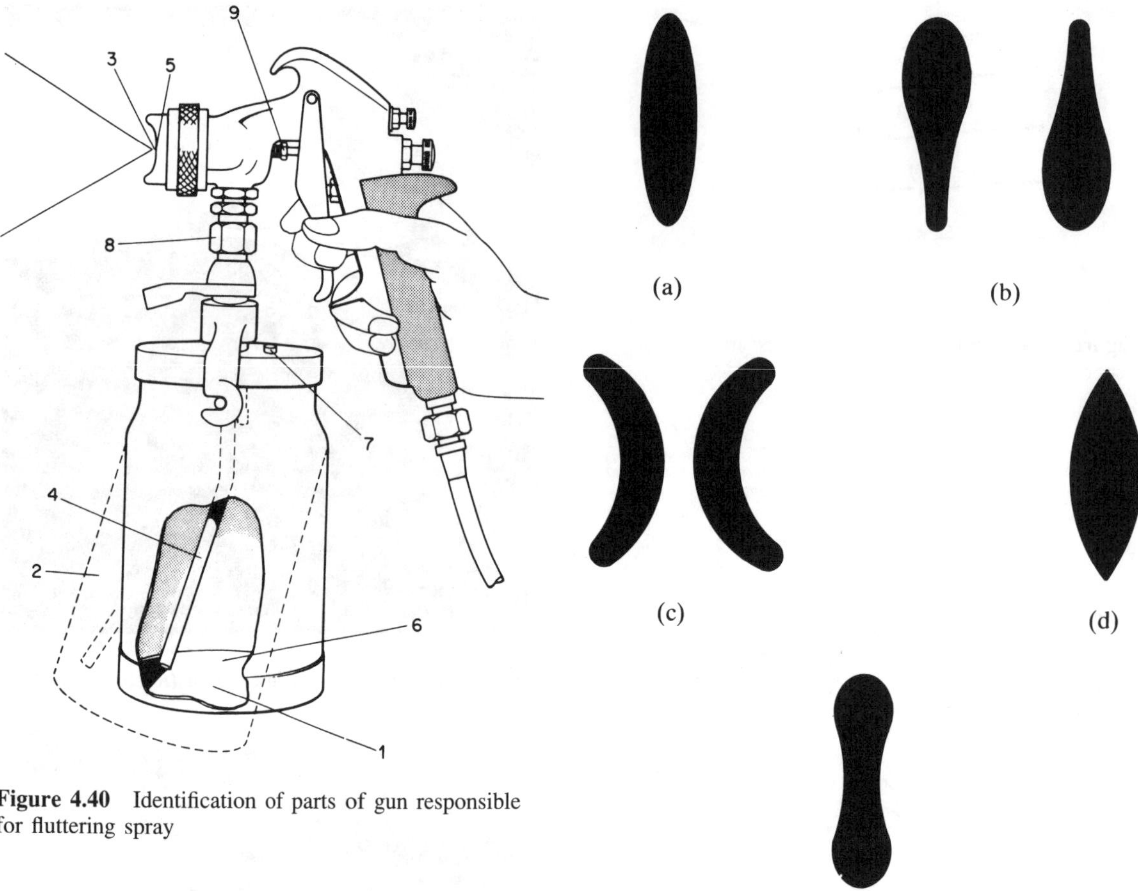

Figure 4.40 Identification of parts of gun responsible for fluttering spray

Figure 4.41 Faulty spray patterns: (a) correct pattern (b) top or bottom heavy (c) right or left sided (d) heavy centred (e) split

1 Insufficient paint in the cup or pressure feed tank so that the end of the fluid tube is uncovered.
2 Tilting the cup of a suction-feed gun at an excessive angle so that the fluid tube does not dip below the surface of the paint.
3 Some obstruction in the fluid passageway which must be removed.
4 Fluid tube loose or cracked or resting on the bottom of the paint container.
5 A loose fluid tip on the spray gun.
6 Too heavy a material for suction feed.
7 A clogged air vent in the cup lid.
8 Loose nut coupling the suction feed cup or fluid hose to the spray gun.
9 Loose fluid needle packing nut or dry packing.

4.11.2 Faulty spray patterns

The normal spray pattern produced by a correctly adjusted spray gun is shown in Figure 4.41a, and defective spray patterns can develop from the following causes:

1 Top or bottom heavy pattern (Figure 4.41b) caused by:
  (a) Horn holes in air cap partially blocked.
  (b) Obstruction on top or bottom of fluid tip.
  (c) Dirt on air cap seat or fluid tip seat.
2 Heavy right or left side pattern (Figure 4.41c) caused by:
  (a) Right or left side horn hole in air cap partially clogged.
  (b) Dirt on right or left side of fluid tip.
3 Heavy centre pattern (see Figure 4.41d) caused by:
  (a) Too low a setting of the spreader adjustment valve on the gun.
  (b) Atomizing air pressure which is too low or paint which is too thick.
  (c) With pressure feed, too high a fluid pressure or

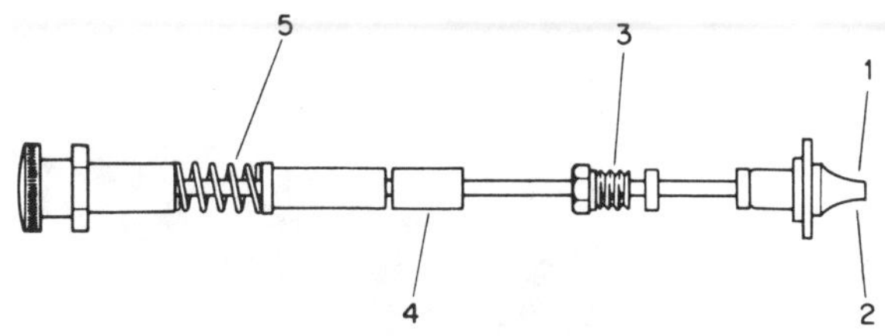

Figure 4.42 Fluid needle assembly

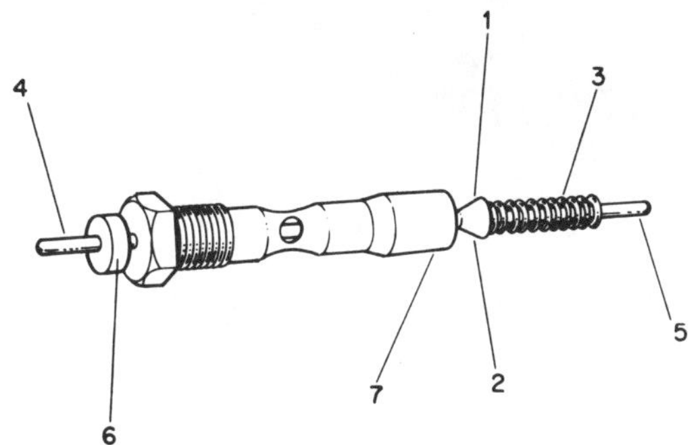

Figure 4.43 Air valve assembly

a flow of paint which exceeds the normal capacity of the air cap.

(d) The wrong size fluid tip for the paint being sprayed.

4 Split spray pattern (Figure 4.41e) caused by the atomizing air and fluid flow not being properly balanced.

To correct defects 1 and 2 (top or bottom heavy pattern, or heavy right or left side pattern) determine whether the obstruction is in the air cap by spraying a test pattern; then rotate the air cap half a turn and spray another test. If the defect is inverted the obstruction is obviously in the air cap, which should be cleaned as previously instructed. If the defect has not changed its position, the obstruction is on the fluid tip. When cleaning the fluid tip, check for fine burr on the tip, which can be removed with P1200 wet-or-dry sandpaper. To rectify defects 3 and 4 (heavy centre pattern, or split spray pattern), if the adjustments are unbalanced readjust the atomizing air pressure, fluid pressure, and spray width control setting until the correct pattern is obtained.

4.11.3 Spray fog

If there is an excessive mist or spray fog, it is caused by:

1 Too thin a paint.
2 Over-atomization, due to using too high an atomizing air pressure for the volume of paint flowing.
3 Improper use of the gun, such as making incorrect strokes or holding the gun too far from the surface.

4.11.4 Paint leakage from gun

Paint leakage from the front of the spray gun is caused by the fluid needle not seating properly (Figure 4.42). This is due to the following (numbers correspond to those on the figure):

1 Worn or damaged fluid tip or needle.
2 Lumps of dried paint or dirt lodged in the fluid tip.
3 Fluid needle packing nut screwed up too tightly.
4 Broken fluid needle spring.
5 Wrong size needle for the fluid tip.

4.11.5 Faulty packing

Paint leakage from the fluid needle packing nut is caused by a loose packing nut or dry fluid needle packing. The packing can be lubricated with a drop or two of light oil, but fitting new packing is strongly advised. Tighten the packing nut with the fingers only to prevent leakage but not so tight as to bind the needle.

4.11.6 Air leakage from gun

Compressed air leakage from the front of the gun (Figure 4.43) is caused by the following (numbers correspond to those on figure):

1 Dirt on the air valve or air valve seating.
2 Worn or damaged air valve or air valve seating.
3 Broken air valve spring.
4 Sticking valve stem due to lack of lubrication.
5 Bent valve stem.
6 Air valve packing nut screwed too tightly.
7 Air valve gasket damaged.

4.11.7 Oil in air line

If the air compressor pumps oil into the air line, it can have the following causes:

1 The strainer on the air intake is clogged with dirt.
2 The intake valve is clogged.
3 There is too much oil in the crank case.
4 The piston rings are worn.

4.11.8 Compressor overheating

An overheated air compressor is caused by:

1 No oil in the crankcase.
2 Oil which is too heavy.
3 Valves which are sticking, or dirty and covered with carbon.
4 Insufficient air circulating round an air-cooled compressor due to it being placed too close to a wall or in a confined space.
5 Cylinder block and head being coated with a thick deposit of paint or dirt.
6 Air inlet strainer clogged.

4.12 SANDING AND POLISHING MACHINES

Sanding and polishing by hand can prove to be both laborious and expensive. Unfortunately there are many parts of vehicle surfaces where there is no alternative but to carry out these processes without the aid of power tools. However, surfaces such as the roof, bonnet, boot lid and parts of the doors and wings of cars can be rubbed down or polished more economically and efficiently with power tools. Damaged areas of paintwork can be rubbed down very quickly with these machines, but final feather edging is best done by hand. Two types of machine are favoured by the refinishing painter, the rotary sander and the orbital sander. Both are obtainable as either compressed air operated, or electrically driven machines (Figures 4.44, 4.45). In addition, orbital sanders are available which operate with compressed air and can be connected to a water supply so that the worked surface is continually washed with clean water while rubbing down takes place.

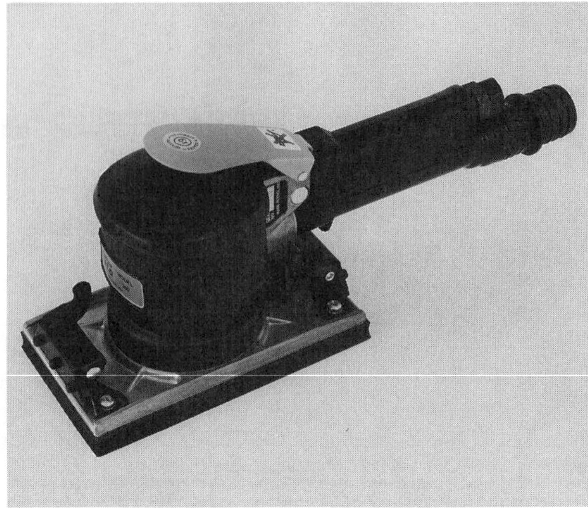

Figure 4.44 Air driven orbital sander with dust extraction (*Desoutter Automotive Lid*)

Figure 4.45 Electrically driven palm grip orbital sander with dust extraction (*Black and Decker Ltd*)

4.12.1 Sanding processes

In order to produce smooth, glossy finishes the substrates and undercoats must be levelled down without leaving deep scratches. Non-sand primer surfaces have virtually eliminated many of the problems associated with scratch swelling, but they must be applied over satisfactorily prepared surfaces. Where repair work has been carried out which includes the use of polyester fillers and/or surfacers, the repaired area must be carefully sanded prior to applying the non-sand coatings.

The wet sanding of high-build undercoats and fillers has long been the accepted method of levelling these materials in order to avoid the creation of excessive dust. In addition, this method of sanding helps to clean the surface but also presents certain problems. Water penetration behind window and windscreen rubbers, door checks etc. can lead to lengthy drying-out times. All undercoats are slightly porous and, if the moisture from the sanding process is not completely dried out, problems of micro-blistering and faulty adhesion may result.

The two main objections to dry sanding have always been that the abrasive paper clogs up and the process creates excessive dust in the workshop. However, clogging up of the abrasive paper is no longer a big problem since the introduction of coated abrasives. These abrasive papers have a coating over the grit particles of a material which is based on zinc stearate. This coating allows the sanded-off residue to be shaken from the paper to produce a fairly clean

Figure 4.46 Festo orbital sander (*Minden Industrial Ltd*)

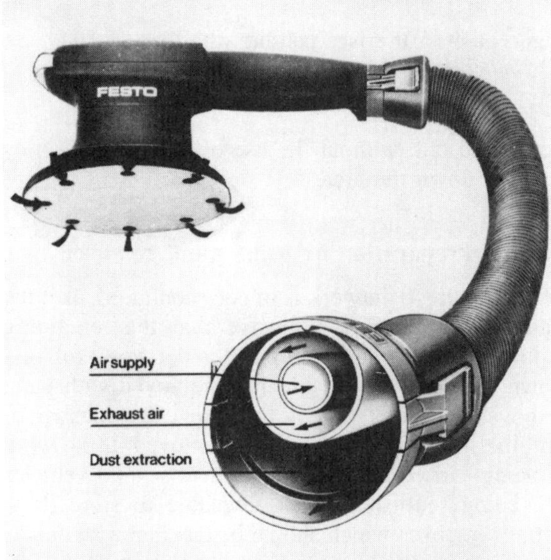

Figure 4.47 Festo random orbital sander (*Minden Industrial Ltd*)

sanding surface, thus creating more mileage from each sheet or disc of the abrasive paper. Coated abrasive papers are available in production paper grades (e.g. P40 to P120) and also in finer grades of lubricoat papers (for example P150 to P600).

It should be noted that, following sanding down with these coated abrasives, the surface should be cleaned with a proprietary spirit cleaner to remove particles of the stearate coating from the surface before applying paint. Failure to do so may result in the appearance of 'fish eyes'.

Dry sanding of painted surfaces, and surfaces to be painted, has increased in popularity as a result of the development of sanding tools which incorporate dust extraction methods. These dust extractors can be in the form of either a vacuum bag attached to the sanding tool, or a large vacuum dust collecting unit remote from the tool and to which two or more sanders can be connected. An added bonus with the remote machines is that they can also be used as a vacuum cleaner for the workshop. The sanding tools have a pad with eight or more holes in them to which are attached sanding discs with similar holes. In operation, the dust created by sanding is drawn through the holes and deposited in the collecting unit or bag.

Sanding machines available include types which are dual acting; that is, they can be set to either rotary or orbital (eccentric) actions (see Figures 4.46, 4.47). The rotary action is in the region of 420 rev/min, while the eccentric action moves at 12 000 strokes/min. The machines can be either electric or driven by compressed air. The rotary action is suitable for removing old paint films and surface rust and, when fitted with a polishing head, the machine can be used for final polishing paintwork. The eccentric action is suitable for feather edging, flatting surfaces prior to painting, and levelling surfacers and fillers. The sanding discs can be either self-adhesive (synthetic resin adhesive) or of a type which has a looped velvet reverse side which is simply pressed on to a special pad. These pads themselves are available as hard, soft and supersoft types depending on the type of work for which they are required.

The Festo sanding systems for body repair and paint shops can be of the mobile type which has the extraction unit mounted on to castors (see Figure 4.48), or of the fixed position type. In the latter type a boom, which can be moved over the car body, has all the facilities that the worker requires, for example compressed air supply hoses, extraction hoses, and electrical supply for both 240 V and 110 V tools.

4.12.2 Polishing machines

There are many types of machine suitable for polishing, and they all operate with a common circular movement. To be suitable for polishing, however, the machine must not rotate with too many revolutions per minute as scorching of the film will easily result. The pad is covered with a lambswool disc or foam pad which must be kept clean at all times (Figure 4.49). Remember that polishing is the final process; should grit be picked up on the disc, scratches will result which can ruin the finish on a vehicle. Polishing machines are best used with liquid polishes and should be used with a sweeping movement. Even with slower revolving machines, polishing in one spot for

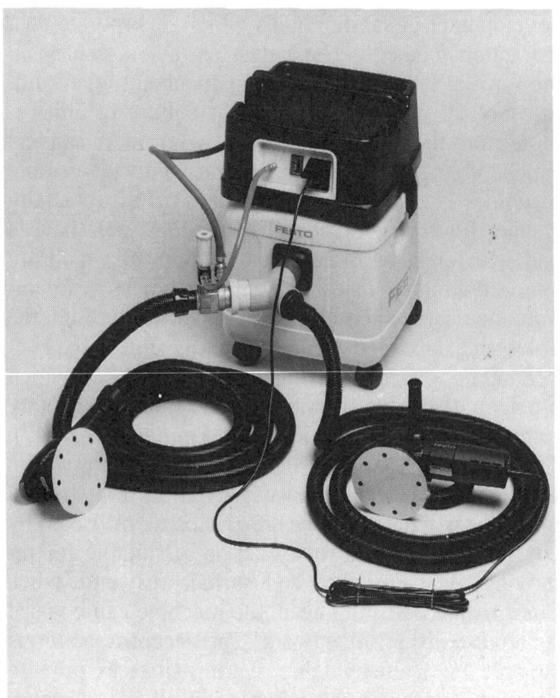

Figure 4.48 Festo dust extractor (*Minden Industrial Ltd*)

too long can cause scorching and blistering of the paint film.

4.13 PREPARATION OF A MOTOR VEHICLE FOR REPAINTING

Two methods of preparing a vehicle for repainting are open to the painter: (a) completely strip the existing finish down to the metal substrate, or (b) prepare the surface by rubbing down with abrasives. The main point to consider when deciding which process to use is the condition of the existing finish allied to the extent of damage in the case of a local repair. Which process to choose can only be decided upon by a close inspection of the vehicle.

On a motor car that has been involved in a collision and received damage to its front wing, but not serious enough to warrant replacement, the panel beater will carry out his work first, bringing the wing as near as possible to its original shapes. The vehicle then becomes the responsibility of the painter, who will carry out a close inspection of the condition of the paintwork. Should the paint show signs of poor adhesion to the substrate, it is advisable to strip it off completely using paint remover. However, if the damaged area is not extensive and the paintwork on the rest of the wing is in sound condition, the work can be carried out, without the use of paint remover, by rubbing down the area.

Figure 4.49 Air driven polisher with foam pad (*Desoutter Automotive Ltd*)

4.13.1 Preparation by using paint remover

If the existing paintwork is in poor condition, or if the damaged area is so extensive that the remaining paintwork on the damaged panel is not worth rubbing down, the existing finish can be removed with paint remover. The first task is to remove any flashes or chrome strips that can easily be removed. If the work is being carried out in a shop where several vehicles are being refinished, it is advisable to store these fittings in boxes which should be labelled with details of the car such as customer's name, registration number and type of car. On a complete refinishing job it is also advisable to label each part to indicate from which side of the vehicle it was removed. Should it prove to be too difficult or impractical to remove any parts, these should be masked completely. Though it is not always necessary to mask the surrounding areas when stripping, it is advisable to do so because if any paint remover accidentally came into contact with an adjoining door or bonnet, damage to these areas would occur. However, if masking paper is applied it should be of a fairly stout nature and laid in two or three thicknesses to ensure that any paint remover accidentally deposited on it does not immediately penetrate to the paintwork beneath.

Paint removers were formerly of a highly flammable nature and constituted a fire risk, but now non-flammable types are available and are widely used.

As well as removing fire risks they also evaporate at a much slower speed; hence they remain wet for a longer period and so have better penetration properties. Though they are quite easy to use, some precautions should be taken when working with these paint removers. They can be a cause of dermatitis by penetrating the pores of the skin, dissolving the natural oils beneath and so leaving the pores open to attack by bacteria. A simple precaution is to apply a barrier cream to the hands before commencing paint stripping, and to rinse off with water any remover which comes into contact with the skin. Strong rubber gloves are very useful when using liquid paint removers. The operator should not smoke when using paint remover, even though it is not flammable, as it gives off a strong vapour which, on coming into contact with a naked light, becomes toxic in nature. This toxic vapour if inhaled into the lungs will cause dizziness, nausea, and even vomiting. Adequate ventilation should be provided in any shop where paint remover is used, for the same reasons.

A fairly liberal coat of stripper should be applied to the surface and allowed sufficient time to penetrate through the various layers of paint. The softened paint is then removed with a stripping knife and the surface washed down with a generous quantity of water. Both the laying on and the washing off of the stripper can be done with old or cheap paint brushes. To assist the removal of paint at awkward places, a wire brush is an invaluable tool. Very often a layer of scum or a fine stain of primer is left on the surface but this can be quite easily removed with steel wool and water. It is imperative that all traces of paint remover be thoroughly rinsed off, otherwise it will cause damage to any subsequent paint system applied to the job.

4.13.2 Preparation by using abrasive paper

Before the rubbing down of a local repair commences, the whole panel should be thoroughly washed down with a detergent to remove all traces of wax polish. Nowadays nearly all wax polishes contain silicones which can create 'pinholing' in any paint applied over them. Water-miscible cleaning solutions, which are normally used for degreasing, are suitable for wax removal. Thorough rinsing with water and drying should follow. The grade of wet-or-dry paper chosen to rub down the damaged area depends very greatly on the degree of damage, but normally 180 grade is suitable. The paper should be folded and torn so that one-quarter of the length can be removed (Figure 4.50). This is then attached to a rubber rubbing block and the damaged area abraded using a liberal quantity of water to keep the work clean so that continual inspection can take place. The rubbing down is normally carried out with a forward and backward motion until the paint is removed and a feather edge achieved. Should paint remain in recessed areas, it can be removed with a wire brush, coarse steel wool, or abrasive paper without the rubbing block.

Figure 4.50 Method of fixing abrasive paper into rubbing block

As an alternative to rubbing down by hand, power tools can be used to speed up the process. These can be either electric or driven by compressed air. Those which are electrically driven are used dry, but compressed air tools can be used either wet or dry. The latter types are generally favoured for lighter work where damage to the paintwork is not extensive, but the electric tools are most suitable where severe rusting of the metal has occurred.

4.13.3 Treatment with phosphating liquid

Following the removal of the paint by either of the methods described, the bare metal should be treated with a proprietary phosphating liquid. These liquids (there are several brands available) have an acid content, mainly phosphoric, which etches the metal and completes the derusting of it. In addition to etching the metal, phosphating solutions deposit a layer of iron or zinc phosphates on to the metal which inhibits corrosion. These liquids should be used according to the manufacturer's instructions, but are generally diluted in the ratio 1:1 with water, applied with an old paintbrush and washed off with water after about 15 minutes. The liquid, being of acid content, should not be allowed to come into contact with the skin, eyes or clothing of the operator, and rubber gloves should be worn to protect the hands. In the event of its accidentally doing so, the affected part should be rinsed thoroughly in running water. Following drying of the bare metal with an air duster and wash leather, the area is now ready for repainting.

4.14 FINISHING AND REFINISHING PROCESSES

There are four main car refinishing paints at the refinisher's disposal: cellulose synthetic (half-hour enamel) paint; acrylic resin paint; low-bake synthetic paint; and two-pack paints. Cellulose synthetic and low-bake synthetics are used as repair materials over the high-bake synthetics applied by car manufacturers; in addition, of course, they can be used for complete resprays. Acrylics are best used on repair work over an original high-bake acrylic finish, but can also be used as a refinishing material on complete resprays. To determine whether a paint is suitable for repairing a particular job, a flat area should be chosen and a wet coat of the paint sprayed on to it. If wrinkling or lifting occurs this is proof that the solvent is too strong and will probably lift the existing finish.

The following paint systems are typical of those carried out in refinishing shops. They do not follow any one particular paint manufacturer's specifications but are intended as a guide to the use of the various paint types. When using a particular brand of paint, the operator should always follow the maker's instructions as to viscosities, drying times and temperatures. The processes outlined are based on the assumption that the vehicle has been prepared for refinishing as described in Section 4.13, followed by the necessary masking up.

4.14.1 Coach finish

This is the traditional material, used in coach painting workshops for many years, but it has lost favour to more modern materials mainly because of its lengthy drying time. However, in paint shops where spraying equipment is not available, this material still has a use. It is essential that the workshop should be kept clean, otherwise dirt in the finish is a certainty owing to the prolonged period of paint film wetness.

The finishing material can be sprayed either hot or cold and can also be brushed. It is suitable for application on all vehicle construction materials including wood. Undercoats may also be sprayed hot when thinned 6:1 with white spirit. The paint system is of the simplest type. For example, for metal surfaces the procedure is as follows:

1 Prepare the surface.
2 Carry out any necessary masking.
3 Wipe the surface down with a proprietary spirit wipe or a mixture of methylated spirit and water.
4 Apply one coat of self-etching primer, either spraying or brushing type. Allow to dry and then apply ground coats as soon as possible.
5 Apply two coats of ground coat:
   (a) If cold sprayed, allow 2 hours between coats.
   (b) If hot sprayed, heat to 60°C and allow 1 hour between coats.
   (c) When brushed, allow to dry for 6 hours between coats.
6 Should stopping up be necessary, this should be done after the first ground coat which should, of course, be allowed to through-dry completely. A synthetic resin stopper should be used and, when hard-dry, wet flatted with P360 wet-or-dry paper.
7 Should sanding down of the ground coats be necessary, allow to dry for at least 6 hours and dry-sand using P400 lubricoat paper. Dust and tack off. If sanding down is not necessary, the finishing material can be applied following the flash-off times listed in point 5.
8 Apply finishing enamel:
   (a) When brushed, no thinning is required.
   (b) When hot sprayed, heat to 60°C without thinner.
   (c) When cold sprayed, thin 6:1 with white spirit or preferably with a thinner supplied by the paint manufacturer.

   Apply two coats as follows:
   (a) Brushed: allow at least 6 hours between coats, the time depending on workshop temperature.
   (b) Hot sprayed: one light coat; allow 30 minutes and apply a double-header coating.
   (c) Cold sprayed: one light coat; allow 60 to 90 minutes before applying a double-header coat, but take care not to apply too wet a coating otherwise runs will result.

   Spraying pressures should be in the order of 3–4 bars (45–60 psi). Spray gun fluid tips should be of the smaller types (say 1.25 mm diameter) to avoid flooding the job. Spray gun distance from job is best about 200 mm.
9 Remove masking while the paint is at the tacky stage. This will allow the edges to settle down into place.
10 After overnight hardening, transfers, lining and lettering may be applied.

4.14.2 Cellulose synthetic (half-hour enamel): complete respray from bare metal

The process summary is as follows:

1 Spray a thin coat of etching primer, mixed 1:1 with activator, thinned to 20–25 seconds viscosity.
2 Allow to dry for 15 minutes.
3 Spray cellulose primer surfacer, thinned 1:1, to 19–22 seconds viscosity. Apply three full coats, allowing 10–15 minutes between coats.
4 Allow to dry for 1 hour.
5 Wet flat with P400 or P600 wet-or-dry paper.
6 Apply cellulose stopper where necessary, allowing 15–20 minutes between layers.
7 Allow to dry for 1–1½ hours.

8 Wet flat stopper with 320 wet-or-dry paper.
9 Spray cellulose primer surfacer to stopped-up areas, and flat with P600 grade paper.
10 Blow off vehicle with air gun and tack off.
11 Spray finishing material thinned 1:1 to a viscosity of 21–23 seconds. Apply one coat and allow to dry for 15–30 minutes. Apply second coat.
12 Allow overnight drying.
13 Wet flat with P800 grade paper, dry with air gun, tack off.
14 Spray double-header coat, thinned as before.

Alterations to this standard method can be made where necessary. For example, an alternative to using cellulose primer surfacer at stage 3 would be to use a synthetic resin primer surfacer which has better build qualities and may be preferable on rough surfaces. It is thinned 4:1 giving a viscosity of 26–30 seconds. Spray two or three coats, allowing a flash-off period of 30 minutes between coats, and the surface will appear completely matt. Allow the final coat to dry for 4 to 6 hours before dry scuffing, or to dry overnight before wet flatting with P600 or P400 grade paper. Cellulose stopper cannot be applied between coats of this material, and when used after the third coat must be oversprayed with cellulose primer filler. Cellulose finishing paints can be applied directly over synthetic resin primer surfacers, but if drying conditions are less than ideal a certain amount of lifting of the surfacer could be experienced. This is caused by the strong solvent used in cellulose finishes penetrating the still soft synthetic primer surfacer and acting as a paint remover. Therefore when adverse drying conditions prevail, such as low temperature, dampness or high humidity, it is advisable to spray a coat of sealer or cellulose surfacer over the job before applying the cellulose-type finish.

At stage 11 in the process, an alternative method is to apply one coat of the finishing paint, allow to dry for 15 to 30 minutes, then spray a double-header coating. This is a quicker method than that mentioned in the process summary, but does not produce the same standard of finish.

It should be noted that with the exception of tacking off and the actual application of the paint, all the process stages should be done outside the spray booth. Cleanliness within the booth or spray room is essential if high-class finishes are to be obtained. Rubbing down, even though done with water, will leave a scum on the workshop floor, which when dried out will leave behind a powdery residue. When disturbed this will cause air contamination. Burnishing and polishing cloths also cause contamination of the atmosphere, and these processes should be carried out in a part of the workshop away from the spray booth. These polishing cloths must not be put down on the floor or dirty work benches, for if grit should be picked up on them, damage to the new paint finish will occur.

As to the correct air pressure to use when spraying, this will vary according to the type of spray gun being used, the paint viscosity and the type of paint. For cellulose synthetic paints an air pressure of about 4 bars (60 psi) is normal, though this can be adjusted by the spray painter to suit his own requirements. The golden rule is to use the lowest air pressure that will give satisfactory atomization.

Stage 6 in the process summary refers to the application of cellulose stopper. It should be borne in mind that this material dries by the evaporation of the solvent content, and if it is applied in heavy layers the inside of the coating will remain wet for a considerable period. In addition to this, when drying does eventually take place the stopper will contract and sink. A much more satisfactory repair will be obtained if the material is applied in thin layers with a suitable drying period between applications. Plastic spreaders are to be preferred to metal knives when laying on the stopper, as these cause less damage to surrounding areas of paintwork. The lid must always be firmly replaced on the tin or tube of the stopper when not in use, otherwise the stopper will harden in the container and cannot be satisfactorily resoftened.

A coat of sealer could be added to the process, where costs permit, between stages 10 and 11. This would reduce absorption of the finishing coats by the surfacer and promote a high gloss.

Following the finishing coats, the vehicle may, if required, be burnished and polished; the method for this is described in Section 4.15.

Finally, it should be borne in mind that the use of a guide coat at each of the stages of sanding down is invaluable for locating surface defects.

4.14.3 Cellulose synthetic (half-hour enamel): refinishing over an existing finish

The majority of cars or vans that are refinished today are first patch repaired, followed by refinishing of the entire body. Only when the paintwork is in very bad condition is it considered necessary to strip it all off and it is usually only essential to cut back to bare metal those parts where the coating is damaged or corrosion has set in. However, it is not always necessary to repaint the whole car or van. A typical example would be that of a comparatively new car with paintwork in excellent condition but with local damage such as would be sustained in a collision. In this case it would only be necessary to repaint the repaired areas.

Following careful inspection of the vehicle, it can be brought up to the painting stage by using the method described in Section 4.13.2. A process summary would be:

1 Apply etch primer to bare metal parts only.
2 Spray one or two coats of cellulose primer surfacer, allowing 10–15 minutes between coats, on to the etch primer.
3 Apply cellulose stopper.
4 Rub down stopper with 360 paper.
5 Spray sufficient coats of primer surfacer or filler to bring the repaired area up to the level of the surrounding surface; use air pressure of 3 bars (45 psi).
6 Wet flat these areas with 360 paper.
7 Apply three full coats of primer surfacer to the entire body. Alternatively, synthetic resin primer could be used, providing more build to the system and allowing for easier flatting at stage 8.
8 Wet flat with P600 grade paper.
9 Apply cellulose stopper where necessary.
10 Rub down stopper with P400 grade paper, and spray locally with cellulose surfacer. Wet flat these areas with P600 paper.
11 Dry off and tack off.
12 Spray finishing coats as described in Section 4.14.2.

4.14.4 Cellulose synthetic (half-hour enamel): local repair

1 Rub down damaged area with P180 grade paper (wet) to obtain a feather edge. Wet flat surrounding paintwork with P400–P600 grade paper. Alternatively, dry sand with a dual-acting sanding tool using P80–P150 grade paper.
2 Treat exposed metal with phosphating solution.
3 Apply thin coat of etch primer to bare metal only; alternatively, spray one coat of primer surfacer directly on to bare metal, thinned 1:1.
4 Fill up defects with cellulose stopper in thin layers.
5 Wet flat stopper with 320–360 grade paper.
6 Spray sufficient coats of primer surfacer or filler (thinned 1:1) at 3 bars (45 psi) to bring repair up to level of surrounding surface.
7 Wet flat with P600 grade paper.
8 Burnish surrounding area with rubbing compound and a damp cloth to ensure a good colour match and better blending in.
9 Tack off.
10 Spray colour coats, thinned 1:1 to a viscosity of 19–23 seconds, at a pressure of 3 bars (45 psi). Spray several coats lightly until a good colour match is achieved. Allow to dry hard.
11 Wet flat with P800 grade paper, dry off and tack off.
12 Overspray the repair with a mixture of 75 per cent thinner, 25 per cent colour, carrying the spraying beyond the edge of the repair to obtain a soft blend.
13 Allow to dry hard (preferably overnight) and wet flat with P200 grade paper.
14 Burnish and polish.

Any masking off that may be required should be done between stages 2 and 3. A coat of sealer may be applied just prior to the colour coats to provide better hold-out, thus obtaining a smoother finish. Synthetic resin primers have not been included in this process as they have a tendency to peel back from the edges when rubbing down takes place. Overspray from the spray gun can create unnecessary work when carrying out local paint repairs, but this can be restricted by using a narrower fan pattern than that used for spraying a whole panel. When this is done, a higher volume of paint will be applied, increasing the risk of runs, and so the fluid needle adjusting screw should be turned to the right until a satisfactory volume of paint issues from the fluid nozzle.

Occasionally the spray painter may be called upon to repaint a motor car on which the paintwork is in excellent condition, the customer simply desiring a change of colour. In this case he may only require to wet flat the surface with P600 grade paper using a weak solution of water-miscible cleaning solution to remove any wax polish. Then following thorough drying off and masking up, three coats of half-hour enamel will produce quite a good finish which can be further improved by flatting, burnishing and polishing.

4.14.5 Acrylic lacquer

Though cellulose synthetic finishes are best left to dry in their own time to obtain best results, acrylic lacquer can be force dried without damage to the paint film, with a consequent speeding up of the process. This can be done with infrared lamps or in a heated booth (though not in excess of 50°C). The filler materials possess better build and flow-out than the cellulose-based materials, thus providing better surfaces for the finishing coats. The spraying viscosity of the finishing enamel is more critical than the half-hour enamel, and only the thinner recommended by the paint manufacturer must be used. Solvent evaporation from the wet paint film is governed to some extent by the workroom temperature, and should this be below 15°C a special quick repair thinner should be used. This thinner evaporates very quickly but tends to produce a low gloss which will require burnishing and polishing. However, should the ambient temperature be 15°C or above, a good hard glossy finish can be obtained straight from the gun which does not require polishing. As acrylic primer fillers have good adhesion properties the use of etch primer, though recommended, is not essential provided that the metal substrate has been properly prepared and treated with a phosphating solution.

4.14.6 Acrylic lacquer: complete respray from pretreated bare metal

1. Spray one coat of acrylic primer filler thinned to a viscosity of 21 to 23 seconds at 25°C. Allow 5 to 10 minutes to flash off.
2. Spray two coats of primer filler thinned 1:1, 26 to 29 seconds at 25°C. Allow 5 to 10 minutes between coats and 1 to 2 hours after second coat.
3. Apply cellulose stopper where necessary in thin layers, allowing 15 to 20 minutes between layers.
4. Wet flat stopper with 320–P400 grade paper, dry off and tack off.
5. Spray stopper locally with primer surfacer, allow 5 to 10 minutes, then apply a full coat over the entire surface. Leave to dry for 1 to 2 hours.
6. Wet flat with P600 grade paper, rinse, dry off and tack off.
7. Apply one wet coating of acrylic sealer, allow to dry for 30 minutes, denib and dry. Do not wet flat. The lacquer coats should be applied within 2 hours. Tack off before spraying the enamel.
8. Spray lacquer coats as necessary using one light coat followed by a double header. Thin the enamel 2:3 to a viscosity of 16 to 19 seconds at 25°C. Use the appropriate thinner only. Should more than one coating of the lacquer be considered necessary, allow a flash-off time of 5 to 10 minutes between coats. The lacquer is touch dry after 15 minutes and can be safely handled after 1 to 2 hours, depending on room temperature. If necessary it can be burnished and polished after overnight drying, though this can be done after 4 to 6 hours if the enamel is force dried. The air pressure used when applying acrylic lacquer is between 3 bars (45 psi) and 4 bars (60 psi) depending on the make of spray gun employed.

4.14.7 Acrylic lacquer: complete respray over an existing finish

The method chosen here must obviously depend on the condition of the paint film. If the surface requires filling and stopping up, it should be wet flatted with 280 grade paper using a water-miscible solution or liquid detergent. After rinsing and drying off, the system described for bare metal can then be used. If filling and stopping is not considered necessary and the paint is sound, it can be wet flatted with P600 grade paper and a solution of liquid detergent. Following rinsing and drying off, tack off and apply the sealer and lacquer coats as described.

4.14.8 Acrylic lacquer: local repair

Acrylic lacquer can be used to repair high- and low-bake enamels, but is not recommended for repairs to half-hour enamels or nitrocellulose-based air drying finishes. Nor is it suitable for use on wood or the repair of synthetic coach finishes.

A typical system for a local or spot repair is as follows:

1. Degrease with a solution of liquid detergent.
2. Wet flat damage area with 180 grade paper and feather edge.
3. Treat bare metal with metal conditioner or phosphating liquid, rinse and dry off.
4. Spot prime with acrylic primer filler.
5. Stop up with cellulose stopper where required.
6. Wet flat with 280 grade paper and rubbing-down block. Finish off with 320 grade paper, rinse and dry off.
7. Spray in with acrylic primer filler sufficient coats to level up the surface.
8. Wet flat with P600 grade paper, rinse, dry off and tack off.
9. Spray over the repair with acrylic sealer.
10. Denib and dry. Wet flat around the edge of repair with P800–P1000 grade paper. Burnish surrounding panel.
11. Apply acrylic finish, thinned with quick-drying repair thinner, in light coats. Finish off with a double-header coat to obtain a smooth finish.
12. When dry, wet flat with P1200 grade paper, burnish and polish.

Should the damage to the panel be too severe for satisfactory or economical levelling up with cellulose stopper, the two-pack polyester resin stopper described in Section 4.5.2 could be used. This is applied to the bare metal prior to the paint system. It is best rubbed down and, after dusting off and tacking off, coated with primer filler.

The use of a sealer coat at stage 9 may be eliminated if the surface is carefully prepared. Sealers are supplied ready for use, and it is difficult to spray them without leaving an edge which is difficult to remove. Providing that the original finish is flatted

50–75 mm beyond the repair area, a satisfactory job can be produced by spraying the finishing coats on to the filler to slightly beyond its edge. Care should be taken not to overlap the colour on to unflatted enamel.

4.14.9 Metallic finishes

Practically without exception, metallic colours being applied by car manufacturers in Britain are based on acrylic resins. However, they present problems not experienced when applying straight colours, which are caused by the metallic particle content. Without delving too deeply into the realms of paint technology, a metallic finishing paint could be described as a tinted, semi-transparent varnish containing finely ground metallic particles such as aluminium, bronze and copper. Polished aluminium flakes, because of their silvery metal appearance, are the pigments most widely used. Because they are lacking in opacity, these paints are generally applied in several layers to achieve the desired effect of colour depth and an even distribution of the metal flakes. The best method is to apply a single coat followed by wet on wet (double-header) coats. The coverage may vary from colour to colour. Apply as many coats as may be necessary.

Though car manufacturers favour acrylic-based metallic paints, they are, however, available in cellulose synthetic, slower drying types of spraying synthetics, and, most widely used nowadays, two-pack synthetics. They are not suitable for brush application. It is not necessary to outline a complete paint system for metallic finishes; this section deals with the application of the finishing coats. The actual spraying of these has a great influence on the finished appearance as regards colour. Spray gun technique is very important. The gun should be held at right angles to the surface and at a distance of 150–200 mm approximately. If the distance between the gun and the painted surface varies, dark and light patches will result. A 50 per cent overlap of gun stroke is essential to obtain an overall even colour and texture.

When too wet a coating is applied, the metallic flakes move freely within the wet film, and when solvent evaporation takes place they are generally in a fairly upright position. This tends to darken the final effect, as light does not reflect too well from the flakes in this position (Figure 4.51). Opacity is also reduced and when sinkage takes place as a result of solvent evaporation, the particles tend to stick through the top skin of the paint film, causing the finish to have a seedy appearance.

Too dry an application will create a dusty effect, too pale a colour when viewed head on, and too deep a colour when viewed from the side. This is because the metallic flakes have a tendency to lie parallel to the surface when a dry coating is applied.

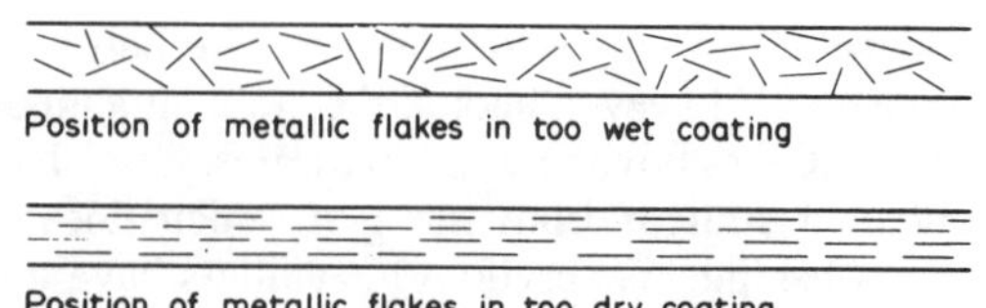

Figure 4.51 Coatings too wet and too dry

When repairing a metallic finish, two options are available to the refinisher. One is to extend the repair to a natural break in the car body lines and refinish the entire panel. Using this method there is always the danger of ending up with a mismatch when the masking is removed. A better method is to apply the colour coats to the whole panel, remove the masking and then recoat but extend the colour further still, using a fade-out technique. Overspray thus created should be overcoated with a clear blend-in material to protect the overspray when burnishing is done. When using this second method with the base-coat-and-clear materials, it is best to coat the panel with the clear coating first and, while this is still wet, to apply the colour; this should then be overcoated with further coats of clear.

There are too many variables involved to make it possible to lay down a hard and fast system in obtaining a perfect match on repair work. No matter how much technical data is available, in the final analysis it becomes a system of trial and error, with the spray painter trying out various spray gun set-ups, varying the air pressure, speed of gun stroke, and distance from the surface. The spraying viscosities vary according to the type of material being used but a rough guide is as follows: acrylic 16–19 seconds, cellulose synthetic 19–23 seconds, synthetic 23–28 seconds.

The damaged area should be levelled up as previously described, and the whole of the surrounding panel wet flatted with P400 grade paper. Following drying and tacking off, the panel should be sprayed with primer surfacer or filler and flatted with P600 grade paper. After drying and tacking, the colour coats are applied. The function of this final coat of surfacer is to equalize solvent penetration from the finishing material. Surfacers and fillers are more porous than enamels, and when the surrounding panel is sprayed with the finishing colour only the repaired area can be detected by a slight variation in colour and texture. Where costs permit, a coat of sealer could be applied prior to the finishing colour.

Light burnishing and polishing of metallic finishes can be done after overnight drying in the case of acrylic and cellulose-based materials. With the slower drying synthetics this process is best left for about four days to allow for complete solvent evaporation and to give the paint film time to harden off. When using two-pack materials, burnishing can be carried out after 16 hours at a workshop temperature of 15–20°C.

4.14.10 Pearl finishes

Pearl finishes are obtained by using pearl or mica pigments. These are translucent laminer particles which by varying their composition can selectively reflect parts of the visible spectrum of light. This property of the pigment enables a wide range of spectacular effects in paint, from the well-known mother-of-pearl appearance to the brighter and more pronounced flip/flop effect colours.

Pearl tinters are used to match OEM colours and offer the capacity of producing a personal touch by applying a custom finish using the same application methods as base-coat/clear-coat processes. There are two different types of pearl finish which are referred to as three-coat and two-coat finishes. The three-coat system consists of a ground coat, followed by a pearl coat, which is then overcovered with a clear coat. Two-coat pearl finishes are similar to conventional base-coat/clear-coat metallics, but the formula includes mica instead of aluminium flake, as well as other pigmented bases.

Method of application

1 On original coloured paint

Degrease and sand the surface with P400 or P500 paper

Apply two coats of white

Sand with P800 paper

Sanding might remove primer at edges. If so apply two or three coats of white thinned 1:1 with relevant base-coat thinner to the damaged areas. Do not sand the base.

Apply two coats followed by a mist coat of your chosen pearl base. Wait for 15–20 minutes before applying clear coat. (One light coat followed by two wet coats.)

2 On original white paint

Sand to eliminate orange peel and apply two coats of white

Apply two normal coats and one mist coat of your chosen pearl base

Finish with clear (one light coat followed by two wet coats).

Note:
The iridescent effect appears more or less pronounced according to:

(a) the quantity of pearl base in the formulation, and
(b) the choice of pigmented bases in the formulation. The deeper the colour of the pigmented bases, the more the pearl effect will appear, primarily with interference pearls.

4.14.11 Quick air drying synthetics

This type of material offers special advantages for low-cost resprays and rapid finishing of commercial vehicles. The main advantage is that it is suitable for application over all types of existing finishes, and is suitable for hot or cold spray application.

The rapid respray system is as follows:

1 Clean off all traces of traffic dirt, grease, wax and silicone polishes by using water-miscible cleaning solution.
2 Feather edge all damaged areas including parts damaged by stone chips (generally on sills). Treat bare metal with derusting liquid.
3 Spot prime any bare metal areas with cellulose-based primer surfacer, and allow to dry for half to one hour. Stopping, if necessary, is best done before priming with two-pack polyester stopper.
4 Wet flat the whole vehicle with P600 grade paper, using liquid detergent or water-miscible solution as a lubricant and to remove wax. Rinse, dry and tack off.
5 Thin the enamel to a viscosity of 23–26 seconds (4:1). Spray one coat and allow to flash off for 10 minutes, then apply a double-header coating. The finish may also be applied hot, in which case it requires no thinning. The paint is heated to a temperature of 70–80°C, and one coat will be sufficient to give the required film thickness. The flash-off times between coats recommended by the paint manufacturer must be strictly observed. If too great a time lapse is allowed, then problems of lifting or wrinkling may be encountered.

4.14.12 Low-bake enamels

These are modified high-bake enamels rather similar in composition to those used by motor vehicle manufacturers. Though developed primarily for the refinishing trade, they are in fact used by vehicle manufacturers in the rectification of damaged or faultily finished vehicles. One of the problems encountered by the refinisher has always been to obtain a close match to the original finish. When exposed to sunlight some colours have a tendency to fade, while a certain amount of discolouration takes place in others; some whites, for instance, tend to yellow with time. In addition to this, some pigments combine very well with a synthetic medium but not so well with cellulose-based vehicles. As the latter have always been the most widely used materials in refinishing, colour matching thus presents problems. These would have been fewer if refinishing shops were equipped with equipment similar to that of the vehicle makers, allowing them to use the same finishing materials. This, of course, would require a tremendous capital outlay which would be well beyond the reach of the refinisher, and would in all probability never be recovered.

In an effort to improve refinishing techniques, to reduce drying times and to obtain faster production, low-bake enamels (i.e. paints that cure at lower temperatures than the original high-bake finish) were developed. Being similar in structure to the high-bake material, these paints offer a closer match in colour and texture than do cellulose synthetic materials. Burnishing and polishing is seldom required as these paints provide a good gloss from the gun.

Equipment

An obvious requirement for the drying of low-bake enamels is a stoving oven, and various types are available. A stoving oven can be acquired which can be connected to an existing spray booth, or where workshop space is limited a combined spray booth and low-bake oven can be installed (Figure 4.52). The obvious limitation of a combined spray booth and low-bake oven is that no further paint spraying can be done during the stoving schedule, in addition to which valuable working time can be lost while waiting for the booth to return to a comfortable working temperature.

Where workshop space permits, a better proposition would be to install a low-bake oven which can be joined up to an existing spray booth with sliding shutter doors to seal the two areas (Figure 4.53). This can be further improved upon by having entrance/exit doors at each end of the unit which would provide a flow-line system of painting and stoving (Figure 4.54), thus reducing time wastage on vehicle movement. The type of installation must obviously be governed by the size and shape of the workshop (Figures 4.55–4.59). A further variation is shown in Figure 4.60, in which the vehicle being painted is moved sideways on rails and bogies.

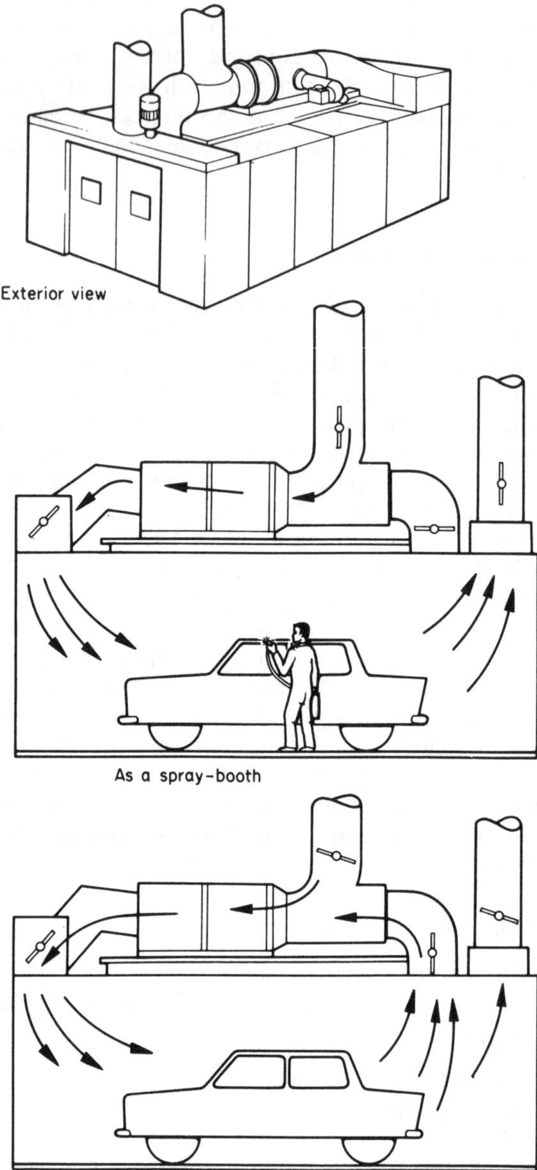

Figure 4.52 Spray booth combined with low-bake oven

Apart from the stoving oven, the usual spray painting equipment used in refinishing shops is required.

Preparatory work

Deep indentations in the vehicle body can be filled with polyester stoppers, which cure very quickly and have good adhesion properties when baked. Following rubbing down and cleaning off, the surface can be sprayed with synthetic resin primer surfacer if it is of a rough nature. As previously stated, these primers

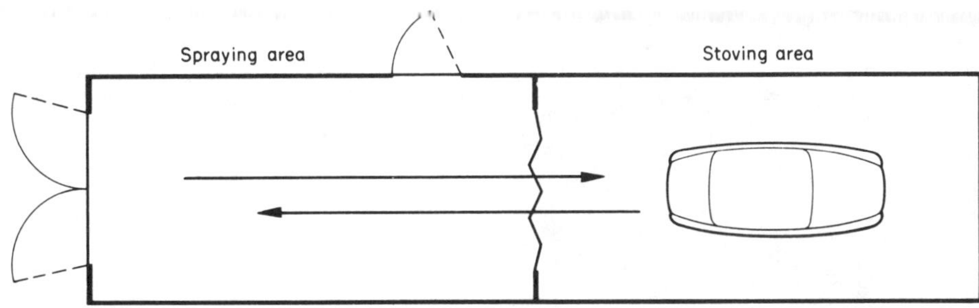

Figure 4.53 Low-bake oven joined to spray booth

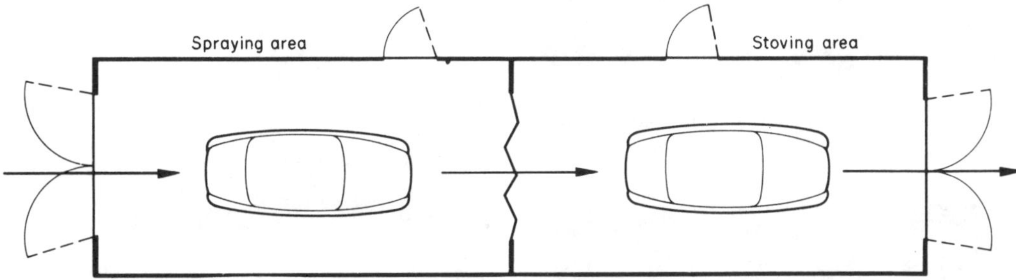

Figure 4.54 Oven joined to spray booth, with through doors for flow line

Figure 4.55 Spraybake exterior unit, height 3 m (*Spraybake Ltd*)

Figure 4.56 Multiple installation showing drive through and paint mix room (integral) between units (*Spraybake Ltd*)

have exceptional filling properties, but when used with low-bake finishes the stoving schedule (1–1$\frac{1}{2}$ hours at 82–93°C) can prove to be too long to be economical. Where the surface is not too rough, a better method would be to use cellulose primer surfacer thinned 1:1. Slight imperfections should be levelled with cellulose stopper, which must be overcoated with surfacer. The surfacer can then be stoved for 30 minutes at 70–80°C. A third material finding favour is the polyester primer filler described in Section 4.5.1, which obviates the need for stopper on small imperfections and provides a single coat build-up to the finishing stage. As previously stated, several coats can be applied wet on wet without flash-off periods, and force dried for 15 minutes at about 75°C. A further advantage is that polyester stopper can be used, if necessary, between

Figure 4.57 Multiple installation showing side loading doors and track, and housing IRT arch (*Spraybake Ltd*)

Figure 4.58 Corner installation showing two low-bake units, height 2.5 m (*Spraybake Ltd*)

coats of polyester primer filler. This material should be overcoated with cellulose primer surfacer before applying the finishing material. Whichever of the above materials is chosen, the final coating must be rubbed down wet, and following drying and tacking off, the colour coats can then be applied.

Spraying

The finishing material is thinned 7:2 and sprayed in one single coat followed by a double-header coat, with a 15-minute flash-off between coats. Another flash-off period of 15 minutes should be allowed before stoving.

Preparation for stoving

It is current practice in the motor industry to repair areas damaged during assembly with low-bake enamels

Figure 4.59 Interior shot of Green Booth designed to meet EPA regulations (*Spraybake Ltd*)

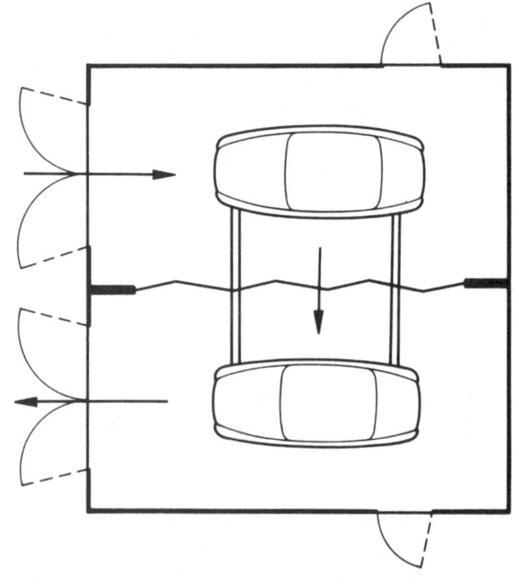

Figure 4.60 Joined oven and spray booth, with sideways flow

and then to stove the whole body fully trimmed. The temperature inside the vehicle does not rise sufficiently to damage the trim provided that the doors and windows are properly closed. Before stoving a vehicle refinished with low-bake enamel, check that doors and windows are closed, remove all exterior plastic fittings, increase tyre pressure by 0.5 bar (5 psi), check that the petrol tank is not too full and, if duotone work is being done, remove any masking. It is not necessary to remove the battery.

Stoving
The usual stoving schedule is 30–40 minutes at 80–100°C, following which the body should be allowed to cool for about 45 minutes (longer if necessary) before further masking up for duotone work is carried out. Certain parts of the car will be sheltered from the heat, and consequently paint applied to these parts will not be fully cured. A converter liquid is available from most manufacturers of low-bake paints which can be added to the enamel for brushing in door edges and insides of boot lids and bonnets, or these areas can be sprayed if the viscosity is reduced to 22 seconds with the appropriate thinner. Any areas touched in thus will dry quite hard at the normal paint shop temperature.

Finishing
Polishing of low-bake enamels is not normally necessary as a good gloss from the gun is obtainable, but it should be borne in mind that these paints remain more open, i.e. wetter, than more conventional materials during the flash-off period prior to stoving. For this reason, a clean workshop is a necessity; otherwise dust may settle on the still wet film which will necessitate flatting and burnishing to obtain a smooth finish. The nibs should be removed with P800 grade wet-or-dry paper and the flatted area brought up to a high gloss with burnishing compound, followed with a good-quality car polish. Low-bake finishing can be carried out over both stoving and nitrocellulose finishes provided they are in a sound condition. Red or maroon cellulose finishes which might bleed are best coated with a sealer before commencing spraying of surfacer coatings.

Developments
Apart from the cost of the stoving equipment, one of the reasons why low-bake finishes were slow to find popularity with refinishers was that it meant having to increase the already extensive stock of paints normally carried in refinishing shops. With this in mind, paint manufacturers have developed a type of thinner which, when added to their half-hour enamel (nitrocellulose based), will convert it to a low-bake cellulose enamel. Over 2000 car colours are available in this material, and so colour matching does not present too many problems. The cellulose stoving enamel should be thinned with the converter thinner to a viscosity of 23–26 seconds with a BS B4 flow cup. Spray either three single coats or one single and one double-header coat at 4–5 bars (60–75 psi), allowing 15 minutes flash-off between coats and another 15 minutes before stoving.

The stoving schedule recommended is 40 minutes at 91–93°C, during which the car body attains a metal temperature of 80°C for 30 minutes. This temperature is not high enough to cause damage to the car interior but is high enough to effect a complete cure of the paint film.

Advantages
Low-bake refinishing, though costly to set up initially, undoubtedly increases the potential productivity of refinishing shops. Faster drying times are obtained at almost every stage of the job, with the result that vehicles are refinished much more quickly than when using air-drying materials. This leads to faster delivery of customers' vehicles, which is a very good selling point, and as the finished article can be moved out of the working space after cooling of the body, freer movement within the workshop is obtained. Used properly, low-bake finishes lead to a higher turnover of work, which in turn increases the need for good organization within the workshop (see Section 4.18).

4.14.13 Short-wave infrared paint drying

Infrared drying lamps have been used in finishing shops for many years to force dry localized areas of paintwork and to accelerate the curing of two-pack polyester stoppers and fillers. A more recent development is the short-wave infrared heating module which, instead of easily breakable lamps, has heating elements mounted into an aluminium cassette.

One firm specializing in this type of equipment is Infrarödteknik AB, whose IRT 100 unit has a reflector coated with a thin layer of gold to give maximum reflectivity and long life. In addition it is equipped with a fan for solvent vapour removal, and a control panel which guarantees efficient control of the drying process according to the type of paint being used. The unit is mounted on a stand with a flexible support lever which makes it possible to locate the heating module in any position without locking devices (see Figure 4.61). A variation on this type of heater is a vertical heater mounted on castors for force drying of doors, quarter panels and so on.

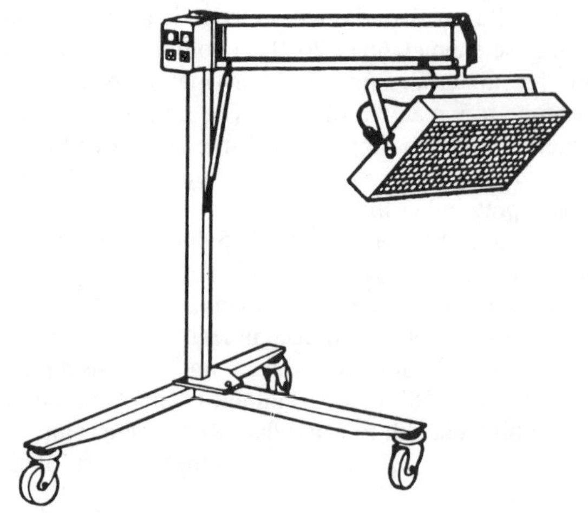

Figure 4.61 Short-wave infrared heater IRT 100 (*Infrarödteknik AB – Stanners*)

A further development is the infrared drying arch, designed to be mounted inside an existing spray booth. The arch consists of a number of heating modules mounted on to a steel frame in the form of an arch. This arch is stored at the rear end of the booth and is sheltered from overspray. Following the spraying process, it is moved quickly to the vehicle and then moves along the length of the vehicle at a slower speed. After the stoving of the vehicle, it then switches off automatically and moves back to the rear of the booth (see Figure 4.62). Being computer controlled, the arch is capable of providing the refinisher with the option of drying a complete vehicle or one or more panels at various positions on the vehicle body. It also caters for the various types of paint and colours.

Short-wave infrared units require no preheating, which makes for fairly high savings where energy costs are concerned. They also cool in seconds, which means the vehicle can be moved out of the booth almost immediately. When used to force dry two-pack materials, the drying times claimed are quite extraordinary: for example, a complete respray can be dried in 5 to 10 minutes, and minor touch-ups in 1 to 2 minutes. Single panels finished in a material such as Bergers Standox 2K are baked in 4 to 5 minutes, while primers and fillers bake even faster. With drying times like this, the savings on energy costs and the gain in turnover of work are obvious.

4.14.14 Two-pack paint system

There are many variations available to the refinisher with these materials, and the paint manufacturers' literature must be referred to. It is inadvisable to mix one manufacturer's materials with another's.

Although it is common practice in some workshops to use cellulose-based primer surfacers under two-pack finishes, better build and intercoat adhesion is obtained when two-pack undercoats are used. These undercoats are multipurpose materials which can be used as primers or non-sand surfacers. Self-etch 2K primer surfacers are available for use on bare, prepared substrates such as steel, aluminium and glass fibre, and can even be applied over vehicle manufacturers' finishes. Transparent adhesion primers are available which can be applied on to manufacturers' finishes which have merely been cleaned with a cleaning spirit and a Scotchbrite pad. These primers can then be overcoated with the finishing material after 15 minutes as a wet-on-wet process. As previously stated, there are many variations of materials with these products,

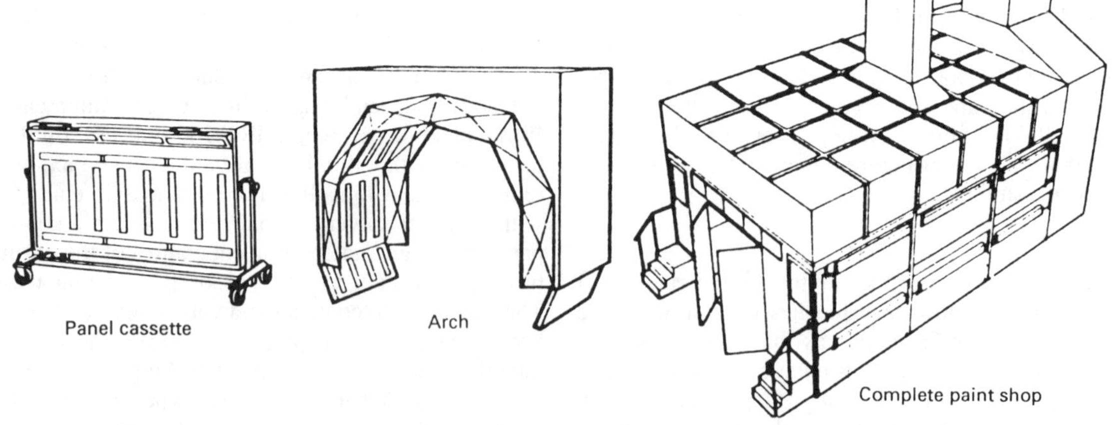

Figure 4.62 Infrared cassette, arch and paint shop (*Infrarödteknik AB – Stanners*)

and so there is no basic system or process which can be described. A variety of thinners are available for use at various workshop temperatures. Various hardeners may be used depending on workshop temperature and humidity.

Two-pack paints are available as straight colours, metallics and base-coat-and-clear systems. To give some idea of the range of choice in two-pack base-coat-and-clear materials, the following list will be helpful. Each of the clear coatings contains an isocyanate hardener:

1 Base coat: acrylic/polyurethane synthetic
  Clear coat: two pack (2K)
2 Base coat: acrylic synthetic
  Clear coat: 2K modified acrylic synthetic
3 Base coat: acrylic lacquer
  Clear coat: 2K modified acrylic synthetic
4 Base coat: cellulose
  Clear coat: 2K acrylic synthetic or polyester/acrylic.

It must be stressed that the time lapse between applying the base coat and the clear coatings recommended by the paint manufacturer must be strictly observed if intercoat adhesion problems are to be avoided.

Figure 4.63 Burnishing using rubbing compound and a sponge polishing pad (*Farécla Products Ltd*)

4.15 BURNISHING, POLISHING AND FINAL DETAIL WORK

Though brief reference to burnishing and polishing has been made earlier, these subjects are important enough to warrant fuller description. Even though most car refinishing paints are nowadays formulated to provide a good gloss from the gun, the final appearance of the vehicle can be further enhanced by careful burnishing and polishing. Burnishing helps to smooth out the surface while imparting a fuller gloss and revealing depth to the colour (Figure 4.63). Polishing, if carefully done, will improve the lustre still further and provide a protective coating over the paint film.

4.15.1 Burnishing

Where the final coat of enamel contains small particles of foreign matter, it is best flatted with P1200 grade wet-or-dry paper with soap as a lubricant. The soap must be rubbed into the paper and not into the painted surface. Care must be taken not to rub through the colour coats, especially where projections exist in the vehicle construction.

Following washing down, the car or van is now ready for burnishing. The mutton cloth (or any suitable soft cloth) is first wetted in clean water and wrung out. A small quantity of burnishing compound is applied to the cloth and rubbing can commence, working in straight lines over a small area. A fairly firm pressure should be applied at first, but as the area shows sign of glossiness the pressure should be reduced. The friction caused by rubbing will generate heat, and when this takes place the cloth must be turned over and the burnishing continued until a smooth glossy surface is achieved. Some of the colour will come away on the cloth but, provided that a sufficient coating thickness has been applied, this is unimportant. The cloths used for burnishing should be washed out periodically in clean water to remove the build-up of pigment which will hinder the action of the cloth.

Machines can be used for burnishing large areas of panel using liquid burnishing materials. These liquids, which contain milder abrasives, reduce the risk of swirls appearing on the surface as would be the case should burnishing compounds be used with machines. Hand burnishing, however, will produce the better finish should costs permit. When using machines for burnishing, the precautions set out in Section 4.12.2 should be observed. Following the burnishing process comes the final polishing.

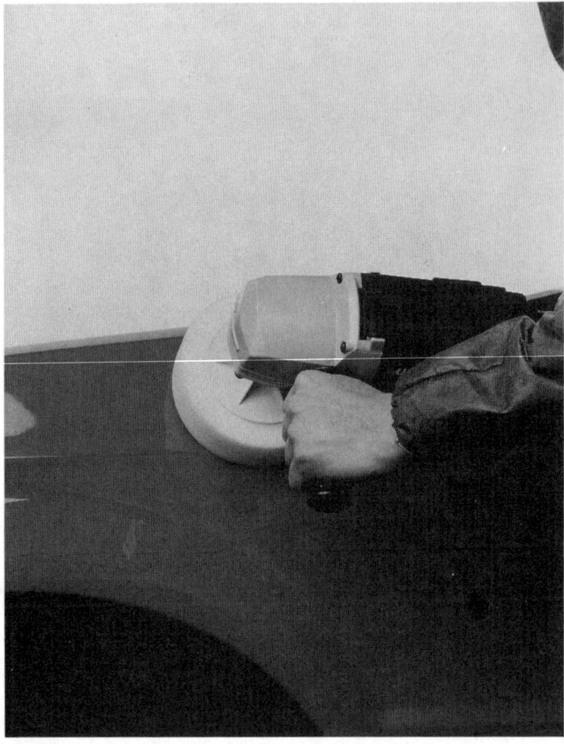

Figure 4.64 Rotary polisher fitted with a sponge polishing pad (*Farécla Products Ltd*)

4.15.2 Polishing

This is where the rotary-type polisher fitted with a lambswool pad (Figure 4.64) comes into its own. These machines are best used with liquid car polishes, but are also very effective when used with the more viscous conventional polishes. The polish is spread lightly over the surface with soft clean cloth, following which the finish is brought up to a high lustre with the machine. Heavy pressure is not required when polishing with a machine. Some parts of the vehicle body are obviously not suitable for machine polishing and must be done by hand. This is best done with soft yellow dusters rather than mutton cloth, which can leave striations on the surface and spoil the finished effect. The same care and precautions should be practised when polishing as when burnishing (see Section 4.12.2). The importance of working cleanly cannot be overemphasized.

4.15.3 Final detail work

Attention to detail when completing a respray can make the difference between a professionally done job and the amateur one. The sequence of operations is usually as follows:

1 Remove overspray from windows with a razor-blade scraper and wash leather.
2 Clean chrome bumper bars with rubbing compound.
3 Clean and refit all parts which have been removed for the painting process.
4 Vacuum clean inside of car.
5 Paint tyres with tyre black. Rubber glazing strips can also be painted with this material.
6 Finally carry out any lining work, using either lining tape, fine lining brushes or roller-type lining tools. The type of paint used is usually of a synthetic resin nature.

4.15.4 Corrosion

Rust is a corrosion, known chemically as iron oxide, which occurs on the surface of iron and most of its alloys, especially steel, when they are exposed to air and moisture.

As an electrochemical reaction, corrosion requires three components for it to proceed:

1 A reactive steel surface.
2 Atmospheric oxygen.
3 Water or atmospheric moisture.

When the three come together a spontaneous reaction takes place which is called corrosion.

Oxidation of steel, or any iron alloy, increases also when electrical conductivity of moisture on the metal surface increases. When an excess of ionizable salts (for example sodium chloride) dissolves in water to form electrolyte. Chlorides further increase the corrosion rate by virtue of the high mobility of their chloride ions, which diffuse rapidly throughout the corrosion deposit keeping it loose and preventing it from becoming hard enough to slow or halt the reaction of the corrosion. Deicing salts used on roads in winter time thus accelerate the continuous corrosion at a greater rate than normal. Therefore it is essential that vehicles are washed regularly, especially in the winter, to remove the salt deposits which accelerate the corrosion.

Types of corrosion

General corrosion

This generally occurs in vehicles having areas of steel which have become unprotected, often in wheel arches where gravel thrown up by the wheels has worn away the protective coatings of underseal compounds. This type of corrosion also occurs in chassis members, pillars, wing edges, or any area that can be penetrated by moisture. If spaces of this kind are also filled with dirt from the road, the corrosion will often be combined with deposit attack. Ambient atmospheric conditions commonly influence the speed with which general corrosion occurs.

General corrosion depends on the following factors:

(a) the length of time the surface stays wet
(b) the relative humidity of the air
(c) the air temperature
(d) the level of air pollution (especially of sulphur dioxide and chloride)

The most important of these factors is the length of time the surface is wet, for example from rain or damp snow. Thus the speed of corrosion on a dry surface outdoors at 90 per cent humidity is less than 1 per cent of what it is on a surface dampened by rain. The chassis member surfaces filled with road dirt and water remain damp for a very long period, even after other parts of the car are dry. Floor and boot areas inside the car often rust because they are subjected to long periods of dampness. The water that succeeds in penetrating these areas through leaking windows and faulty seam sealing often collects under the floor coverings of the vehicle. Such moisture can be retained for long periods of time, thus increasing the rate of corrosion. Similar corrosion damage can also occur in doors where clogging of the drainage holes at the bottom of the door cause a considerable amount of water to be trapped there, resulting in the eventual rusting through at the bottom edge. On a surface covered with road dirt that contains salt, corrosion can proceed at a lower level of humidity than on a clean surface.

The presence of air pollutants (acid rain) especially sulphur dioxide and chlorides, effect the speed of atmospheric corrosion. The higher sulphur dioxide content in urban atmospheres results in heavy erosion of steel and zinc-coated steel.

Crevice corrosion

Crevice corrosion can occur when constricted gaps are filled with water. This is very evident on body panels that are welded together, e.g. vehicle front wing, invariably forming narrow gaps between the two components. It is practically impossible to avoid gaps when car body panels are spot welded together. If the gap is not filled with some form of coating or sealing compound rust will develop, eventually causing the joint to fail. Sometimes corrosion can be aggravated by the gap between the two components being filled with a porous material which will retain moisture and create crevice corrosion.

Pitting corrosion

Pitting corrosion can occur when paintwork is damaged, particularly by loose stone chippings thrown up from the road and hitting the paintwork on the car and leaving minute damaged areas. The exposed metal surface forms an anode when moisture is present and corrosion occurs which, if not treated, can eventually cause a hole. The worst pitting corrosion is normally found along the front edge of bonnets, the panel beneath the bumper, sill panels and the lower edge of doors. It is also a problem for brake pipes which, if not inspected, can have serious implications as a result of pitting corrosion.

Galvanic corrosion

Another form of corrosion found on car bodies is known as galvanic corrosion. This happens when two dissimilar metals are in close contact with each other as is often the case between the steel body panel and a stainless steel trim or an aluminium fitting connected to a steel surface. Under such circumstances a difference in potential exists between the two metals and an electrolytic cell is set up whenever moisture covers the surface between these two materials. One of the metals is then eroded by the other.

Fatigue corrosion

Fatigue corrosion occurs in and near spring mounting brackets, especially coil spring mounting plates in the front of the vehicle. This weakens the structure and combined with shock and vibration eventually results in cracks in these areas. Periodic inspection of these areas on older vehicles is essential to preempt collapse of these sections.

4.16 RUST-PROOFING

A vehicle body contains many recessed areas and box sections which are difficult, and in some cases impossible, to paint properly with a spray gun and conventional paints. The only time that they receive a coat of paint is in the electropriming process carried out by the vehicle manufacturer. All primers are porous, and this material is no exception. Should moisture be allowed to penetrate the primer, it will eventually break down and corrosion of the substrate will result. It is estimated that 90 per cent of corrosion on vehicle bodies is of the 'inside-out' type.

In an effort to prolong the life of a car body and provide an additional selling point, most manufacturers now carry out a rust-proofing process as part of the finished product or as a chargeable extra. A warranty is provided to the customer, but usually on condition that he or she has the vehicle checked for signs of corrosion at regular intervals and is prepared to pay for any rectification work that may be required.

The material used to treat these cavity areas is based on wax dissolved in a solvent with rust inhibitors added. A compound of this type has good capillary attraction and can filtrate into joints, seams and otherwise inaccessible areas. Different types of wax coatings are available, some classed as penetrants and others as heavy-duty coatings. There are several firms worldwide which specialize in these types of materials and the equipment for applying them. One such company is Tuff Kote Dinol (TKD).

Measures employed by car manufacturers in an effort to combat corrosion include the use of precoated steel; washing and spraying the assembled body; and immersion in anti-rust primer, which is then baked. Spot-welded seams, which would be prime sites for corrosion, are sealed; and wheel arches, along with the underbody, are protected with anti-chip coatings. After a further baking period the car body is painted and some of the critical cavities can be flooded with hot wax. However, a car is subjected to a great deal of stress throughout its life; it is scratched, scraped and banged, and panels may flex, seams move and joints vibrate. If untreated metal becomes exposed, rust will get in; therefore cars need to be treated regularly, and so rustproofing becomes a continuous process.

The basic anti-corrosion tool is the spray gun with its four attachments. The rigid and flexible lances are best for all enclosed areas; the hook nozzle is designed for more accurate, directional work; while the fan spray tool is for coating open underbody surfaces.

4.16.1 Spray nozzles

Rigid lance (1100 mm)

This produces a 360 degrees spherical spray pattern at right angles to the lance, combined with a forward and backward directed spray that allows all surfaces – front, back, sides, top and bottom – to be coated in one single sweep or stroke. It is highly effective in places where straight structures such as doors, tailgate panels and long channels exist, or where the operator needs to control the position of the lance, such as enclosed front or rear wing box sections.

It is advisable to place the tip of the lance into the section being processed and to fog spray on both inward and outward strokes. Do not force into access holes. Test penetration before spraying to ensure adequate clearance for the lance.

Flexible lance (1100 mm)

This is basically a flexible version of the rigid lance. It produces a 360 degrees spherical spray at right angles to the lance combined with a straight ahead jet. It is highly effective in long narrow sections which would otherwise be inaccessible or awkwardly positioned and where its flexibility enables it to operate even where there are bends and restrictions which would prohibit the use of the rigid lance.

The main use of the flexible lance is for treating sills, underbody box sections, strengtheners and pillars. Its limitations are in narrow sections where it cannot gain entry or in very large box sections where it is desirable to control the position of the tip of the lances, such as in doors.

It is advisable to work the lance into the desired position and to fog on the outward stroke. In larger sections, retract the lance slowly so that one sweep will sufficiently cover all surfaces.

To extend the life of the nylon tube, the sides of the lance should be held away from the edges of the hole as it is entered or withdrawn. To eliminate snagging on the edge, always hold the exposed part of the lance at right angles to the hole. Feed the lance into the section gradually, keeping thumb and forefinger around the tube, close to the hole. Use the same technique when withdrawing the lance. The tube will go slack just before the tip emerges from the opening. This reduces the risk of accidentally spraying the interior trim and fittings by withdrawing the lance too rapidly.

Hook nozzle

This produces a highly atomized forward-directed full-cone jet which gives a powerful long range and, at the same time, good dispersion. Its value lies in its long range and directional capability, so that the product can be directed precisely where pointed.

This makes it suitable for treating narrow box sections such as door pillars, boot and bonnet lid reinforcements, and areas with difficult access such as door hinge areas, head and tail lamp housings, wing supports, door sills and underbody box sections, reinforcements, and suspension mountings. It may also be used for spray coating the wheel arches and underside of the vehicle.

The thumb should always be positioned on the machined flat portion of the neck to direct the nozzle where required. When spraying into narrow sections, the nozzle should enter the section with the spray directed towards the surface immediately opposite the opening into the section. As soon as the trigger is applied, the nozzle should be moved in an arc until the spray is being directed towards the end of the section. This should be repeated in the opposite direction. Where seam penetration is required, the nozzle should be specifically directed toward the seam.

Straight nozzle
This produces a forward spray with a restricted spray width. Ideally this nozzle should only be used with underbody sealants on to easily accessible panels. More obscure areas such as ledges within wheel arches should be processed with the hook nozzle.

Rules for nozzle selection
To minimize operator error when processing enclosed sections, always use the flexible or rigid lances first. Only use the hook nozzle where openings are insufficient for access with the 360 degree tooling. An exception would be where a very narrow section is to be processed and it would be preferable to spray with a powerful directional jet, such as a bonnet or boot lid strengthener.

4.16.2 Materials

Various anti-rust corrosion compounds include heavy-duty waxes and sealants for wings, wheel arches and underneath the vehicles; high-performance cavity waxes and penetrants for all enclosed areas, box sections, doors and pillars; special engine compartment waxes; and waxes for special purposes. These compounds are waxes and inhibitors dissolved in a solvent. They are applied as a liquid, but solidify as the solvent evaporates. For ease of application make sure the materials and the vehicle to be treated are at room temperature (about 15°C). This is particularly important in winter.

4.16.3 Process

To assemble the gun, put a can of wax into the pressure pot, connect the head to the air supply, and regulate the pressure. In general, work with pressures of 5–8 bars (75–120 psi). Attach the appropriate tool and then spray. To change materials, release the pressure and replace the can. Try not to let the gun get too dirty because, although it is robust, dust and dirt will clog the nozzle.

The principle of rust-proofing is to prevent the atmosphere and corrosive substances from attacking the metal of the car body by applying an impermeable layer to the metal. In this case it is a layer of a penetrant or a sealant. Applying these materials is straightforward enough, but it does need care and attention.

Engine compartment
When protecting the engine compartment, the only preparation necessary is thoroughly to clean and dry the engine unit and compartment. Assemble the gun and load the engine compartment wax, which is designed to withstand high temperatures. The hook nozzle is best here, as the spray from the flexible lance would go everywhere and cover everything with wax. A face mask is necessary when working in confined areas or on open surfaces.

Putting your thumb on the flat section at the base of the hook will help to direct the spray. Beware of one piece of metal getting in the way of another, and by obscuring it prevent it from getting fully coated. The gun is two stage: the first pressure delivers just air, and the second delivers both air and material. The quantity of material is proportional to the distance you pull back the trigger. Match the speed of your sweep to the amount of spray to avoid build-up of wax. Work in short bursts rather than spraying continuously, and always remember to hold the gun upright.

Using the flexible lance, spray inside the front cross member and any other reinforcing sections. Unprotected cavities are prone to rust. Watch out for the join between sheets of metal, as dirt and moisture can gather here and start rust. Use the hook spray, since it is directional. Always wipe away any surplus anti-corrosion fluid as you go. Make sure you do inside all the reinforcements. Spray with reduced pressure, otherwise anti-corrosion fluid will go everywhere. Work from the bottom up, so that anti-corrosion fluid will not drip on you. When you rust-proof the chassis legs you should use a penetrating fluid and the flexible lance for all enclosed areas.

Pillars
Whenever possible remove the courtesy input switch for access, otherwise extra holes may have to be made. Spray as far up the pillar as possible, then down to the sills. Finally extend and secure the webbing on inertia reel seat belts and spray the surrounding areas with great care. Spray into the D-post. Pay attention to fully coating the wheel arch seam area.

Door panels and rear quarter panels
Use the existing openings to spray into door panels, or drill extra access holes as required. Take care to thoroughly protect all door hinges. Remove the trim where necessary to gain access to the rear quarter panel. Reach over the rear wheel arch and spray coat the wheel arch seam, the lower rear wing, around the tail lights and into all the inaccessible areas. Remove the existing plugs for access to the tailgate panel. Spray up beside the window of the tailgate; if unprotected this is particularly vulnerable to rust.

Sill panels and underbody sections
It is most important to achieve thorough coverage of the sills. Use the manufacturer's existing access

hole, or open extra access holes. Look out for double sill sections, which must both be treated. Treat the underbody box sections and cavities in a systematic manner. Underbody seams, joints, brackets and attachments must all be coated with the fluid. Thoroughly apply a coating of rust-proofing fluid to the underside of the car, paying special attention to exposed seams and joints well away from the spray.

Wheel arches and underbody

Before applying a coat of sealant or heavy-duty wax, first remove any loose or flaking materials. When spraying, avoid blocking the manufacturer's drain holes; they are a vital part of anti-rust protection. If they do get blocked, clear out the surplus.

The wheel arches must be done carefully (Figure 4.65). Apply an extra coating to all forward-facing surfaces. To give the best possible abrasion resistance, pay particular attention to all joints and recesses. These are the places where mud can easily accumulate. When spraying the chassis legs try not to get any overspray in the engine compartment. Using the hook nozzle, deposit a generous coating of sealant all the way round. Make sure that the ledges are fully coated; not all of these are visible from underneath, so check them when the vehicle is back on the ground.

Spray the underbody from every direction to prevent shadow areas (Figure 4.66). Check your work as you go. Use the fan spray nozzle for rapid coverage of large open areas. Use the hook nozzle to get a good finish around the edges and to reach into awkward corners, then wipe away any excess sealant before it sets.

Clean any excess rust-proofing fluids off the car body with the recommended solvent, and remember always to clean the gun thoroughly after using the sealant or other heavy-bodied products.

In some situations you *must* rust-proof: for instance, when fitting a new body panel which is supplied untreated or when you do service bodywork repairs. But rust can also develop as a result of any accidents. Check the vehicle when it comes in for signs of external corrosion or paint damage; if neglected it risks forfeiting the protection of any body warranty given.

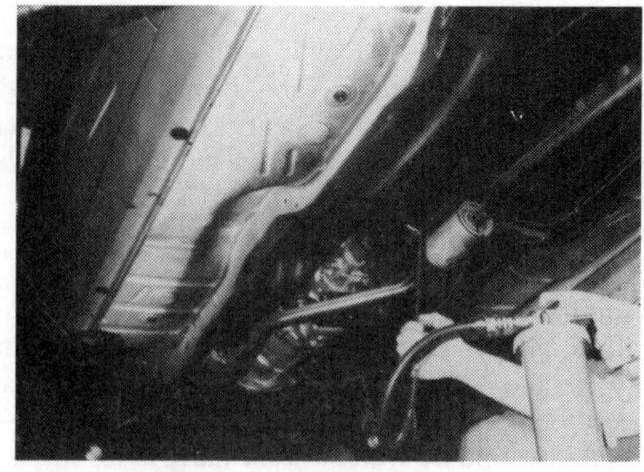

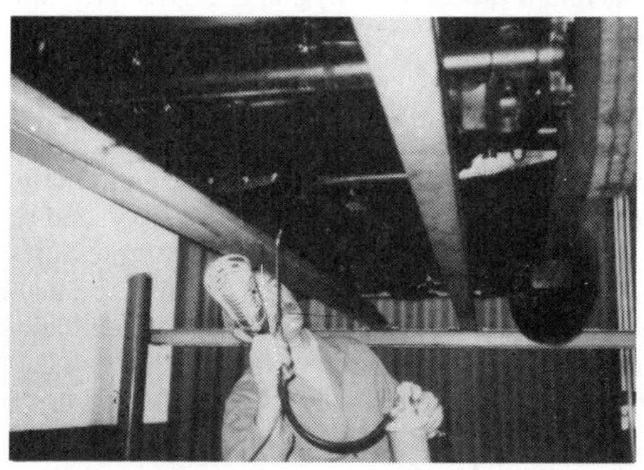

Figure 4.65 Sealing wheel arches and floor pan (*Tuff Kote Dinol*)

4.17 COMPARISON OF HOT AND COLD SPRAYING

The true value of a paint film is the amount and character of the solids deposited that remain after drying or curing. Generally it is not possible to apply paint by spray as supplied by the paint manufacturers. To obtain the necessary consistency for spray application and good flow-out, thinners are normally used. The function of a solvent is merely to reduce the viscosity of the paint to assist its application to the work.

A paint at a viscosity of 35 seconds BS B4 flow cup for conventional spray application requires a high air pressure to produce good atomization (Figure 4.67). The high pressure and the expanding compressed air

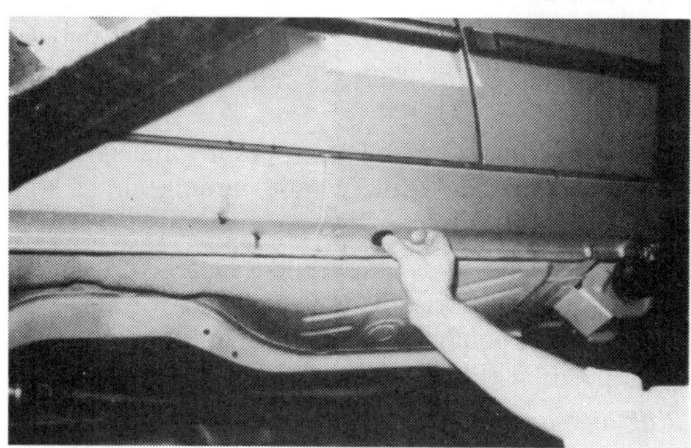

Figure 4.66 Sealing underbody (*Tuff Kote Dinol*)

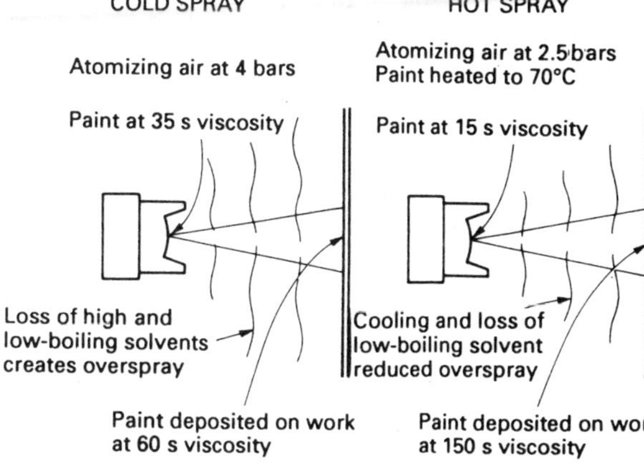

Figure 4.67 Comparison of cold spraying and hot spraying

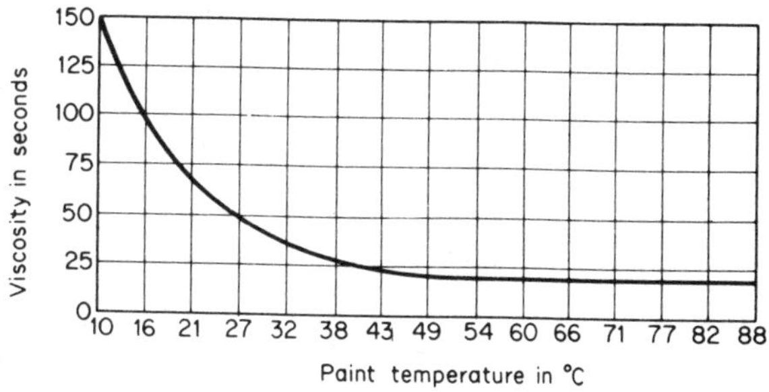

Figure 4.68 Typical viscosity curve

evaporate a lot of the solvent and carry off some of the finely atomized paint as overspray. This is a recognized loss, and therefore less solids are deposited per coat. The basic principle of hot spray is to reduce the initial viscosity of the paint by heat, in contrast to cold spray where the viscosity is reduced by the addition of a thinner which subsequently evaporates. The underlying principle of a spray system involves splitting up the fluid paint into tiny droplets. This atomization is only possible if the viscosity of the liquid is sufficiently low. In conventional cold spray application this low viscosity is achieved by the addition of a solvent, whereas in hot spraying the necessary reduction in viscosity is achieved by heat. Heating of enamels etc. to 60–80°C generally results in the viscosity becoming one-third to one-quarter that at ambient temperature. Further heating does not usually cause any further fall (Figure 4.68).

The viscosity of paints can vary in the paint shop owing to changes in the atmospheric and shop temperatures throughout the working day. This can be a problem to the painter, as work can be subject to uneven build and runs and sags can occur in spite of gun adjustments. Even if wide temperature changes do occur in hot spray, the atomization and spray pattern is unaffected because the paint heater delivers the fluid at a controlled temperature and therefore the viscosity remains constant. At higher temperatures temperature viscosity is negligible, but there is a sharp variation within normal atmospheric temperature range (Figure 4.69).

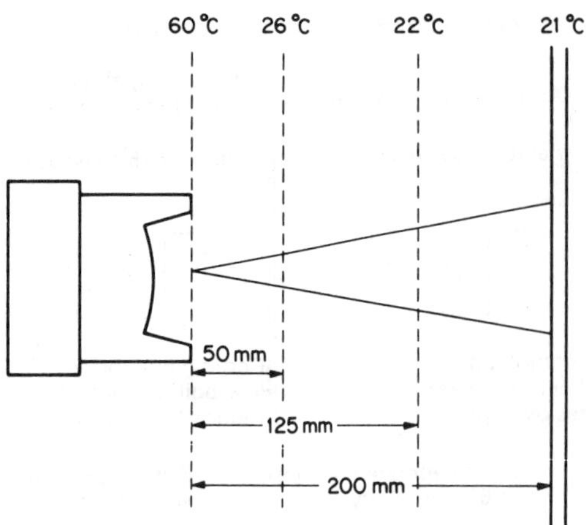

Figure 4.69 Rapid drop in paint temperature from fluid tip to job surface

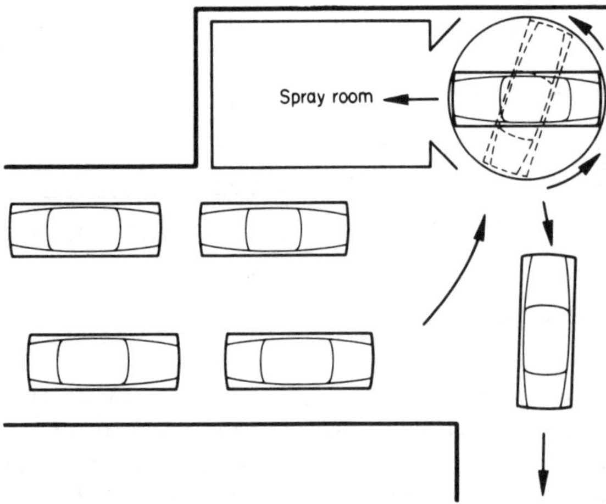

Figure 4.70 Suggested use of turntable

In hot spray application, low boiling solvents are released more quickly during atomization because of the heat. These are therefore not present in the applied liquid film to cause chilling and resulting condensation, called 'blushing', which occurs with certain paints on applications under humid conditions. The effect is further checked because the material does not reach the low temperature when chilled by the evaporating solvent at the spray tip of the gun.

4.18 MOVEMENT OF VEHICLE IN THE PAINT SHOP

In any refinishing shop, the movement of vehicles is of primary importance. Much time and effort may be wasted in moving vehicles from one point to another as dictated by the finishing process: This is particularly true in those shops where expensive air conditioned spray booth and/or spray booth ovens have been installed. To recoup the high capital outlay in such units it is essential that they should be used at full capacity for the whole of the working day. The simplest spraying/stoving unit, i.e. the combined spray booth and oven, is capable of handling six to seven vehicles per working day of eight hours, provided it is kept in use over the lunch hour. Furthermore, with soaring building costs and high rateable values, coupled in some cases with restricted space for expansion, it is important that floor space be kept to a minimum. While it is not possible to achieve a factory-type flow line for refinishing jobs that may vary tremendously in their requirements, for example size, extent of repair and amount of work to be carried out, the desired continuous supply of vehicles to the spray booth may be achieved. Careful consideration of the problems of vehicle movement will enable this supply to be obtained with the minimum of time, labour and space.

There are several methods of moving vehicles within the refinish shop:

Under own power This method is the most economical of labour, but requires the maximum of space. Also it gives rise to additional dirt, air pollution and fire hazards.

Manually Space requirements are as above but demand on labour is heavier. Three men may be required to move an average car.

Mobile hydraulic jacks These permit some saving of space in that they allow a tight turning circle both front and rear.

Turntable This gives an excellent means of utilizing what would otherwise be a dead corner, or may permit the spray booth to be sited in an otherwise impossible position (Figure 4.70).

Rail and bogie system The vehicle is moved along railway-like tracks with bogies under each of its four wheels. This offers the greatest advantages of all sideways movement of the vehicles. Normally four sets of parallel rails are involved. The first pair constitutes the track on which the vehicles are first received into the paint shop, and along which the preparatory work is usually carried out, that is discing, flatting and stopping. The second pair is kept clear to allow movement of selected bodies to the spraying and/or stoving position (Figure 4.71). Vehicles are either driven or manhandled from one track to another. It is possible to have a layout utilizing only one track, but move-

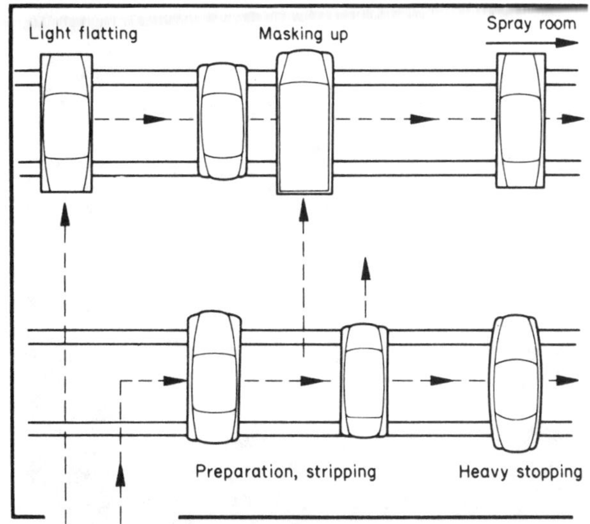

Figure 4.71 Lateral movement of vehicles using two sets of tracks

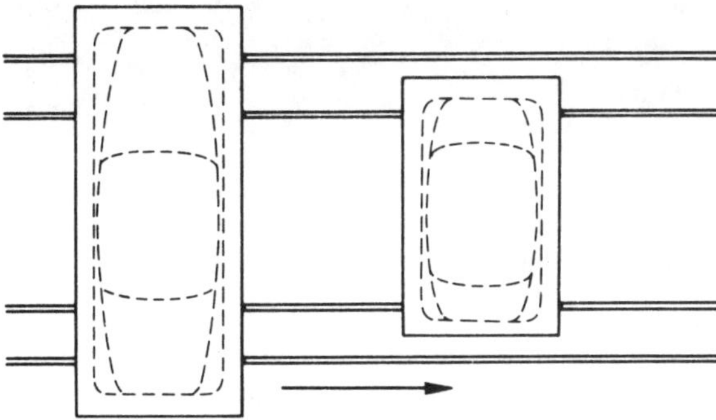

Figure 4.72 Two-track rails requiring long bogies

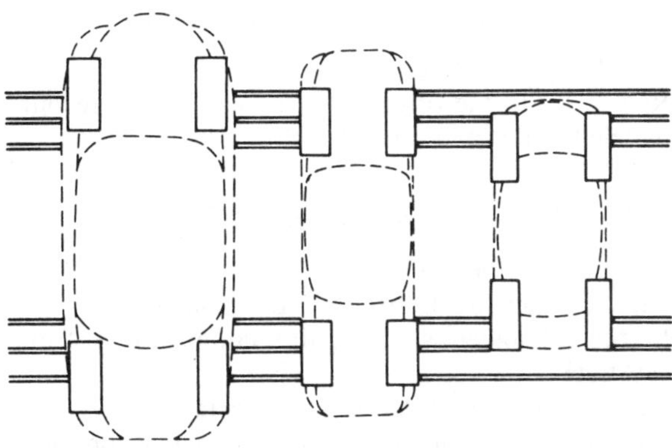

Figure 4.73 Smaller bogies can be used with three-track rails

ments are more restricted than with two tracks. Two types of track are available: (a) those with sets of two rails, and (b) those with sets of three rails. With type (a), bogies long enough to accommodate vehicles of differing wheel bases must be used (Figure 4.72). Type (b) permits the use of smaller, lighter bogies which are on selected pairs of rails from each set (Figure 4.73).

4.19 COMMON SPRAY PAINTING DEFECTS

4.19.1 Paint poblems

Complaints about faulty paintwork are a problem for everyone concerned. They mean spending extra time and money on each job. There are many reasons for faults in spraying.

Establishing the precise causes of faulty paint finishes often involves complicated and costly investigations. Many faults, particularly in refinishing, only crop up in isolated instances.

As a rough classification, the reasons for faulty paintwork can be subdivided into three main groups:

1 Incorrect procedures during preparation in the spray shop.
2 Adverse operating conditions during spraying.
3 Wrong treatment and care of the paintwork.

The most common faults are inadequate or incorrect preparation of the underlying surface and insufficient care in handling the actual paint materials. There are also environmental issues: corrosive industrial atmospheres, salty sea air, strong UV radiation and heavy pollution have a direct influence on the paint finish. Working conditions when spraying are important too, for example relative humidity, temperatures, drying times and film thicknesses.

4.19.2 The most frequent types of defect.

Damage caused by external action:

1 Stone chipping.
2 Corrosion.
3 Acid damage.

Damage caused by procedural errors or poor-quality materials:

1 Orange peel.
2 Scratch opening.
3 Sinkage.
4 Showthrough of stopper or filler contours.

Figure 4.74 Blistering (ICI Autocolor)

5 Cloudiness.
6 Popping.
7 Peeling of primer or filler.
8 Colour changes due to excess peroxide in stopper.
9 Poor opacity.
10 Poor interfilm adhesion.
11 Poor colour matching with solid colours and metallics. (This is why fading out is recommended.)

4.19.3 Bleaching

Recognizing the defect:
Yellowing of the surface corresponding to areas of filler in the substrate.

Possible causes:
Excessive quantities of peroxide used in the filler.

Rectifying the defect:
Rub down the effected area to the surface of the filler, seal with isolator or epoxy and repaint.

4.19.4 Bleeding

Recognizing the defect:
Discolouration of the top coat, either in the form of a halo, or, in severe cases, a complete colour change. This defect usually only occurs when spraying over red or maroon paint.

Possible causes:
Absorption of pigment from the underlying paint, dissolved by the solvents of the new coat.

Rectifying the defect:
Rub down to the original finish, seal with a recommended sealer and repaint.

4.19.5 Blistering (see Figure 4.74)

Recognizing the defect:
Air- or water-filled blisters in the paint film.
In the advanced stages paint comes away from substrate.
Size may vary with the weathering factors.
Long periods of humidity and wet weather promote the phenomenon.
In extreme cases rusting occurs.

Possible causes:
Spraying over rust.
Undried previous materials sanded too soon.
High relative humidity or condensation generated by variations in temperature during the spraying process.
Condensation during spraying.
Residues of sanding dust.
Sanding water not dried off.
Top coat too thin.

Rectifying the defect:
Having determined the extent of the damage, sand down the areas to sound substrate and respray.

4.19.6 Blooming

Recognizing the defect:
A milky white haze or mist formed on the surface of the paint film.

Possible causes:
Moisture condensing on, and being trapped in, the wet film. This may be due to:

(a) spraying during cold, wet or humid weather
(b) using too fast or poor quality thinners
(c) compressed air pressure too high, and/or poor spray gun set-up
(d) fanning compressed air onto the film to speed up solvent release
(e) draughty paint shop, or inadequate heating and/or air movement.

Rectifying the defect:
Slight blooming may be removed by the use of polishing compound after the paint film has hardened, or by spraying the affected area with non-blooming thinners.
In more severe cases rub down the surface and repaint using the correct grade of thinner or non-blooming thinner.
If these remedies fail to correct the fault, raise the temperature of the paint shop by 5–10 degrees, avoid all direct draughts, flat and repaint the affected area.

4.19.7 Blowing/air trapping

Recognizing the defect:
Large rounded air bubbles or blisters, usually occurring in the areas of seams and boxed in corners, or over heavily filled or plastic surfaces.

Possible causes:
Air trapped beneath the paint expands resulting in the detachment of the paint film from the substrate. This frequently results from:

(a) poor application of filler, stopper or primer, resulting in entrapped air
(b) poor feather edging of chipped film
(c) bridging of seams and boxed corners by the paint film
(d) porosity and air pockets in the primer due to inferior or insufficient thinner, the compressed air pressure too high, or dry spraying
(e) failure to repair and seal substrate correctly, especially when spraying GRP
(f) excessive application of heat during drying

Rectifying the defect:
The paint must be removed to the depth of the bubble, any underlying defect remedied and the area repainted.

4.19.8 Chalking

Recognizing the defect:
A chalky dusting or powdering at the paint surface, often associated with old, weathered paint film.

Possible causes:
Precipitation of elements within the paint. This may be due to:

(a) incompatible or defective materials within the paint
(b) degradation of the binding agent
(c) degradation of pigment

Rectifying the defect:
Flat, compound and polish the surface to restore the gloss.
In severe cases repaint the top coat.

4.19.9 Checking/Crazing (see Figure 4.75)

Recognizing the defect:
Small cracks resembling a spider's web running in all directions.

Possible causes:
Very high film thicknesses.
Top coat not cross-linked by hardener or not adequately cross-linked.
Exposure to UV.

Rectifying the defect:
Sand the cracked layers away down to the intact undercoat or strip completely.
Respray the system with primer/top coat.

4.19.10 Cloudiness/mottling (see Figure 4.76)

Recognizing the defect:
This phenomenon is only possible with metallic finishes.

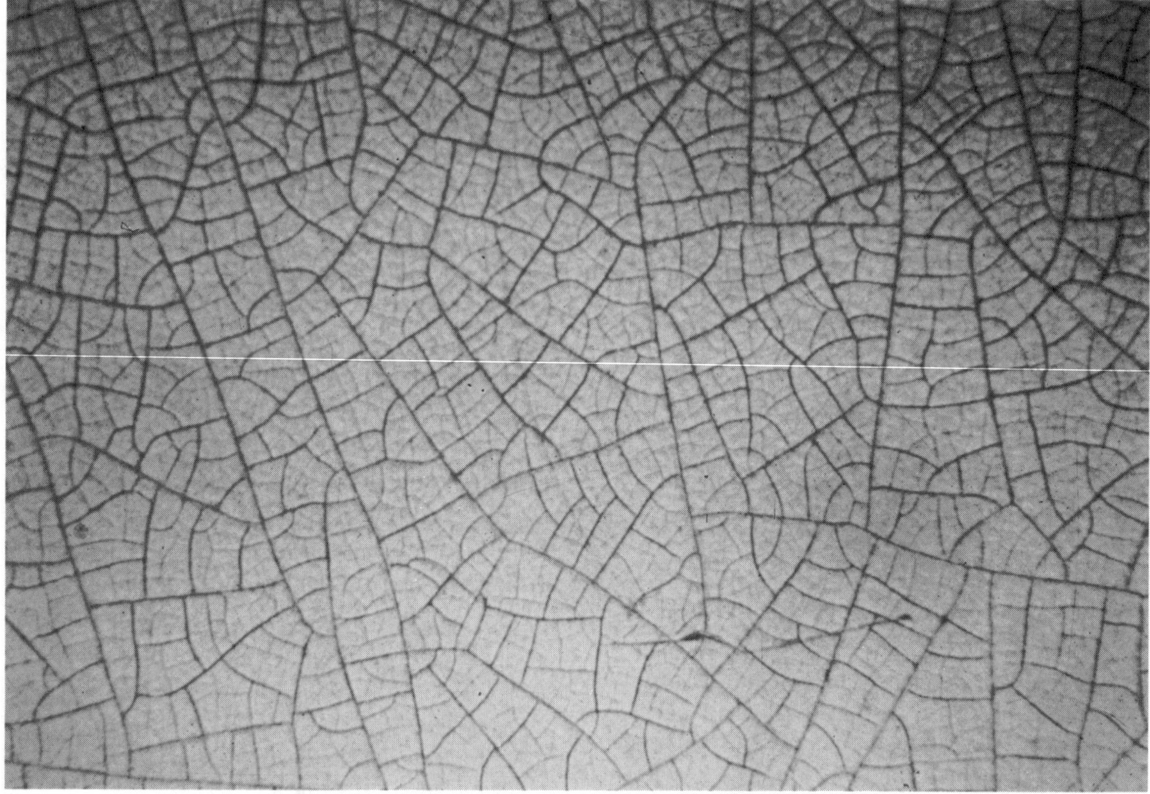

Figure 4.75 Checking/Crazing (ICI Autocolor)

The strength of the colour varies from light to dark; hence it detracts from a uniform appearance.

Possible causes:
Different float-out or orientation of the flakes.
Different film thicknesses.
Paint viscosity too high or too low.
Unsuitable solvents.
Flash-off times between the individual coats too short.

Rectifying the defect:
After appropriate flash-off time, mist-coat to cover by applying repeated thin coats.
If drying has already taken place, sand and

- spray, bearing in mind the viscosity
- observe flash-off times between the spraying operations
- mist-coat
- spray uniform layers

4.19.11 Cracking (see Figure 4.77)

Recognizing the defect:
Cracks of different sizes running in all directions. In extreme cases the undercoat may be visible.

Possible causes:
Very high film thicknesses.
Repeated spraying.
Inadequate drying process.
Undercoats not fully hardened.
Temperature fluctuations.

Rectifying the defect:
Sand the cracked layers down to the intact undercoat or strip completely.
Repaint the system with primer/top coat.

4.19.12 Dirt/seeds/bits (see Figure 4.78)

Recognizing the defect:
Inclusion of contaminant particles of different sizes and forms in top coat or primers.
Look of finish impaired.

Possible causes:
The surface to be painted was not sufficiently well cleaned.
Paint materials not filtered.
Spraying in an unclean area.

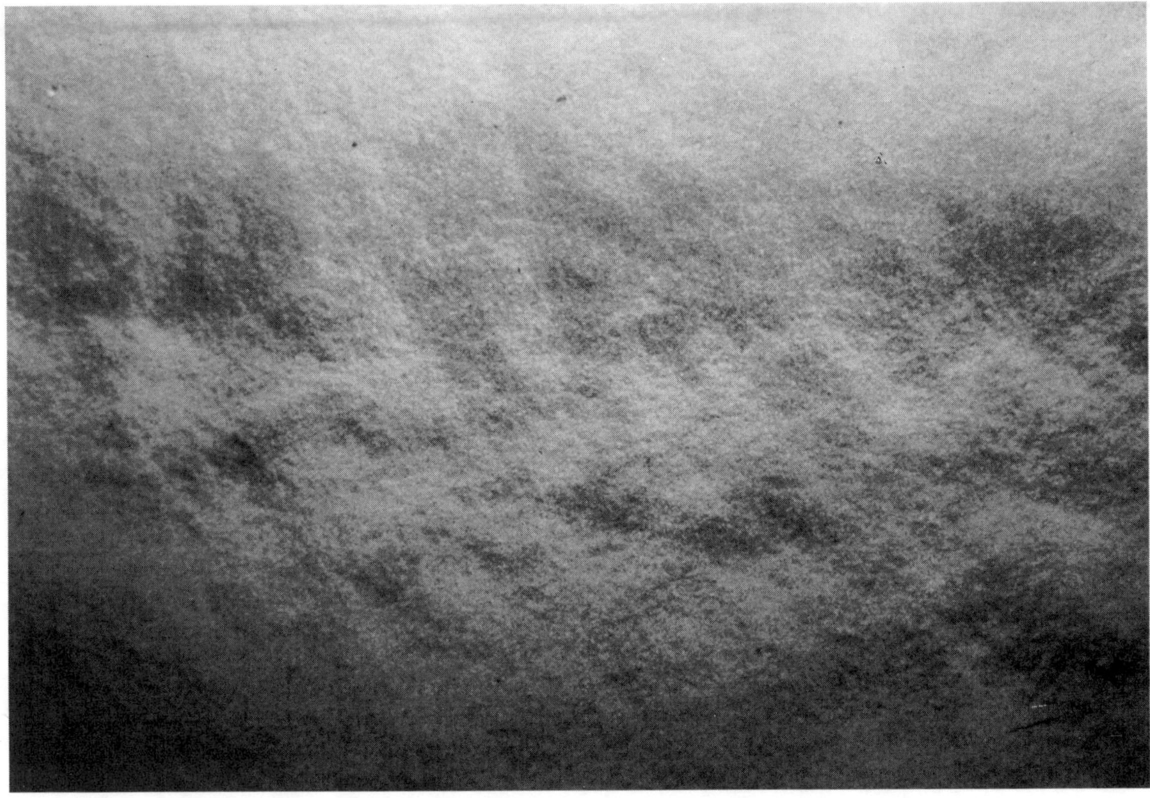

Figure 4.76 Cloudiness/Mottling (ICI Autocolor)

Filters dirty.
Unsuitable clothing being worn. (Should be lint free.)

Rectifying the defect:
Once the finish has cured, odd specks should be sanded out with P1200 abrasive paper and polished again with a suitable solvent-free abrasive or polishing compound.
If a considerable area is involved, sand and repaint.

4.19.13 Dry spray

Recognizing the defect:
A granular or coarse textured finish with no gloss.

Possible causes:
Paint being deposited on the surface in a powdery condition.

(a) Viscosity of paint too high, use of incorrect or poor quality thinners.
(b) Poor spraying technique, dirty spray gun, compressed air pressure too high, gun held too far from the surface during spraying.
(c) Spraying in draughts or in a high velocity air flow.

Rectifying the defect:
Flat, compound and polish.
If the texture is too coarse for this to correct the defect, rub down the top coat and repaint.
Metallic finishes must always be rubbed down and repainted.

4.19.14 Dull finish/abnormal loss of gloss

Recognizing the defect:
Although apparently smooth and evenly applied the surface lacks shine.

Possible causes:
Microscopic roughness of the surface, which may result from:

(a) poor hold out of primer, or the application of top coat over primer which is not thoroughly dry
(b) poor quality or incorrect thinners, or the use of additives in the paint
(c) incorrectly prepared or poorly applied paint
(d) application over a poor substrate
(e) excessively slow drying due to high humidity or low temperature

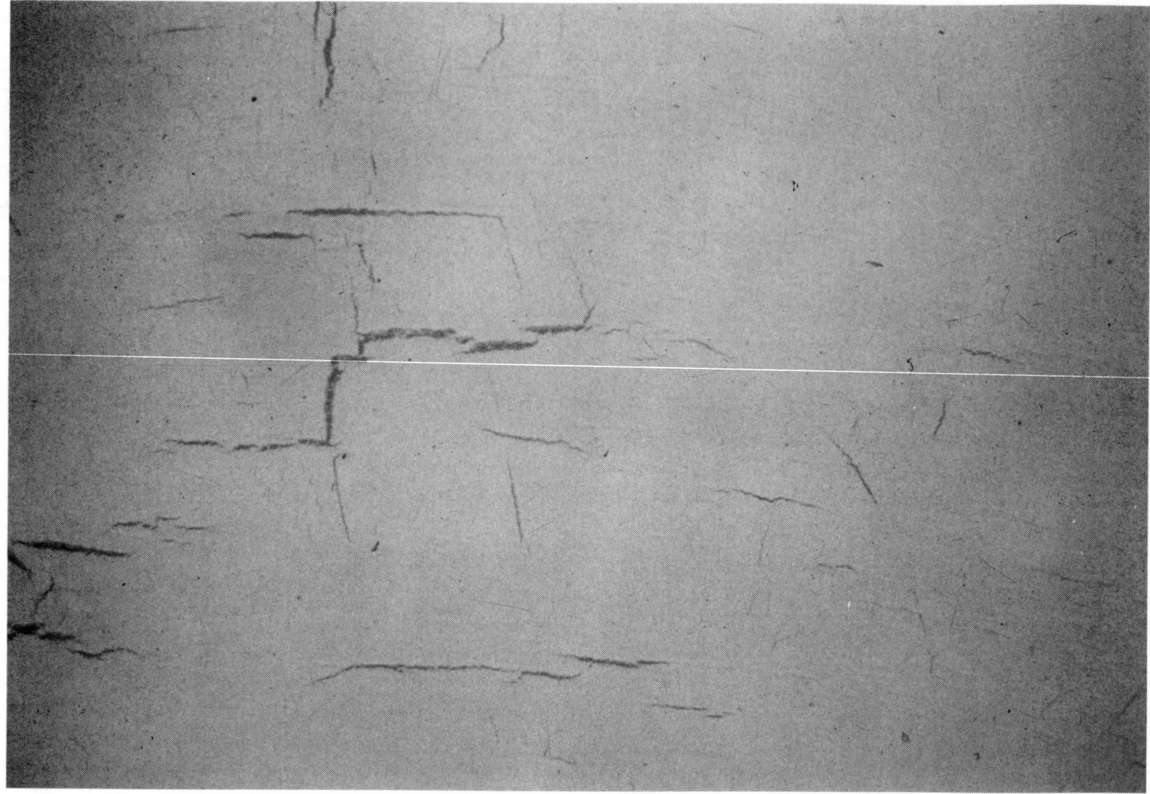

Figure 4.77 Cracking (ICI Autocolor)

(f) solvent fumes or exhaust gases attacking the surface
(g) surface contamination by wax, grease, oil, soap or water
(h) the use of strong detergents or cleaners on a newly painted surface, compounding too soon after painting or using compound which is too coarse

Rectifying the defect:
Normally the shine may be restored by rubbing down with abrasive compound and polishing.
If the dulling is too severe for this to yield satisfactory results, rub down the top coat and repaint.

4.19.15 Feather edge lifting (see Figure 4.79)

Recognizing the defect:
When used for recoating, two-coat finishes push up at points where rubthroughs have occurred and filler or primer have not been applied.

Possible causes:
Old finish not thoroughly cured.
Base coat not isolated or applied too thickly.
Mixed system not developed together causing reaction.
Unsuitable solvents.

Rectifying the defects:
Cure the defective area thoroughly; force dry (at temperature) if possible.
Sand out with fine-grade abrasive paper.
Respray paint finish.
Where the defect covers a large area, it is advisable to strip and repaint the finish.

4.19.16 Metamerism

Recognizing the defect:
Colours which are perceived as identical show differences under different light sources.

Possible causes:
Use of paints containing pigments of different types (a green, for example, may be formulated from yellow and blue or straight from green).
Use of unsuitable paints for respraying.

Figure 4.78 Dirt/Seeds/Bits (ICI Autocolor)

Rectifying the defect:
Spray over with paint based on the same pigmentation.

4.19.17 Off-colour/poor colour match

Recognizing the defect:
Adjacent areas exhibit difference in shade. This is most frequently noticeable on adjacent complete panels.

Possible causes:
There is no single cause, the defect may result from a number of factors:

(a) use of differing or incorrect materials
(b) inadequate mixing of the paint
(c) fading due to weathering or exposure
(d) incorrect application
(e) metameric distortion (colour variants in different light)
(f) incorrect colour choice or use of wrong variant of the colour

Rectifying the defect:
Flat down the surface and repaint using the correct colour and variant.

4.19.18 Orange peel (see Figure 4.80)

Recognizing the defect:
Pronounced pebbling of the sprayed paint.
Orange-peel surface profile.

Possible causes:
Paint viscosity too high.
Thinner too fast.
Fan too large on spray gun.
Spray booth temperature too high (30°C).
Air pressure incorrect.

Rectifying the defect:
Sand off and respray.
Sand and polish small areas.

4.19.19 Overspray

Recognizing the defect:
Area of granular paint particles adhering to, or partly absorbed in, the surface of the film.

Possible causes:
Spray dust deposited on the surface. This results from:

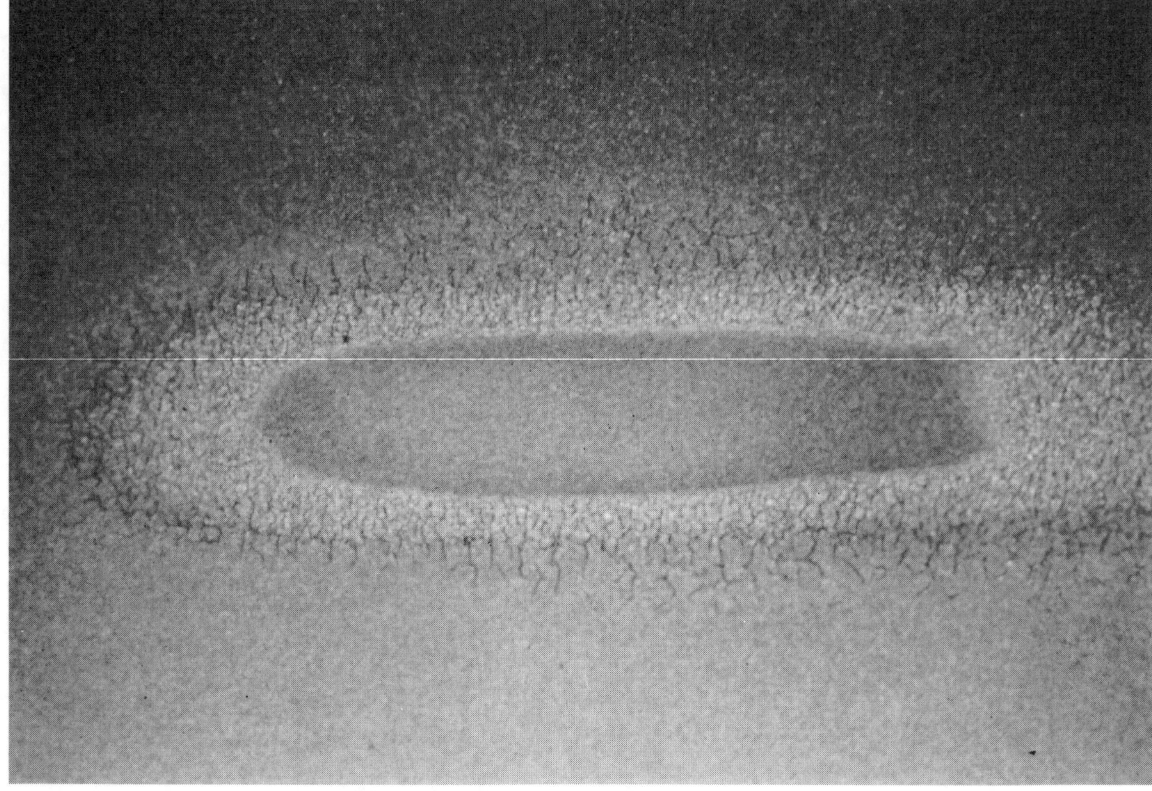

Figure 4.79 Feather edge lifting (ICI Autocolor)

(a) poor masking
(b) paint from a subsequent application settling on the surface
(c) compressed air pressure too high
(d) inadequate extraction or ventilation

Rectifying the defect:
Rub down with abrasive compound and polish.

With plastics:

(a) unsuitable adhesion promoter used
(b) paints not plasticized or insufficiently plasticized

Rectifying the defect:
Sand down to intact film or strip to substrate.
Respray.

4.19.20 Peeling/flaking (see Figures 4.81a, 4.81b)

Recognizing the defect:
There is no interfilm adhesion between coats or the substrate. The loss of adhesion is often not apparent until after mechanical stress has been applied.
From crosshatch tests it is possible to see from which coat the loss of adhesion originates.
Loss of adhesion or peeling on plastic parts.

Possible causes:
The undercoat was not free of grease, dust or moisture at the time of spraying.
Paints not developed as a system.
Undercoats not sanded.

4.19.21 Peroxide specks in metallic finishes (see Figure 4.82)

Recognizing the defect:
Deviant, different coloured specks over the entire finish, chiefly in silver and blue metallics; does not show up until fully cured.

Possible causes:
Excess hardener (peroxide) in the polyester stopper (more than 3 per cent of hardener paste may produce this type of defect).

Rectifying the defect:
Seal off with polyester or epoxy filler.
Respray.

Figure 4.80 Orange peel (ICI Autocolor)

4.19.22 Pinholing

Recognizing the defect:
Small cavities, generally less than 1 mm in diameter, occurring over stopper, filler or GRP substrate.

Possible causes:
Absorption of paint into holes in the substrate this is due to:

(a) air inclusions in GRP resin
(b) inadequate preparation and sealing of the substrate
(c) poor quality filler or stopper
(d) poor mixing of filler, poor application of filler or stopper

Rectifying the defect:
Rub down the affected area to the primer, stop any pinholes, spot prime and flat until the surface is smooth, then repaint.

4.19.23 Popping (see Figure 4.83)

Recognizing the defect:
This phenomenon occurs particularly when two-pack acrylics are being used. This is true both of primers and top coats.

Possible causes:
The bubbles are caused by solvents or air trapped during the drying process, especially when the film thicknesses are high.
Not ideally matched solvent combinations in the paint system.
Non-linear relationship between cross-linking and solvent evaporation.
Incorrect flash-off times.
Poor spray booth conditions.

Rectifying the defect:
After cooling-down, spray again without sanding.

4.19.24 Runs/sags

Recognizing the defect:
Well-defined local thickening of the paint film in the form of a wavy line or shallow rounded ridges, normally confined to sharply sloping or vertical surfaces.

Possible causes:
Slumping of the paint film due to:

(a) excess thickness of application
(b) air pressure too low

Figure 4.81a Peeling (ICI Autocolor)

(c) fan width too narrow
(d) spray gun too close to the surface or moving too slowly
(e) use of poor quality or incorrect thinner
(f) incorrect viscosity of the paint
(g) air or surface temperature too low
(h) contamination of the underlying surface

Rectifying the defect:
Allow the paint to harden thoroughly, rub down excess paint, flat, compound and polish.
In severe cases it may be necessary to rub down and repaint the surface.

4.19.25 Scratch opening (see Figure 4.84)

Recognizing the defect:
Look of finish spoiled by scratch lines in the paintwork.

Possible causes:
Scratches, originating in the underlying metal or stopper, are not filled by the primer and filler.
Final sanding was too coarse.
Topcoat sprayed too thin.
Extra swelling on primers which are too fresh, or use of unsuitable solvents.
Old finish not fully cured.

Rectifying the defect:
Dry out thoroughly and respray.

4.19.26 Shrivel (see Figure 4.85)

Recognizing the defect:
One or more layers of the primer or top coat pull up.

Possible causes:
Old finish not hardened through.
Paint systems not adapted to one another (for example nitrocellulose or acrylic undercoats or top coat paint used over synthetics).
Unsuitable solvents.

Rectifying the defect:
Harden defective area thoroughly; force dry (at temperature) if possible.
Sand off with fine-grade abrasive paper.

Figure 4.81b Flaking (ICI Autocolor)

Repaint paint finish.
Where the defect covers a large area, it is advisable to strip and repaint the finish.

4.19.27 Silicone cratering (cissing/fish eyes) (see Figure 4.86)

Recognizing the defect:
Markings in the final coats, generally circular and pinhead in size and occurring at isolated points or over large areas.

Possible causes:
Contamination by, or residues of, grease, wax, oil, or polish and above all substances containing silicone on the undercoat.
Spray air supply contains oil or condensation.
Contaminated air is being brought in with the air supply.
Equipment (spray gun) is greasy.
Polishes or sprays containing silicone being used in or near the spray booth.
Working apparel contaminated with polishing dust containing silicone.

Rectifying the defect:
The defect cannot be rectified or avoided until the reason for it has been cleared up.
Where the reasons lie with the undercoat, it is advisable for the area due for spraying to be cleaned beforehand with a silicone remover.
Where individual craters are observed after the first thin pass of top coat, tighten up the top coat thoroughly and apply a number of coats (sprayed dry).
Remedy other factors caused by spray air, equipment, polishes etc.

4.19.28 Sinkage (see Figure 4.87)

Recognizing the defect:
Ringing of stopper or filler and the top coat loses gloss.

Possible causes:
Undercoat not fully cured.
Stopper or filler recoated too soon.
Unsuitable solvents used.
Drying temperature too high.

Figure 4.82 Peroxide specks in metallic finishes (ICI Autocolor)

Rectifying the defect:
After curing or drying, sand down the defective area thoroughly and spray again.

4.19.29 Slow drying/softness

Recognizing the defect:
The paint film requires an excessive drying period, or fails to harden thoroughly.

Possible causes:
Slow evaporation of solvent from the paint. This may be due to:

(a) excessive thickness of the paint film
(b) poor atmospheric conditions during spraying or drying, coldness, humidity, lack of air movement
(c) insufficient drying time between coats
(d) insufficient, poor quality or incorrect thinners

Rectifying the defect:
Move the vehicle to a warm, well-ventilated area. Low heat may be applied to improve drying, but care must be exercised to avoid wrinkling.

4.19.30 Spotting/colour change due to chemical factors (see Figure 4.88)

Recognizing the defect:
Light or dark spots, mainly on horizontal surfaces, chiefly affecting synthetics.

Possible causes:
Exposure to acid, alkaline or other chemicals.
Industrial fall-out; sulphur dioxide combined with rain.
Cement or lime deposits.
Exposure to brake fluid or organic acids.
Corrosive substances of animal or plant origin (for example bird droppings).

Rectifying the defect:
Depending on the seriousness of the damage, it may be possible to polish out isolated spots.
Large areas of damage may need respraying.
Where exposure to brake fluid has occurred or damage has been done by acids, sand down the areas affected, and respray.

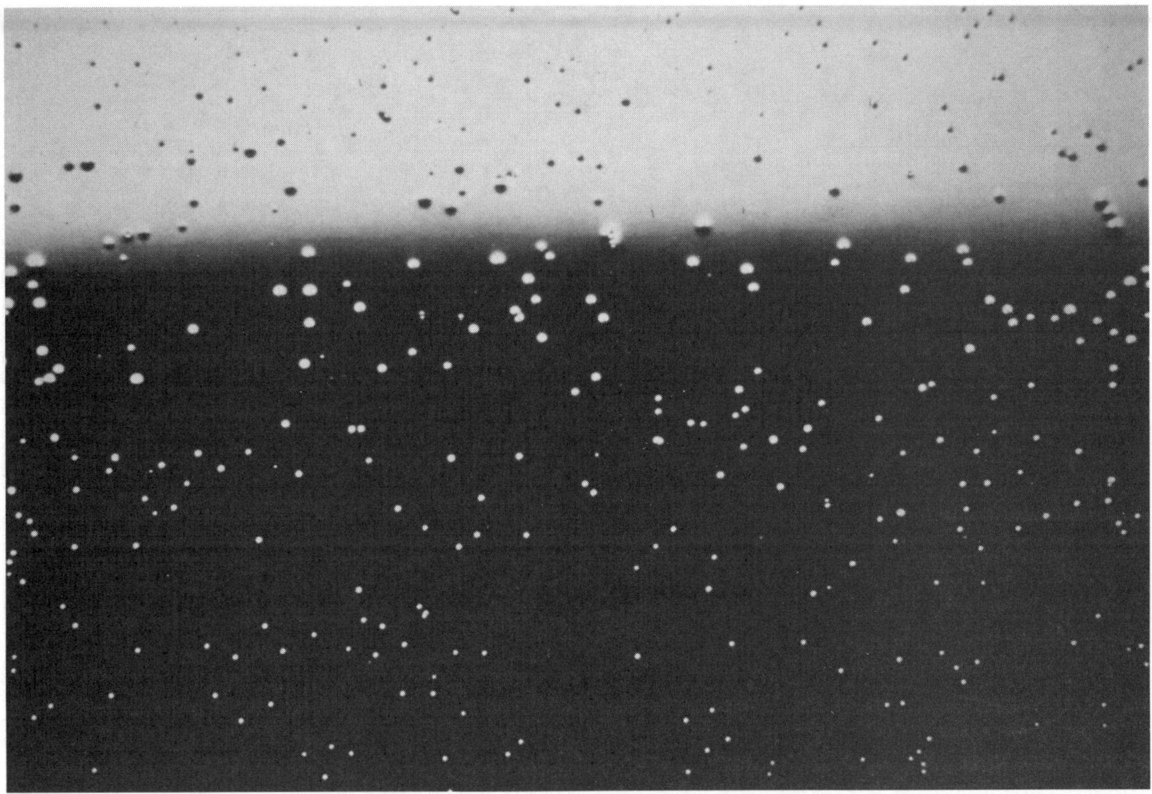

Figure 4.83 Popping (ICI Autocolor)

4.19.31 Stone chipping or mechanical damage (see Figure 4.89)

Recognizing the defect:
Mechanical damage or stone chipping extending through to the underlying metal will very soon produce rust.

Possible causes:
Damage due to external factors going through to the filler, primer or underlying metal.

Rectifying the defect:
Sand out or blast clean.
Use primer giving anti-corrosion protection.
Apply top coat.

4.19.32 Water marking and spotting (see Figure 4.90)

Recognizing the defect:
Marking due to a slight depression and whitish discolouration in the paint surface.

Possible causes:
Finish was exposed to water before it was fully hardened.
The phenomenon is generally found only on horizontal surfaces.

Rectifying the defect:
Where required, faint, isolated spots or marks should be sanded out with P1000/1200 grit wet-and-dry abrasive paper and polished.
Large and heavy spots or marks should be sanded out and resprayed.

4.20 COLOUR MIXING AND MATCHING

4.20.1 Colour mixing systems

Surveys have shown that between them 40 car manufacturers offer approximately 600 colours on current models. These are new colours only, and the total number of colours that the refinisher may have to match can run into thousands.

The development of a new car colour can be broken down into four stages:

1 Initiated by clothing and other fashions, new colours are reproduced in a range of paints. The car

Figure 4.84 Scratch opening (ICI Autocolor)

body stylist then selects a colour by trial and error, to suit a new car model.
2 Paint manufacturers then reproduce this colour and carry out the usual tests for colour stability, durability etc.
3 The vehicle manufacturer paints a number of car bodies with the colour to ensure that the paint is suitable for mass production application.
4 Finally, the paint manufacturers are required to produce a coloured paint to suit the vehicle refinisher and to provide colour mixing formulae as well as basic colours and tinters.

There are two colour mixing systems available to the refinisher:

Gravimetric This system involves the use of a set of scales. The base colour and tinters are mixed according to their weight, the quantity of each being obtained from a given formula.

Volumetric The various quantities of each ingredient are added by volume. An adjustable measuring rod is the main tool involved (see Figure 4.91).

In both systems the materials required to mix and match colours are base colours, full-strength tinters, and reduced-strength tinters. The last are necessary in order to provide the refinisher with a measurable quantity. For example, instead of adding 0.5 grams of blue-black full-strength tinter, he can more easily add 5 grams of reduced-strength tinter.

A complete set of colour formulae is provided by the paint supplier and is updated at regular intervals. Colour samples are also provided which are added to as new colours are produced. There can be several slight variations of a colour as vehicles leave the production line, and these can provide the refinisher with real problems when trying to match a colour. To solve these problems, paint manufacturers produce samples of these colour variants and literature with hints on how to match them.

Colour mixing systems should not be installed in the workshop if at all possible. They constitute a fire hazard, and can collect dust and overspray which may affect the accuracy of the scales. A separate colour mixing room should be provided where possible, which has plenty of natural light and also has 'daylight' fluorescent lighting.

It should be fully understood that the vehicle manufacturer uses high-bake materials, whereas the

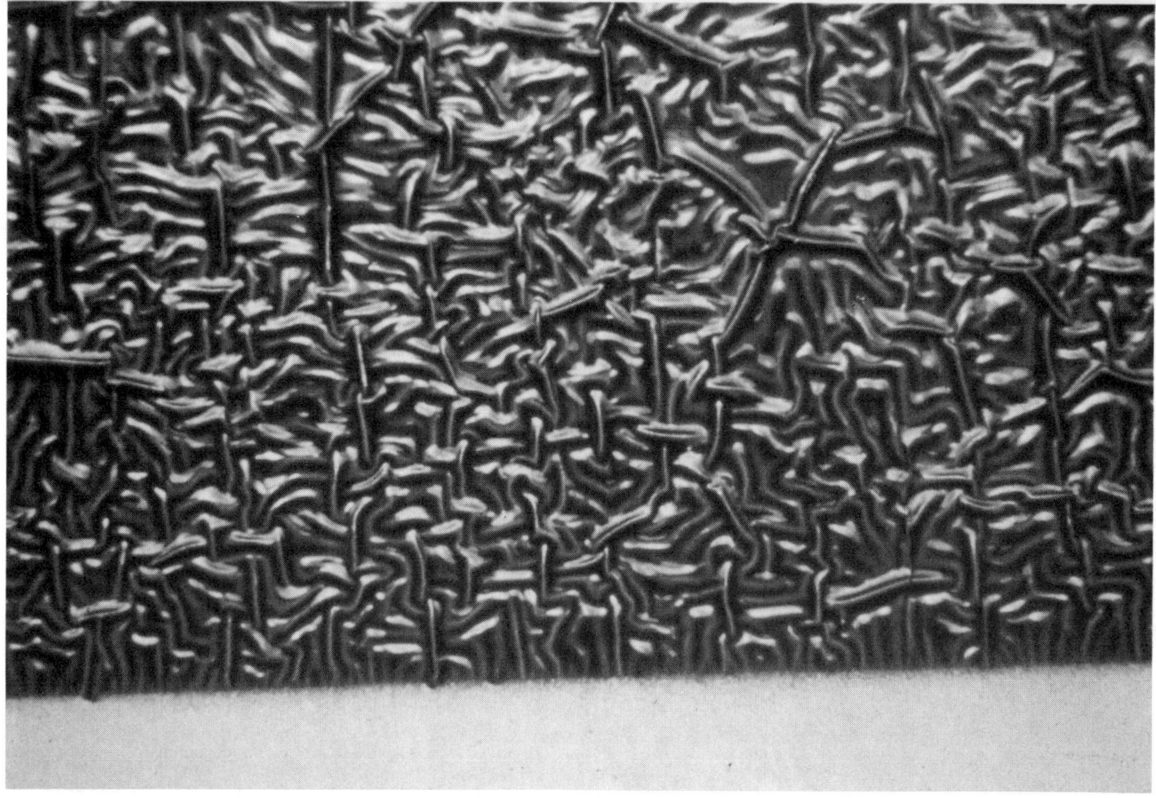

Figure 4.85 Shrinel (ICI Autocolor)

refinisher cannot. In order to match a colour, different pigments may have to be used in the refinishing material in order for them to be compatible with the binder. Consequently, a perfect match may be obtained in daylight conditions, but the colour may alter drastically when viewed under sodium or mercury street lights. There is nothing that the refinisher can do about this, and it is advisable to explain this to the customer when only part of the car, such as a wing or door, is to be painted.

The modern system of supplying the refinisher with colour formulae is in the form of microfiches, which require a microfiche viewer or reader.

4.20.2 Procedure for identifying a particular colour

1 Note the make and year of manufacture of the vehicle.
2 Find the colour identification plate on the vehicle (the positioning of this varies from one vehicle manufacturer to another) and note the colour coding.
3 Select the appropriate microfiche, place it in the viewer and find the mixing formula.
4 Double check that the correct colour has been found by finding the coloured sample in the books provided by the paint manufacturer, and placing this against a cleaned-up part of the car.
5 If this does not match perfectly, check the book on colour variants and consult the manufacturer's hints on tinting.

4.20.3 Hints on colour mixing and matching

Never try to match a colour by dabbing a sample of the colour on to the vehicle and making adjustments. Colours alter during the drying process. Always spray a sample card with the appropriate number of coats and allow it to dry before comparing it with the colour to be matched.

On a volumetric system, never use a dented tin or one with a concaved or convexed bottom to mix the colour in. On a gravimetric system, always remember to place the tin on the scales before zeroing them.

When comparing the painted sample against the vehicle, daylight, preferably from a northern source, is best. Bright sunlight or artificial lighting can sometimes exaggerate or diminish a colour effect.

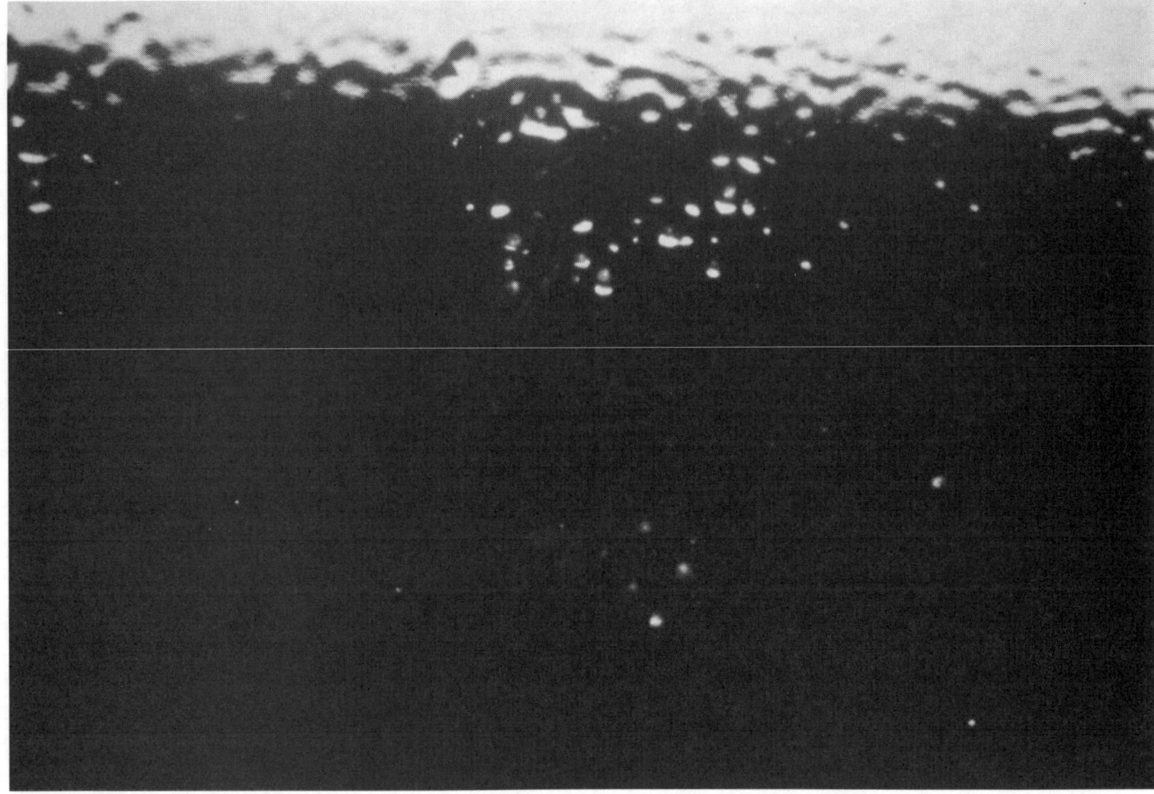

Figure 4.86 Silicone cratering (Cissing/Fish eyes) (ICI Autocolor)

4.21 VALETING OF REPAIRED VEHICLES

After repair work has been completed, the vehicle should be presented to the customer in a first-class, clean condition. All repaired vehicles should be thoroughly inspected for quality control. The repairs should be guaranteed for a specific mileage or period, as required by the Code of Practice for the Motor Industry and by the VBRA Code.

Many insurance companies now make an allowance for valeting within their agreed repair price, and this provides the opportunity for the bodyshop to produce a first-class factory finish to both the exterior and interior of the repair.

4.21.1 Exterior valeting

Engine compartment
One of the first processes to be completed when valeting is the cleaning of the engine compartment. If the exterior of the vehicle body is cleaned prior to the engine compartment, contamination may be transferred from the engine compartment to the exterior, resulting in the recleaning of the exterior.

Before the cleaning of the engine compartment it is advisable to have the engine of the vehicle warm to aid subsequent starting of the vehicle. Then the electrical system must be switched off and sensitive electronic equipment covered with plastic sheeting to prevent the penetration of water which may damage components or impair subsequent starting of the vehicle.

The first stage of the process is to apply an appropriate engine cleaner to remove transport wax, engine oil and road dirt. This must be worked into all of the recesses within the engine compartment using a stiff brush and allowed to stand to dissolve the grease and dirt prior to the use of clean, and preferably warm, water to remove the dirt and solution.

Alternatively a high-pressure steam cleaner may be used, in conjunction with a solution of dewaxing agent to remove oil and dirt from both the engine compartment and engine.

Consideration must be taken about the type of material used during the process. If these materials are entering normal drains a biodegradable material

Figure 4.87 Sinkage (ICI Autocolor)

should be used, or alternatively waste disposal arrangements must be made (see Environmental Protection Act – EPA).

Once this process is complete the engine should be run for a short while to help dry it and the engine compartment off. After this, allow to cool and ensure it is thoroughly dry, then apply a light coat of 'aerosol engine clear coat' to enhance and help to sustain the cleanliness of the engine and engine compartment.

This process will also help to prevent moisture entering certain engine components which could impair engine starting.

Exterior paintwork

The cleaning of the exterior vehicle paintwork will vary from vehicle to vehicle depending on the condition of the paintwork. It is recommended that the vehicle is washed at least once a week, to remove normal traffic film build-up, following the recommended method of using a hose or pressure washer to apply copious amounts of clean water to the surface of the paintwork to remove loose deposits of grit and dirt which could scratch vehicle paintwork during the cleaning process. Take into consideration that when working with water, for example high-pressure washer, appropriate water resistant clothing, including gloves and goggles, should be worn.

The next stage is to remove any dirt that has become attached to the painted surface by using a brush attachment to a pressure washer or manually with a clean sponge and hot water, with the addition of a high active detergent. This solution can be brushed or wiped across the surface starting from the highest point on the roof, working down across horizontal surfaces such as bonnet, boot, doors and eventually finishing with sills and wheels.

Next rinse down with copious amounts of clean water to remove dirt and detergent solution from the exterior of the vehicle. Then dry using an appropriate wash leather. When using a new wash leather first soak it in warm water before use then wring it out. This will increase the absorbent capabilities of the wash leather to remove smears and detergent residue from the paintwork of the vehicle.

Care must be taken when washing the wheels that

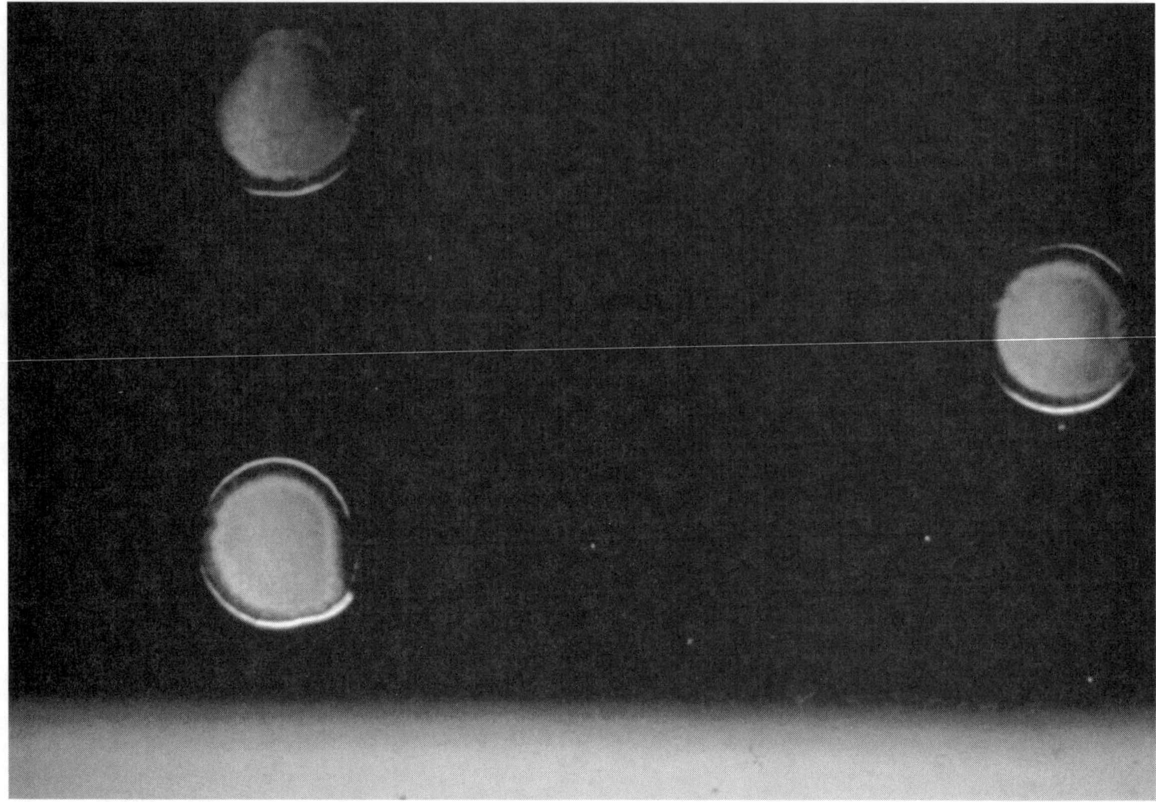

Figure 4.88 Spotting/Colour change due to chemical factors (ICI Autocolor)

all detergent and clean water applied to remove it does not effect the braking system.

Finally the paintwork should be examined with a view to burnishing and polishing if required.

Polishing and burnishing of exterior paintwork
This process must be carried out indoors or out of direct sunlight, as direct sunlight will dry burnishing compounds and polishes prematurely leaving smearing of residues on the surface of the paintwork.

Burnishing is the use of abrasive materials, for example. rubbing compound.

Polishing is the use of soft wax (non-abrasive). A non-silicone wax should be used in order to prevent contamination to new paintwork.

In the situation where the vehicle already has a build-up of several layers of polish, it is advisable to remove these before repolishing. This can be done by wiping down the paintwork with a cloth soaked in white spirit to soften the polish which should then be removed with clean cloths. Work on one area at a time, for example bonnet, boot lid, wing. Should any smearing be present when the work is complete repeat the process on that area.

It is possible that at this stage there may be some discolouration of the paintwork due to traffic film. This can be removed by the use of a burnishing cream which at the same time, will improve the gloss level.

An electric or air driven polishing machine can be used to speed up the work on large panels, though hand burnishing is better on some awkward areas. During the process spread some cream on the job and burnish with the machine. Do not apply any pressure and always keep the machine moving. Lambswool bonnets or sponge heads can be used on the machine for burnishing. When the burnishing is complete on one area clean off residue powder with a clean cloth and inspect. When burnishing by hand use a slightly damp piece of mutton cloth and work with a forward and backward, never circular, action.

Work on one area at a time and check that the work is satisfactory before moving on.

Polishing is done in a similar manner to burnishing, either by hand or machine. In the latter case lamb's wool bonnets are better than sponge heads. Spread

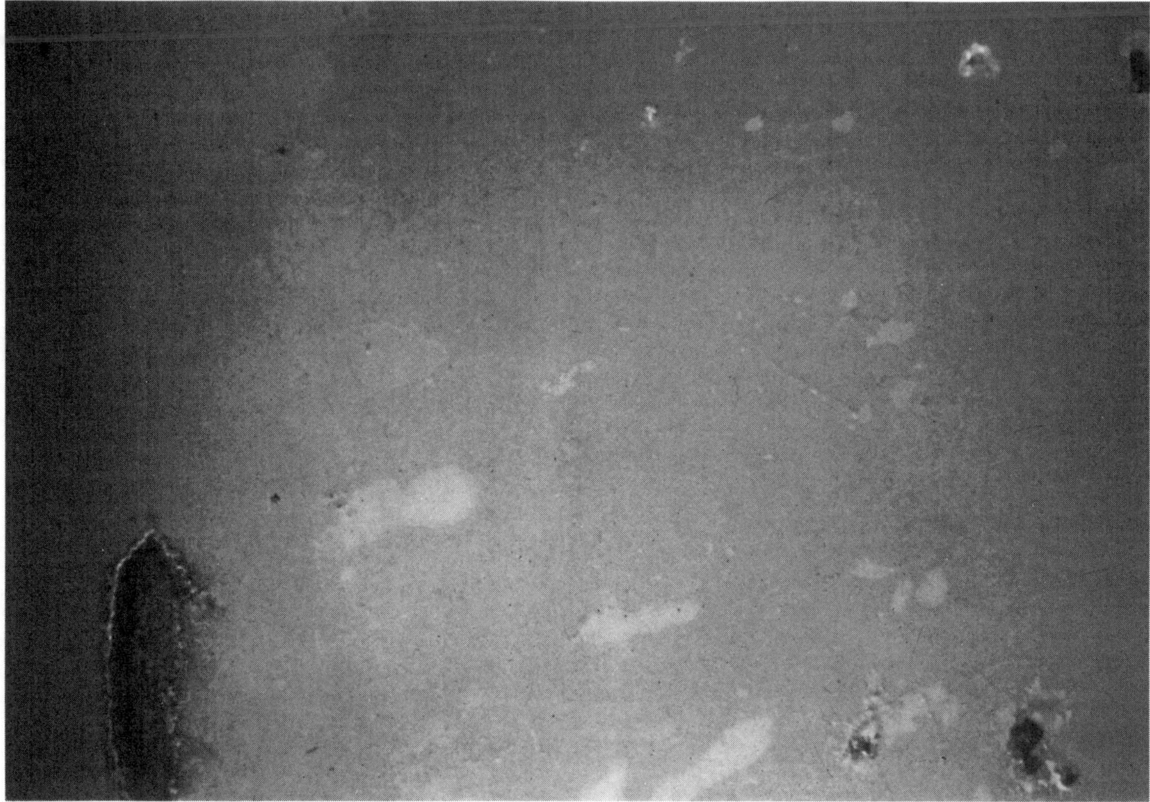

Figure 4.89 Stone chipping or Mechanical damage (ICI Autocolor)

the polish sparingly with a clean cloth before using the machine which should be moved lightly across the surface. Always work cleanly, do not lay cloths or polishing or burnishing heads on the workshop floor to avoid contamination. After use wash the lambswool bonnet in clean soapy water, rinse thoroughly and hang up to dry.

Glass
After cleaning the vehicle paintwork the glass surfaces are the next stage. Glass can be cleaned externally and internally by using a proprietary glass cleaner with clean cloths, one to apply the material and one to remove it. Never use wax polish on windows as this may render the windscreen wipers ineffective. Should overspray be present on the vehicle glass it can be removed with a Stanley-type blade. This can be made easier if the glass is first wiped over with a damp cloth or sponge.

Chrome trims
After a period of time chromed steel can show signs of pitted corrosion. This can be treated with a proprietary rust-removal acid, left for a stated time and then rinsed off with clean water and dried off. The chrome trim can be brightened by rubbing with a cloth and metal polish or rubbing compound which should then be removed with a clean cloth.

If the chrome trims are made from stainless steel and not chromed, acid must not be used and they should only be polished with a rubbing compound.

4.21.2 Interior valeting

Dry vacuuming of passenger compartment
Items of customer's personal belongings should be placed safely and securely in a suitable container marked with the vehicle's registration number and customer's name, and returned to the vehicle when the cleaning has been completed.

The initial operation in this area would be to remove, from the floor of the passenger compartment, any debris or items too large to be picked up using the vacuum cleaner or that might damage the machine. During this operation it is advisable to use suitable gloves, particularly when dealing with broken glass,

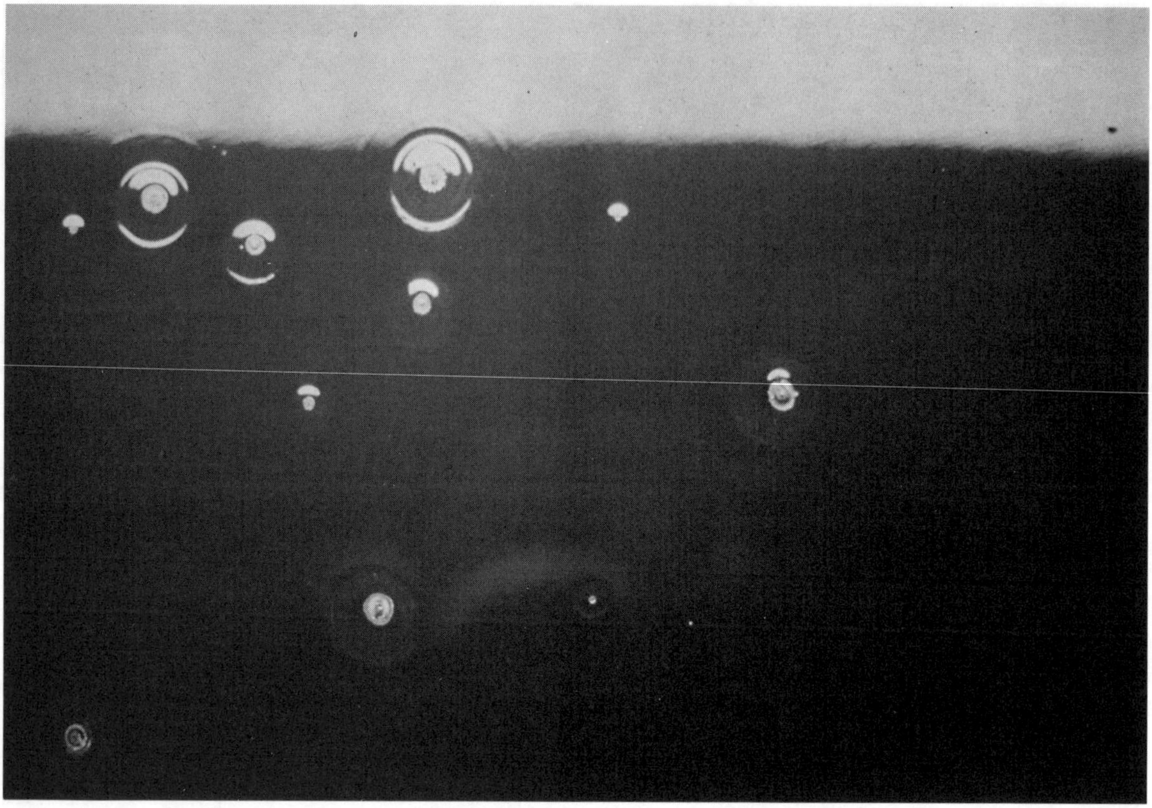

Figure 4.90 Water marking and spotting (ICI Autocolor)

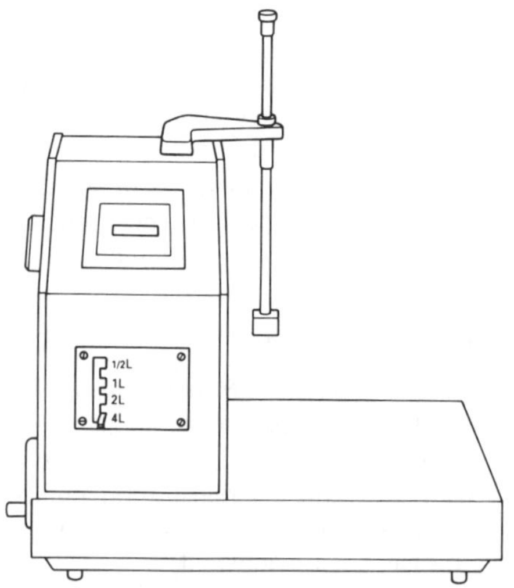

Figure 4.91 Volumetric colour mixing machine

where for instance a windscreen or window has been broken and the glass has to be removed from the interior of the vehicle.

At this stage ashtrays should be emptied and cleaned, and rubber mats removed for cleaning.

Once loose debris has been removed from the vehicle, attention can be focused on the cleaning of the soft trim. This can be done in two ways. First, by the use of a bucket, cloth, water and mild detergent to remove dirt and dust from plastic and vinyl soft trim. Care must be taken not to use excessive amounts of water which may enter electrical components and cause damage. This method has limited success on fabric cloth trim especially with stubborn stains which may require a stronger chemical solution to remove them.

The second method is to use a proprietary dash and trim cleaner which must always be used according to manufacturer's instructions, as some of these materials, because of their strong solvent or chemical action, could damage the trim material.

Glass can be cleaned using the same methods as for the exterior of the vehicle.

Soft fabrics such as the type used on vehicle seats and carpets may be cleaned thoroughly using upholstery cleaning machines. This is a time-consuming job which requires lengthy drying times during which time the vehicle cannot be used. In general soft fabrics if lightly soiled can be cleaned using a vacuum cleaner with various attachments available.

Leather upholstery cleaning must be considered very carefully due to the nature of the material. It can be easily damaged using incorrect materials. The process used for general cleaning is that of wiping items of trim with a solution of warm soapy water to remove light soiling. Use of upholstery cleaners should be avoided on leather as these machines and materials could cause serious damage to the leather trim. Once all areas of light soiling have been removed the affected areas can be dried using a clean dry cloth. Once the leather has dried it must be inspected closely as any cracks in the leather grain or damaged stitching or tears in the material must be reported to the supervisor or customer so the defect can be rectified. To prevent the situation deteriorating after leather has been dried, it should then be treated with a quality leather treatment to prevent the leather drying out which could cause subsequent cracking of the material. Oily or heavily soiled materials will require specialist cleaning by leather experts.

Figure 4.92 Wet bay (*Autoglym*)

Figure 4.93 Dry bay (*Autoglym*)

4.21.3 Premises, equipment and chemicals used in cleaning and valeting

When cleaning and valeting a vehicle it is essential to organize a standard method of working to ensure that nothing is inadvertently missed. The ideal premises and equipment should be set out in separate purpose designed working areas.

The cleaning process should be undertaken in a wet bay with effective drainage, good lighting and adequate working space all round the vehicle when its doors are open. Compressed air and high-pressure cold and hot water should be available (Figure 4.92).

Valeting should be carried out in a well-lit, dry bay with adequate working space all round the vehicle. Compressed air, electrical power points, warm water, a workbench and storage cupboards should be available (Figure 4.93).

Equipment

1 High-pressure hot and cold washer, 70–100 bars (1000–1500 psi), with chemical throughput facility to ensure that engine cleaners and traffic film removers work quickly and effectively.
2 Electric/air polisher, 1500–2000 rpm, with polythene foam and lambswool polishing heads.
3 Fine grade, 100 per cent cotton polishing cloths.
4 Vacuum cleaner and/or shampoo vacuum cleaner with assorted upholstery brushes and crevice tools.
5 Hand-pumped, pressurized sprays for dispensing engine cleaners and traffic film remover.
6 Trigger spray dispensers for interior cleaning, together with wheel, engine and carpet brushes.
7 Good-quality chamois leather, sponges, polishing cloths (100 per cent cotton, knitted stockinette type), steel wool, spatula, glass scraper, buckets, hot air gun (useful for removing PVC stickers, self-adhesive design trims).

There are many manufacturers and suppliers of valeting and cleaning chemicals. The wide range of valeting and chemical cleaners available is broadly divided up as follows:

Exterior detergents and solvents

Interior detergents and solvents

Interior/exterior rubber, PVC, plastic cleaners and conditioners

Glass cleaners, paints, lacquer and protectorants

Paintwork polishes and conditioners

4.21.4 Cleaning and valeting process

Prior to commencing work, check that the vehicle is ready. All body repairs, paintwork and mechanical work should be completed. Be aware of areas which have been newly painted as they may be chemically or water sensitive. Individual chemical products carry their own instructions, and health and safety procedures must be adhered to at all times.

Autoglym recommend the following step-by-step procedure for cleaning and valeting.

Cleaning: wet bay

1 Remove spare wheel, rubber mats and wheel trims if fitted.
2 Protect engine air intake and sensitive electrical equipment (distributor cap, fuse box) with plastic sheet.
3 Apply degreaser to engine and compartment. Brush heavy soiling. Alternatively use hot-pressure washer with appropriate detergent (Figure 4.94).
4 Apply degreaser to door apertures and edges. Brush heavy soiling. Alternatively use hot-pressure washer with appropriate detergent.
5 Clean wheels, trims, spare wheel with alloy cleaner. Brush brake dust deposits. Treat bright metal and motifs. Rinse all items well (Figure 4.95).
6 Pressure wash engine compartment to remove degreaser. Commence with lower areas. Work methodically upwards (Figure 4.96).
7 Pressure wash door apertures and edges to remove degreaser. Carefully angle water jet away from the vehicle interior.
8 Apply traffic film remover to engine compartment. If necessary, sponge or brush to remove grime.
9 Pressure wash engine compartment to rinse away traffic film remover.
10 Pressure wash wheel arches to remove mud and debris. Plain water is normally adequate.

Figure 4.94 Degreasing engine (*Autoglym*)

Figure 4.95 Cleaning wheels (*Autoglym*)

11 Pressure wash and rinse bodywork, grilles, tyres, mudflaps. Pay particular attention to difficult to polish areas behind bumpers (Figure 4.97).
12 Apply traffic film remover to rubber mats. Clean with high-pressure washer.
13 Use air line to dry engine. Remove plastic sheeting. Check engine starting. Use water dispersant if necessary. Run engine to aid drying (Figure 4.98).
14 Use air line and chamois leather to remove excess water from bodywork or trim strips which may trap water (Figure 4.99).

Figure 4.96 Pressure washing engine compartment (*Autoglym*)

Figure 4.97 Cleaning bodywork (*Autoglym*)

Figure 4.98 Drying engine (*Autoglym*)

Figure 4.99 Removing water (*Autoglym*)

Valeting: dry bay

1 Finish engine compartment. Clean hoses, wiring, plastic and paintwork with appropriate dressings.
2 Repaint deteriorated black areas. Use fine jet matt black aerosol or small spray gun.
3 Clear lacquer engine if required to enhance and preserve appearance. Close bonnet.
4 Before interior cleaning, remove all loose carpets, tools, ashtrays and personal items to the bench.
5 Use glass scraper to remove all labels from windows. Residual adhesive can be removed with adhesive remover or water.
6 Remove all plastic labels from bodywork using hot air gun. Residual adhesive can be removed with adhesive remover.
7 Vacuum clean all interior surfaces. Slide front seats forward. Use brushes with vacuum nozzle to clean crevices and air vents (Figure 4.100).
8 Clean luggage compartment first. Use interior cleaner by spraying and wiping clean with cloth rinsed frequently. Check body channels and rubbers.
9 Wash interior. Start with headlining and use interior cleaner. Heavily soiled carpets or seats may require shampoo vacuum treatment.

Figure 4.100 Brushing and vacuuming (*Autoglym*)

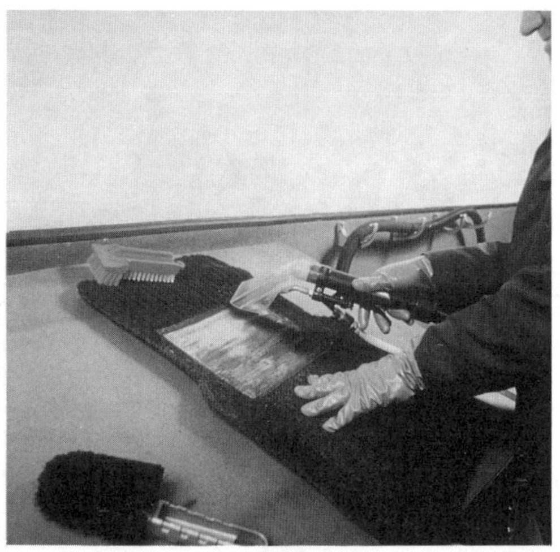

Figure 4.101 Cleaning carpets (*Autoglym*)

10 Removed carpets should be thoroughly brushed, vacuumed and washed on the bench. Use shampoo vacuum if necessary. Allow to dry (Figure 4.101).
11 Tools and jack should be cleaned, and repainted if necessary. Wash out ashtrays.
12 Plastic-coated fibreboard may be painted or stained to cover damage or scrape marks. Check body sides for tar, and top surfaces for industrial fallout. Use appropriate cleaner before polishing.
13 Restore paintwork. Start with roof panel. Use paint renovator with polisher, or appropriate cleaner/polish. Hand polish small areas, corners, edges (Figure 4.102).

Figure 4.102 Polishing (*Autoglym*)

14 Apply protective wax coating by hand, ensuring total coverage of all panels. Leave polish applied at this stage.
15 Clean all external body rubber and plastic mouldings with appropriate dressings.
16 Check all door apertures and rubber seals. Polish door aperture paintwork with clean cloth. Treat rubber seals with appropriate dressing.
17 Check all wheels. If required, clean with steel wool and thinners. Protect tyres with dressing or mask. Respray wheels and clean tyres.
18 Glass cleaning. Lower side windows. Clean top edges completely. Close windows. Polish outside first, then inside. Clean surrounding seals and mirrors.
19 Interior plastic can be dressed to enhance and protect appearance.
20 Replace all carpets, mats, spare wheel, ashtrays, tools, wheel trims.
21 Check interior for remaining imperfections. Check under all seats, glove box, door pockets. Finally vacuum clean. Place protective paper on floor mats.
22 Remove polish. Methodically check all edges, valances, glasses, lights, grilles. Crevices and motifs may be lightly brushed to remove polish.

Final checklist

1 Check exterior mirrors, all glass and surrounding seals. Check reverse of interior mirror.
2 Check polish smears, wheel arch edges, front and rear lower pillars, side sills, door handles, grilles. Touch in stone chips.
3 Check windscreen wiper arms and blades, air grilles.

4 Check all light lenses, motifs, number plates and spot lights.
5 Check door edges and apertures, engine and luggage compartment, body channels and seals.
6 Check control pedals and foot wells, especially under front seats.
7 Check instrument glasses, switches, control levers, interior air vents.
8 Check all ashtrays, glove box, door and seat pockets.

QUESTIONS

1 State three of the advantages to be gained from the method used by vehicle manufacturers for applying the priming coat.
2 State two advantages to be gained by using a guide coat when sanding down a surface.
3 Explain two of the functions of a spray gun air cap.
4 Explain how a suction-fed (or syphon-fed) spray gun operates.
5 Determine the difference between thermoplastic and thermosetting paint coatings.
6 What is the basic difference between those paints classed as lacquers and those classed as enamels?
7 What is the function of a ground coat?
8 Under what circumstances would a bleeder-type spray gun be selected in preference to a non-bleeder?
9 Why are bleeder-type colours a problem to the painter?
10 Explain the working principle of a single-acting, two-stage, piston-type compressor.
11 What is meant by the pot life of a two-pack paint?
12 Give two reasons why cellulose-based paints were replaced by stoving synthetics for mass production finishing on new cars.
13 What type of vehicle finish is liable to become viscid when dry sanded with a sanding tool?
14 Reduced tinters have one-tenth of the tinting strength of a full-strength tinter. Why are they necessary for inclusion in colour mixing schemes?
15 What is the best type of priming paint to use on an aluminium surface?
16 What would be the most likely result of using two-pack polyester stopper as an intercoat stopper in a lacquer-type paint system?
17 State the main disadvantage of a combined spray booth and low-bake oven unit.
18 What is the main health hazard associated with the spraying of two-pack paints?
19 Which method of heat transfer is used in (a) infra-red and (b) low-bake heating units?
20 State two reasons why door handles should not be 'bandaged' when masking up.
21 In which way do thermoplastic vehicle finishes present problems to the vehicle refinisher when carrying out spot repairs?
22 What is an HVLP spray gun?
23 It is estimated that 80–90 per cent of corrosion on motor vehicles is classified as inside-out corrosion. What steps do vehicle manufacturers take to minimize this problem?
24 The spray pattern of a spray gun is top heavy. Describe a simple process to determine whether it is the air cap or fluid tip which is at fault.
25 State what is meant by 'VOCs'.
26 How do HVLP spray guns and VOCs relate to one another?
27 What is the main disadvantage of a combi-unit (spray booth and low-bake oven)?
28 What is the difference between baking and force-drying a paint film?
29 Describe a simple test to determine whether or not an existing paint finish is of the non-convertible type.
30 State three problems which can arise if cellulose stopper is applied too heavily.
31 Describe the paint defect 'blooming'.
32 Describe the cause of the defect 'dry spray'.
33 Explain the cause of 'flaking' or 'peeling' paint.
34 Name any three items of protective clothing which should be worn when using a steam cleaner.
35 Why are biodegradable cleaning materials recommended for the cleaning of engine compartments?

5

Company administration and personnel

5.1 THE COMPANY

5.1.1 Company structures

The company's size will dictate its structure according to the number of employees. Additional factors which also influence the structure are: whether the company is a dealership within a chain of companies, or an independent bodyshop, or a one man owner/employee.

A large company is usually headed by a board of directors and immediately below the board comes the person responsible for the running of the company (managing director).

A smaller company could be run by an owner, director, manager, or one person fulfilling all these roles.

The managing director
From managing director should emerge a structure that would ensure the effective use of all resources and a suitable place for every member of the company. Without such a structure organization will be unsatisfactory and control impossible.

Company structure is usually based on a simple line organization which shows the various staff levels from higher management down to office and workshop. At each level the person in charge is answerable to the figure of authority immediately above him/her, and thus graded levels of authority are laid down.

Below senior management levels come middle managers, then second and third line supervisors who interpret rules and policies. The first-line supervisors oversee rather than perform work, while the second-line supervisors, for example the chargehand, are still performing an operator's job but also have some responsibility. Next are the operators who are skilled tradesmen in their own specialism; then the apprentices trained by the tradesmen; and finally semi-skilled or unskilled operators responsible for their own task.

Many activities are undertaken within vehicle bodywork, which is made up of workshops ranging in size from the one man business, to very large groups employing many hundreds of staff. The average bodyshop is relatively small, between 10 and 30 employees. This is due chiefly to the requirements of the area in which the company is based. (see Figures 5.1a, b, c)

Bodyshops can be classified as:

1 Sole trader
One man business trading under his own name with or without employees.

2 Partnership
Two or more individuals trading under their own names.

3 Firm
A one man business or partnership trading under a name not their own; such a business must be registered. It should be noted that in partnerships each is responsible for all the debts even to the extent of his own private assets.

4 Limited companies – private or public
Shareholders put up the original capital and become risk-takers only up to the amount of their shareholdings. Every business registered under the Companies Act becomes legal, with rights and duties subject to their constitution and the law.

5.1.2 Staff responsibilities

Duties and responsibilities of staff
In order to ensure comprehensive cover of all duties it is essential to define clearly the work of each person in charge. Job descriptions for skilled and unskilled personnel are not usually written down since they know what is expected from them.

Companies of all kinds will have fixed areas of responsibility that consist of clearly defined jobs. The person in charge of each area of responsibility should have a clear knowledge of what is required of him/her in order to prevent the overlap of duties. A written job description in the form of a contract can achieve this.

A job description is usually made out to include the following:

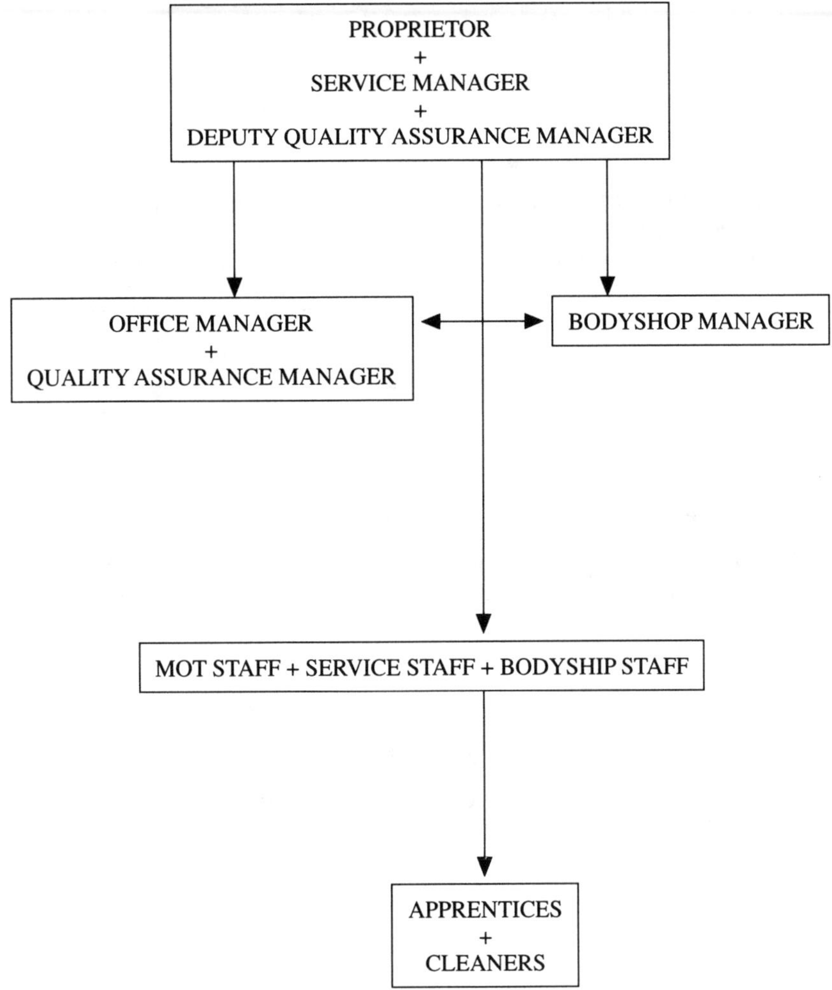

Figure 5.1a Quality assurance organization chart (Parkway Consultancy)

Title of job Job location, main purpose of job.

Duties and responsibilities Limits of authority, scope of job and its responsibility.

Skill and knowledge required Desirable modifications.

Working conditions Dangerous and hazardous conditions should be mentioned.

Economic conditions Rates of pay, bonus payments, holidays, sickness benefits.

Bodyshop manager vehicle repair

1 To be directly responsible to owner/director/board of directors. Also to interpret policy decisions and have them implemented.
2 To achieve an agreed return on the companies investment on its workshop facilities.
3 To promote customer satisfaction by providing good customer care, high quality repairs and workmanship.
4 To supervise all personnel connected with work and to ensure their safe working conditions at all times and also to ensure the company safety policies are applied.
5 To be responsible for all other parts of the workshop.
6 To relate with insurance assessors, other repair establishments, MIRRA and VBRA.
7 To control department costs and supervise all paperwork including invoices, monthly statements and profit and loss accounts.
8 To ensure that apprentice training programmes meet the company requirements and see that apprentices attend a College of Further Education.
9 To hire or dismiss staff according to company policies.

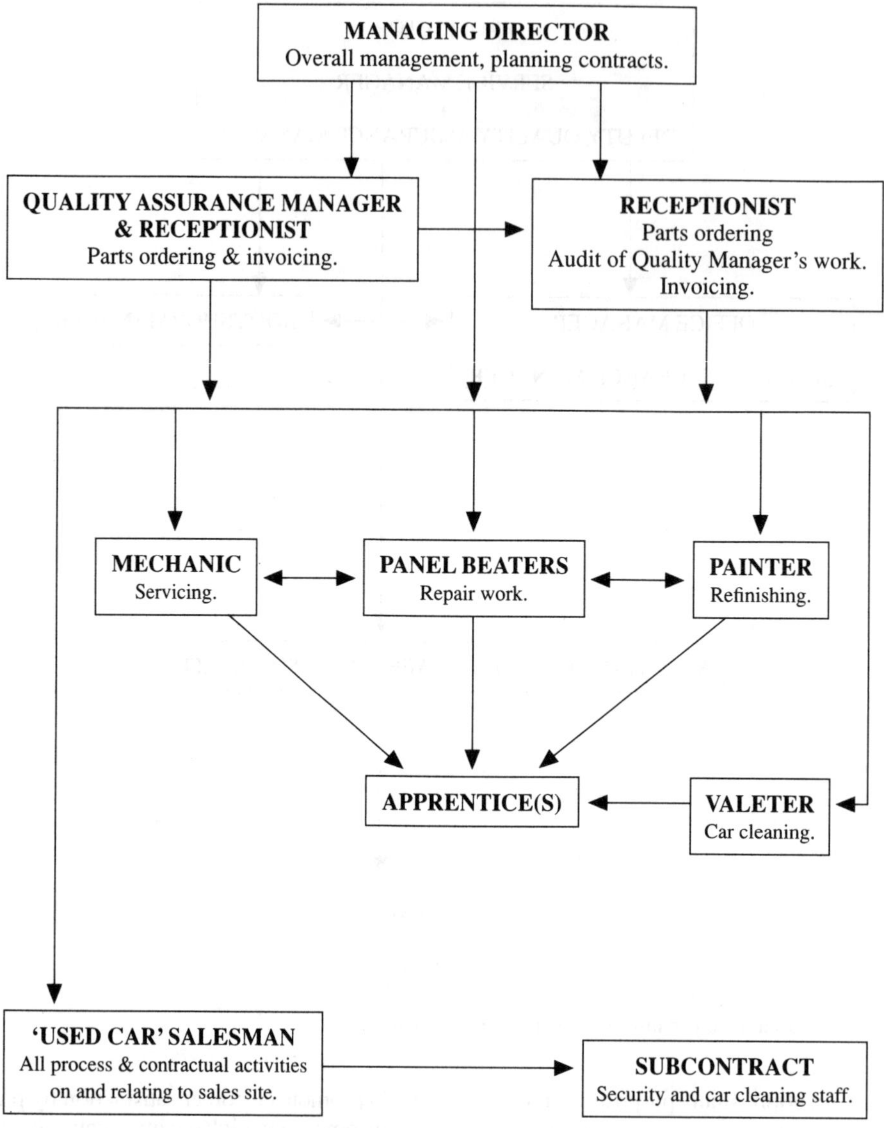

Figure 5.1b Quality assurance organization chart (Parkway Consultancy)

Foreman

The foreman must be a good supervisor and have a first-class practical experience and knowledge of the vehicle body trade, but most important he must have the ability to handle the workforce. His duty is to allocate work to various tradesmen in the workshop, and he is directly responsible to the bodyshop manager for the quantity, quality and safe performance of the work carried out by each employee under his supervision. He is also responsible for team building and providing any training and development needed by the employees to update their individual knowledge and expertise and consequently improve their efficiency and productivity.

Reception engineer/receptionist

It can be the reception engineer or a receptionist who meets the customer and finally hands back the car upon completion of the repair. The attitude and behaviour of the reception engineer must reflect the image of the company. This could be the customer's first contact with the company and it must be a favourable one.

In the case of the *receptionist* he/she is not technically trained and therefore is only there to welcome

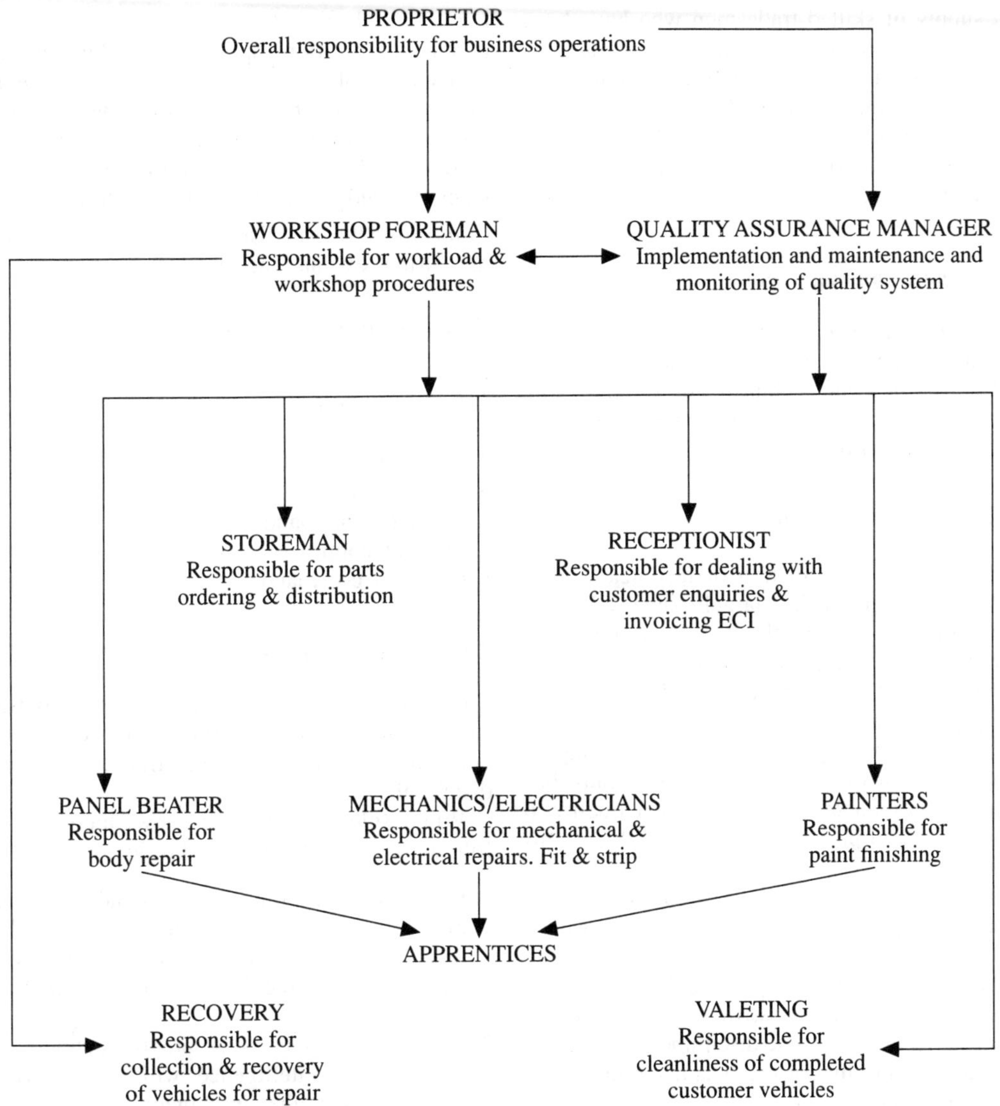

Figure 5.1c Quality assurance organization chart (Parkway Consultancy)

the customer in a friendly, professional manner and take the details of the vehicle.

The *receptionist engineer* must be fully trained both in handling the public and technically, having acquired written qualification and practical expertise.

Skilled vehicle repairers/painters and body fitters
The quality of labour and the high standard of workmanship are two of the most important assets of any bodyshop establishment. It is important that trained personnel are kept up to date with proper training and manufacturers' short courses especially in paint and car repair jig equipment and welding techniques.

Companies wishing to acquire the Quality Standard ISO 9002 (formerly BS 5750) are required to train their operatives to carry out welding tests to an acceptable standard.

Semi-skilled and unskilled personnel
These personnel should have their work checked by the foreman as he is directly responsible for them. They are usually employed on polishing, car valeting and general cleaning duties.

Apprentices
It is essential that a company should have a number of apprentices on the staff so that there will be a

constant supply of skilled tradesmen who have been trained to the company's own high standards. Apprentices are normally taken straight from school and every opportunity must be provided for them to continue their technical training at local colleges through courses such as the NVQ, while also continuing their in-house experience.

5.1.3 Company management

Traditional management structure
Traditional management is based on management by personal experience and technical ability, and quite often lacks any logical method or system or specific managerial qualifications.

Organized management systems (quality procedures)
This is based on a combination of technical ability and managerial skill and has largely superseded the traditional management methods. The principle is based on work study methods which were developed to aid management plan efficiency and productivity into every job within the company.

By operating a management system according to the defined criteria of a standard, the company has then taken all reasonable steps to guarantee quality to their customers, and provided evidence of their commitment to do so.

The assurance of quality is needed to illustrate failures and complaints and their associated costs as far as possible.

Quality Assurance Systems provide the mechanism for ensuring that a business satisfies the needs of its customers by identifying basic disciplines, procedures and criteria to ensure that products and services meet the customer's requirements.

Management functions:
Planning should cover the establishment of objectives to determine the end results; a programme to establish sequence and priorities of action steps to be followed in researching objectives; a schedule to establish the time sequence for programming steps; budgeting to allocate the necessary resources to achieve objectives; establishing procedures to develop and apply standardized working methods; developing policies to make decisions which apply to the problems of the enterprise as a whole.

Organization to develop an organizational structure; delegation, entrusting authority to others and creating accountability; establishing relationships creating conditions for cooperation.

Directing selection, to choose the people for positions in the organization; appraising evaluation performance and capabilities; training and development, helping people to improve their knowledge aptitudes and skill; salary review, to determine a fair day's pay for work carried out; decision making, arriving at conclusions and judgements; dismissing staff.

Controlling the establishing of performance standards, measuring, evaluation, and correcting.

Principles of organization

1 Objective.
2 Specialization.
3 Coordination.
4 Authority.
5 Responsibility.
6 Unity of command.
7 Span of control.
8 Chain of command.

5.1.4 Staff selection

The correct selection of staff is very important because of the Employment Protection Act, which gives employees considerable protection against wrongful dismissal.

When selecting staff for new appointments the following areas should be considered:

1 Physique – General health. Special fitness requirements for the job (for example colour blindness would be a disadvantage for a refinishing operator).
2 Attainments – Special qualifications required for the job.
3 General intelligence – Apprentices should be of at least average intelligence, while managerial posts will demand higher intelligence.
4 Special aptitudes – Special dexterity required for the job.
5 Interests – Any particular constructive hobbies or lively interests that could be potentially useful.
6 Disposition – Would the applicant work well as part of a team?
7 Circumstance – A wide range of factors come under this heading: age, race, mobility, qualifications, experience.

Job evaluation
This is a logical method of working out the relative values of jobs which may be graded according to a particular skill, training or qualifications needed, to physical requirements, to responsibilities involved or to working conditions. However, the establishment of

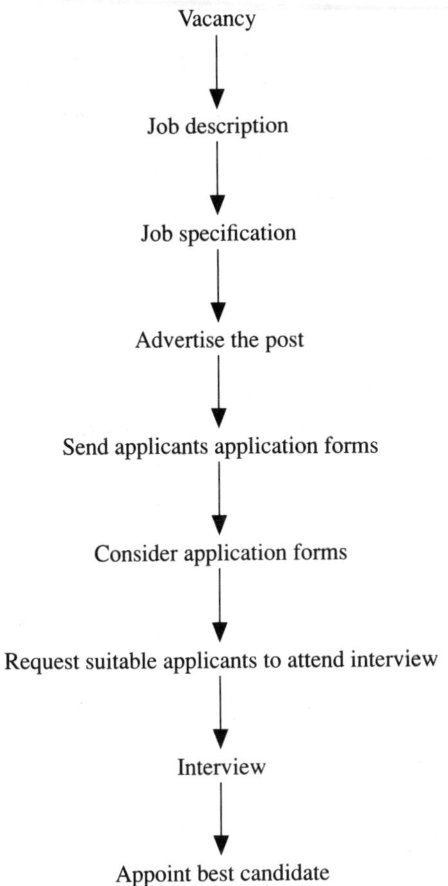

Figure 5.2 Sequence of staff appointments

a fair rate of pay does not usually involve body repair establishments in evaluating jobs which require skill of a specialist practical nature as well as the cooperation of trade union representatives.

Manpower forecasting
This involves attempts to forecast the number and types of personnel that will be required in an organization. The figure obtained will depend on the replacement needs to maintain staff numbers and will be affected by any future technological or organizational developments that call for special skills or qualifications. An effective manpower plan will anticipate such developments and take into consideration the career structures offered by the organization. Manpower planning is a management function and each manager should be familiar with his/her own existing requirements and future needs. People retire or leave for other jobs, new techniques and new equipment come into use, and thus existing staff must be retrained or new staff recruited.

Personnel training
The training of personnel requires a careful study of the objectives related to a particular job before the relevant skills and competences are taught. Continual assessment must be kept throughout the transmission of skills to ensure that the trainee will be able to complete the required work to the standard needed, thus progressive assessment tests should form part of the skill training (see Figure 5.3).

Training and safety are synonymous and the highly skilled worker is usually a safe worker.

5.2 THE CUSTOMER

5.2.1 Types of customer/client

Customer care is important in any company bodyshop if the campany is to retain its customers. Customers must feel satisfied with the service offered from the moment arrive on the premises at reception, where they expect to be treated with respect and listened to with patience and courtesy. Then the work should be completed with skill and to the safety and technical standards laid down by the vehicle manufacturers, ready for collection of the repaired vehicle.

Some companies now assure the quality of their service by achieving ISO 9002, formerly BS 5750, which is a British Standard mark of quality. Customers are also protected by the Motor Manufacturer's Code of Practice and for those bodyshops which are members of the VBRA, their code of practice (see Section 5.3.6).

As customers can be divided into different categories, the relationship with and approach to the customer will depend on this.

Categories can be broadly divided into the following:

Private customers: those whose work is covered by insurance, and those who pay for any work not covered by insurance, will both require the same high standard of customer care and service and quality of work.

Business customers: those who use a company car, who contract all their work, or are part of a dealership which subcontracts work, will expect the same standard of customer care and service as they themselves offer to their clients.

Established customers: those who expect consistency in the quality of work and the maximum of personal attention.

New customers: those who are attracted by the quality of work done and the service offered by the company and therefore require these to meet their expectations.

NAME	START DATE	DATE of BIRTH	EMPLOYEE REF No.
ADDRESS	TELEPHONE	JOB TITLE	HOLIDAY ENTITLEMENT PAYMENT TERMS
QUALIFICATION AND RESPONSIBILITY (see tick box and details overleaf)		NOTICE REQUIRED	
		ASSESSMENT	

TRAINING SCHEDULE	DATES: SCHEDULED/ COMPLETED	EMPLOYEE/ TRAINER	SIGNATURE TRAINEE

Road Test Final Inspect Deliver Move vehicle on company premises Issue stock from stores Handle customer complaints Estimate/Contract/Amendments to Contracts Recovery (Ordering–Controlling) Recovery (Operating) Calibration (In house) Welding Painting Disposal of Scrap	

√ = Allowed to carry out procedure
No mark = Not allowed to carry out procedure
Other work specified:

Figure 5.3 Personnel training record (Parkway Consultancy)

5.2.2 Customer relations

A pleasing approach to customers is essential. Courtesy costs nothing but can mean so much. A smart and clean appearance and ability to instil confidence into customers by demonstrating a good knowledge of operations are essential parts of customer relations. Enquiries should be dealt with promptly and courteously. Visitors to the premises should be offered directions to the appropriate department.

Handling customer complaints
In spite of tight quality control a company can occasionally find a disgruntled customer on their hands

Date: Person taking current details: Job/Ref No:	
Customer and vehicle details	Details of complaint (continue overleaf if required).
Corrective actions (include comment from person responsible for area of complaint).	
(Note what you are going to do to solve the complaint)	
Signature of person responsible for corrective action .... Date scheduled to complete corrective action .... Date 'closed off' .... Signature	

Figure 5.4 Customer complaint and corrective action form (Parkway Consultancy)

with complaints about work carried out. All complaints must be dealt with purposefully and it is the responsibility of senior management to ensure that staff are trained in dealing with complaints (see Figures 5.4 and 5.5).

Staff dealing with complaints should observe the following points:

be a good listener and remain courteous

allow the customer to express his feelings

show genuine interest in the complaint no matter how small

never contradict the customer (remember the customer thinks he is right)

never lose your temper as this can aggravate the situation

To resolve any complaint:

obtain all facts about the work done on job cards, invoices, quality control checks

determine what needs to be done to rectify the complaint

explain to the customer exactly what will be done to remove the cause of the complaint

follow up correction of the fault by contacting the customer in 2 or 3 weeks to ensure that he is now totally satisfied with the work carried out

Where cases cannot be resolved customers can use the conciliatory services of an independent examiner, for example RAC, AA, VBRA or the Institute of Automobile Engineer Assessors, which will endeavour to achieve a settlement based on the facts at their disposal and an examination of the vehicle.

If a genuine interest and concern is demonstrated by the staff involved, the reputation and prestige of the company will be upheld. Quality work in itself is a powerful promotion for the company.

5.3 THE LAW RELATING TO COMPANIES AND CUSTOMERS

5.3.1 Construction and Use Regulations

The Road Vehicles (Construction and Use) Regulations 1986

The motor vehicle construction and use regulations govern the construction and use of motor vehicles on the roads in Great Britain. These regulations are con-

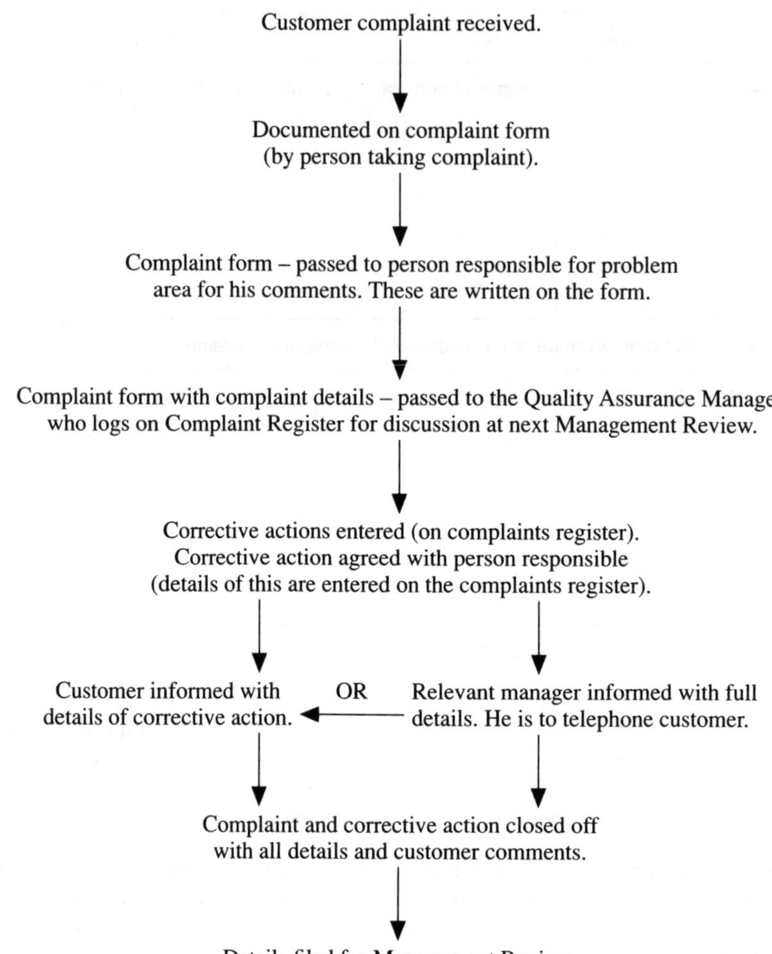

Note: When customer complaint is not settled we accept arbitration by any *mutually* acceptable neutral party (for example AA, RAC, MVRA, IMI, SMMT, RMI, or insurance company assessor).

Figure 5.5 Customer complaint procedure chart (Parkway Consultancy)

stantly being amended and updated. These regulations also require, in some instances, that the vehicles or parts of vehicles should bear a mark confirming that the vehicle conforms to certain specific requirements. Furthermore certain vehicles must be type approved before they can be registered for use on the roads in Great Britain. These requirements are for both private vehicles and for vehicles which carry goods.

Provisions covered by the regulations

A motor vehicle and all parts and accessories shall be at all times in such a condition that no danger is caused or likely to be caused to any person in or on the vehicle or on the road. The vehicle must also be suitable for the purpose for which it is used. The following is a list of the provisions covered by the regulations:

1 Brakes and braking efficiency.
2 Tyre loads and speed ratings.
3 Mixing of tyres.
4 Condition and maintenance of tyres.
5 Steering gear.
6 Windscreen and window glass.
7 Mirrors.
8 Windscreen wipers and washers.
9 Speedometers.
10 Audible warning instruments.
11 Petrol tank.
12 Power to weight ratio.
13 Rear under-run protection and side guards (commercial vehicles).
14 Silencers and smoke emission.
15 Spray suppression devices (commercial vehicles).

16 Towing.
17 Testing and inspection.
18 Ministry plates.

Type approval
Before a passenger car or dual purpose vehicle can be licensed for use on the roads in Great Britain, there must be an approval certificate in force for the vehicle unless it is specifically excluded from the type approval requirements.

'Type approval' is the name given to the system which allows components and vehicles to be tested against prescribed standards. If the tests are successful each component or vehicle manufactured to that specification will carry the type approval.

There are two main type approval systems:

Economic Commission for Europe (ECE)
The ECE agreement on type approval limits itself to setting standards for vehicle components and equipment but not entire vehicles. In the UK a number of ECE regulations have been incorporated into the standards laid down in legislation such as Construction and Use Regulations. The ECE type approved components are identified by a type approval mark which incorporates a circle enclosing an 'E' together with a number, for example (E11)

European Economic Community (EEC)
EEC type approval directives are concerned with vehicles in an assembled state and member states are obliged to incorporate specific standards into their own national legislation. EEC type approval vehicles and vehicle systems are identified by a type approval mark which incorporates the letter 'e' together with a number, for example [e11]

Vehicle Type Approval in the UK
Vehicle type approvals in the UK are based on the EEC system.

Once a vehicle has been type approved the manufacturer or importer will issue a Type Approval Certificate of Conformity (TAC) for vehicles manufactured within the EEC (of which the UK is a member) and which therefore allows vehicles to be used on roads in the UK.

In the case of vehicles manufactured outside the EEC, a Minister's Approval Certificate (MAC) must be obtained.

5.3.2 Road Traffic Act 1988

This Act has very wide ranging powers covering most of the elements surrounding road transport. These include major offences such as driving under the influence of alcohol or drugs to offences concerning construction and use of vehicles. In some cases the Act empowers local authorities to create special legislation tailored to suit the needs of their particular areas i.e. speed limits, restricting access of certain vehicles to congested city centres.
Example 'D' purchased a car from company 'P', then a week later 'D' complained about the vehicle's performance and steering. Company 'P's' foreman stated that there was nothing wrong with the car. An independent inspection of the vehicle showed the steering was faulty and the vehicle was unroadworthy and had been in this condition at the time of the sale.

Company 'P' is guilty under the Road Traffic Act where it is an offence to sell an unroadworthy vehicle, unless its condition is known to the buyer at the time of the sale.

5.3.3 Trade Description Act

The purpose of this Act is to protect the consumer by imposing criminal liability on traders who apply a false description to goods or make misleading statements about services. It does not alter the consumer's contractual rights. To be liable under the Act, the person applying false description must do so within the course of trade or business.

The term 'trade description' means any statement relating to quality, quantity, size, composition, method of manufacture, production, processing, or reconditioning, fitness for purpose, place or date of manufacture, other history including previous ownership of goods.

False description of goods
The act provides that any person who is in the course of trade or business shall, subject to provision of this Act, be guilty of an offence if he/she:

applies a false trade description to any trade or goods

supplies or offers to supply any goods to which a false trade description is applied

Example 1 A motor vehicle which was manufactured in 1990 and first registered in 1993 was later advertised by company 'X' as a 'used 1993 model'.

Company 'X' was guilty under this Act.

Example 2 A second-hand car was advertised for sale by 'D' and described as a 'beautiful car'. Although it looked beautiful, it was mechanically unsound and unroadworthy.

'D' argued that his statement was true and referred only to the appearance of the vehicle. He was guilty because the description was false to a material degree. A reasonable person would have taken the description to refer to the vehicle as a whole, not just its appearance.

5.3.4 Law of Contract

The purpose of a contract is that the contracting parties each undertake to do something, and the failure of either to carry out his/her part of the contract renders him/her liable to a legal action by the other party.

Most simple contracts are oral agreements, for example buying petrol, buying motor vehicle components etc.

A contract may be defined as an agreement which is legally enforceable.

The contract should include four essential components as follows:

1 An agreement between the contracting parties.
2 The provision of consideration to support the contract.
3 An intention to become legally bound by the contract.
4 Tegal capacity held by each of the contracting parties.

The absence of any one of the above will prevent the making of a contract.

A contract cannot be created without the parties reaching an agreement. The idea of agreement is therefore central to an understanding of the Law of Contract.

In the business context there are two legal questions which arise in respect of all transactions. These are whether a legal enforceable agreement has been made and what are the terms upon which it is based. In other words, agreements are arrangements to do, or sometimes refrain from doing, specific things, and these specific things are the terms of the contract.

The law requires that certain contracts should be made in writing because in some cases it provides the consumer with a measure of protection, while in other cases it provides the means of recording their respective obligations, especially in technical transactions.

Contracts that are commonly expressed in writing include, for example, arranging for credit facilities in order to purchase a new car.

5.3.5 Supply of Goods and Services Act 1982

The Act sets out terms which will be implied both into contracts for the supply of services and into contracts for work done and materials supplied.

In a contract for the supply of a service where the supplier is acting in the course of a business, there is an implied understanding that the supplier will carry out the service (a) with reasonable care and skill – in a contract for the supply of a service where the supplier is acting in the course of a business, there is an implied term that the supplier will carry out services with reasonable care and skill; (b) within a reasonable time – where under a contract for the supply of a service by a supplier acting in the course of a business, the time for the service to be carried out is not fixed by the contract, there is an implied term that the supplier will carry out the service within a reasonable time; (c) for a reasonable charge, where no prior agreed payment has been made – where under a contract for the supply of a service, the consideration is not determined by the contract, there is an implied term that the party contracting with the supplier will pay a reasonable charge.

Example 'P' recovered damages for 'D's' unreasonable delay in performing a contract. 'D' took eight weeks to repair 'P's' car when a reasonably competent repairer would have completed the repairs within five weeks. If a reasonable time has elapsed within which the contract should have been performed, the customer is entitled unilaterally to serve a notice fixing a time for performance of the contract (making time of the essence). The new time limit must be reasonable and if the supplier fails to meet it the other party may rescind the contract.

5.3.6 Warranty/guarantees

Both of the above words, warranty and guarantee, are used to refer to the promise given by a manufacturer or retailer when he sells goods or services, that he will meet the cost of correcting defects in those goods or services within a stipulated time or an amount of usage.

With all kinds of work in the vehicle repair industry a warranty against failure due to workmanship must be given for a specific mileage or time period. This should be stated on customers' invoices. All companies should be insured against loss claims arising from such failure. A manufacturer's warranty offers the customer a simple and straightforward way of having manufacturers' faults appearing within certain times put right at little or no cost. Different warranty dates of certain parts must be clearly stated on the warranty form to prevent misunderstanding for the customer.

The useful lifeexpectancy of manufactured products and consumer rights are increasingly being determined by legislation.

The Motor Industry Code of Practice requires that the following inclusions are made in a guarantee:

1 A statement that it does not alter or affect any of the consumer's rights against the retailer under the Sale of Goods Act.
2 The terms of the guarantee must be easily understood especially any inclusions or exclusions.
3 The manufacturers should give advice about what to do if there arises a problem regarding parts and accessories not covered by the guarantee.
4 The manufacturer should allow any unexpired portion of a guarantee to be transferred to a subsequent owner.
5 The manufacturer should make clear that necessary rectification work may be done by any of his franchised dealers whether or not they were the original seller.
6 When a car is off the road for an extended period due to rectification of guarantee faults the manufacturer must operate fair policies to allow the extension of the original time covered.

Code of practice operated by VBRA (Vehicle Body Repair Association)
The code covers the conduct of VBRA members engaged in vehicle body repair and their relationship with their customers. It reflects the high standards adopted by VBRA members and the protection that the customer has in using the facilities offered by a VBRA member. The code's provisions are subject to regular review by VBRA and DGFT (Director General of Fair Trading).

The following gives a summary of this code of practice:

1 To ensure that repair work is carried out at a cost which is fair to both customer and repairer.
2 The repairer shall be responsible for the merchantable quality, fitness for purpose and the description of goods he sells (Sale of Goods Act 1979 and Supply of Goods and Services Act 1982).
3 The repairer shall be responsible to ensure that all repair work is carried out in a proper and workmanlike manner to the safety and technical standards laid down by the vehicle manufacturers.
4 The repairer should provide a written estimate for the cost of a repair if the customer asks for one and it is practicable to provide one. Also, if necessary, the customer's permission for any increase in an agreed estimate.
5 The repairer shall state the estimated time for the repair of a vehicle and shall make every effort to inform the customer if this estimated time cannot be met.
6 The repairer shall supply a detailed invoice including the breakdown of costs, labour charges, parts and materials.
7 The repairer shall provide a written guarantee to exchange or repair any defective parts which need replacement or repair by reason of defective materials, or workmanship during the repair.
8 The guarantee shall be for a period of not less than 12 months or 12 000 miles use, whichever occurs first from the date the repairs were done. This time should be extended to cover any prolonged period a vehicle is off the road due to rectification of faults caused by previous work done during the guarantee period.
9 The repairer should allow the transfer of any unexpired portion of a guarantee to a subsequent owner.

Vehicle manufacturers' guarantees
Some manufacturers give a three-year cosmetic paintwork warranty. This warranty covers eligible vehicles against the eruption, peeling, cracking, discolouration or staining of visible surfaces of all painted body panels as a result of faulty manufacture or materials.

In addition six-year anti-corrosion warranties are given which cover vehicles against perforation due to rust-through of the body panels. The term rust-through means the rusting through from the inside or outside of body panels as a result of faulty manufacture or materials, and does not cover corrosion caused by neglect, accident damage, stone chips or other influences.

QUESTIONS

1 Explain the role and responsibility of the works manager.
2 With the aid of a flowchart show the organizational structure of a typical body repair establishment.
3 Explain the term 'manpower planning'.
4 Explain the term 'job description'.
5 What is meant by the term 'job specification'?
6 What is meant by the term 'technical receptionist'?
7 Explain what is meant by traditional management.
8 Explain the difference between traditional and organized management systems.
9 State four basic management functions.
10 Explain the meaning of the term 'work study'.
11 With the aid of a flowchart explain the sequence for the appointment of new staff.
12 Discuss the need for a personnel training programme.
13 Explain the importance of customer relationships within a bodyshop.

14 Explain the importance of quality control within a bodyshop.
15 What is meant by 'warranty work'?
16 Name four of the provisions covered under the Construction and Use Regulations.
17 Explain what is meant by 'type approval'.
18 Which Act would effect the use of an unroadworthy vehicle on a public highway?
19 What is the purpose of the Trade Description Act?
20 List the four essential elements which should be included in a contract.

Glossary of paint and body repair terms

(Note: Words in italics are specific terms used in NVQ terminology)

Abrasive A substance used for wearing away a surface by rubbing.

Accelerator A constituent of synthetic resin mix which hastens a reaction.

Accessories Optional extras not essential to the running of a vehicle, for example radio, heater.

Accreditation A scheme which identifies the competence skills or knowledge acquired by an individual for which credit can be given.

Acetone A liquid hydrocarbon capable of dissolving 25 times its own volume of acetylene gas at atmospheric pressure.

Acetylene A combustible gas which is mixed with oxygen and used in oxyacetylene welding.

Adhesion The ability to adhere to a surface (the attraction of unlike molecules)

Adhesive A substance that allows two surfaces to adhere together.

Adjustments Necessary alterations to improve tolerances in fit.

Air bags A passive safety restraint system that inflates automatically on vehicle impact to protect the driver.

Air drying Allowing paint to dry at ambient surrounding temperatures without the aid of external heat source.

Air-less spraying A method of spray application in which atomization is achieved by forcing the paint under high pressure through a very small hole in the spray gun cap. The paint instantly expands breaking up into very fine particles.

Alignment The operation of bringing into line two or more specified points on a vehicle structure.

All-metal construction Generally this applies to those body shells of both private cars and light commercial vehicles in which the construction is in the form of steel pressings assembled by welding, thus forming a fabricated unit.

Alloy A mixture of two or more metals with, or without, other metallic or non-metallic elements.

A-post A structural pillar on which the front door is hung.

Assessment Procedure for determining the extent to which an individual has reached a desired level of competence.

Atomize The degree to which the air at the spray gun nozzle breaks up the paint and solvent into fine particles.

Awarding body An examining or validating body.

Backfire In arc welding, a momentary return of gases indicated in the blowpipe by a pop or loud bang, the flame immediately recovering and burning normally at the blowpipe.

Backhand welding Sometimes classified as 'rightward welding'. A technique in which the flame is directed backwards against the completed part of the weld.

Back light A central window in the rear panel of the driving cab, or the rear window of a saloon body.

Base coat and clear system A paint system in which a highly pigmented base coat (which can be either straight colour or metallic) gives the colour effect, while a subsequent clear coat gives gloss and durability.

BC-post Central pillar acting as a central roof and side support between the rear and the front of the car.

Bevel angle The angle of a prepared edge creating a bevel prior to welding.

Billet An oblong piece of metal having a square section.

Binder/paint The ingredients in a paint which hold the pigment particles together, gains adhesion and forms the paint film.

Binder/resin A resin or cementing constituent of a compound.

Bleeding Discolouration of top coat after a new coat has been applied (can occur when spraying over a red colour).

Blowpipe A tool used for welding known as a welding torch.

Body The structured part of a vehicle which encompasses the passenger, engine and luggage compartments.

Body hardware Functional accessories of vehicle body, for example door handles.

Body lock pillar A body pillar that incorporates a lock striker plate.

Body mounting Conventional body mounted on car chassis in composite method of body construction.

Body panels Pressed metal panels, or plastic moulded composite panels, which are fastened together to form the skin of a car body.

Body side moulding The exterior trim moulding fastened to the exterior of the body in a horizontal position.

Body sill The panel directly below the bottom of the doors.

Body spoon A body repairer's hand tool.

Body trim The materials which are used in the interior of the body for lining and upholstery.

Bonnet The metal cover over the engine compartment.

Boot A compartment provided in a car body which takes the luggage and often the spare wheel and fuel tank. It may be at the front or rear of the body depending on the location of the engine.

Boot lid Door covering luggage compartment.

Bottom side The frame member of the base of the body extending along the full length of the main portion of the body.

Brazing A non-fusion process in which the filler metal has a lower melting point than the parent metal(s).

Buckles The resulting distortion of body panels after collision.

Bulkhead A transverse support in a body structure.

Bumping Reshaping metal with a hammer and dolly.

Burnishing To polish or buff a final paint finish, by hand or machine, using a compound or liquid manufactured for this purpose.

Burr The resulting condition left on a metal edge after cutting or filling.

Bursting disc A type of pressure relief device which consists of a disc, usually of metal, which is so held that it confines the pressure of the cylinder under normal conditions. The disc is intended to rupture between limits of overpressure due to abnormal conditions, particularly when the cylinder is exposed to fire.

Butt joint A welded joint in which the ends or edges of two pieces of metal directly face each other.

Calibrate To check irregularities in measuring instruments.

Cant panel The curved section of the roof top running between the comparatively flat top and the rain drip or gutter.

Cantrail The longitudinal framing of the roof at the joint.

Carbon dioxide A heavy colourless and incombustible gas which results from the perfect combustion of carbon.

Carbon fibre An extremely strong, though expensive, reinforcement which can be used in conjunction with fibreglass. It gives increased rigidity to the laminate.

Carbonizing flame An oxyacetylene flame adjustment created by an excess of acetylene over oxygen, resulting in an excess of carbon in the flame.

Case hardening This is the process of hardening the outer case or shell of steel articles, which is accomplished by inducing additional carbon into the case of the steel by a variety of methods.

Catalyst A chemical substance which brings about a chemical change to produce a different substance.

Catalyst dispenser A purpose-designed container for measuring and dispensing liquid catalyst without splashing.

Centre pillar The centre vertical support of a four-door saloon.

Certification The means by which achievement in terms of competence, skills or knowledge are recorded or documented.

Chassis The base frame of a motor vehicle of composite construction to which the body is attached.

Check valve A safety device that controls the passage of gas or air in one direction, in order to prevent the reversal of gas flow and a consequent accident.

Chemical reaction The resulting change when two or more chemical substances are mixed.

Chipping A term used to express the condition in which the paint finish flakes off or chips away from the underneath surface.

Chopped strand mat Chopped strands bonded into a mat to produce a popular economical general-purpose reinforcement.

Chopped strands As the name suggests, glass fibre strands chopped into short (about 12 mm) lengths. They can be used as fillers. Useful for bodywork repairs.

Circuit The path along which electricity flows. When the path is continuous, the circuit is closed and the current flows. When the path is broken, the circuit is open and no current flows.

Clear coat On a painted surface, a top coat that is transparent and allows the colour coat beneath to remain visible.

Client The person or organization employing the services offered, for example employer, sponsor, learner.

Coating A covering material which is used to protect a surface.

Cohesion The attraction of like molecules. Essential in the formation of a cohesive paint film.

Cold curing Generic term for materials which harden at room temperatures, after the addition of catalyst.

Collapsible steering column A safety feature in the form of an energy-absorbing steering column designed to collapse on impact.

Colour This is the visible appearance of reflected light.

Colour retention Good colour retention is said to occur when a paint is exposed to the elements but does not change colour.

Compartment shelf panel The horizontal panel situated between the rear seat back and the back window.

Competence The ability to perform activities to a required standard in an occupation.

Competence element A statement that describes what can be done, an action, behaviour, or outcome which an individual should be able to demonstrate.

Compressive strength The ability of a material to withstand being crushed. It is found by testing a sample to failure: the load applied, divided by the cross-section of the sample, gives the compressive strength.

Compressor A piece of equipment used to compress air, which can be subsequently used to spray paint.

Condensation A change of state from a gas to a liquid caused by temperature or pressure changes. It may also be formed by moisture from the air being deposited on a cool surface.

Conductor Any material or substance that allows current or heat to flow through it.

Contaminants Foreign substances on the surface to be painted or in the paint that would be detrimental to the finish.

Continuous assessment Assessment of competence on every occasion during normal workplace activity. Used for formative assessment and to arrive at a cumulative judgement for final assessment purposes.

Copper acetylide A spontaneously explosive and inflammable substance which forms when acetylene is passed through a copper tube.

Corrosion The wearing away or gradual destruction of a substance, for example rusting of metal.

Coverage The surface area which can be covered by a given amount of paint.

Crazing A paint defect showing minute interlacing cracks on the surface of a finish.

Curing The change of a binder from soluble fusible state to insoluble infusible state by chemical action.

Curing time The time needed for liquid resin to reach a solid state after the catalyst has been added.

Cutting tip A torch especially adapted for cutting.

Cylinder Steel containers used for storage of compressed gases.

Dash panel A panel attached to the front bulkhead assembly and which provides a mounting for all instruments necessary to check the performance of the vehicle.

Degreasing Cleaning a metallic substrate, by removing grease, oil, or other surface contaminates.

Deposited metal Filler metal from a welding rod or electrode which has been melted by a welding process and applied in the form of a joint or built up.

Diagnosis The determination, by examination, of the cause of a problem.

Dinging Straightening damaged metal with spoons, hammers or dollies. In the early days the dingman was the tradesman who worked on completed bodies to remove minor imperfections without injury to the high gloss lacquer or varnish.

Dinging hammer A special hammer used for dinging or removal of dents.

Direct damage Primary damage which results from an impact on the area in actual contact with the object causing the damage.

Dolly block A hand tool, made from special steel, shaped to suit the contour of various panel assemblies and used in conjunction with a planishing hammer to smooth out damaged panel surfaces.

Door skins Outside door panels.

Door trim The interior lining of a door.

Double-header coating One coat sprayed immediately after another

D-post The rear standing pillar providing a shut face for the rear door and forming the rear quarter panel area.

Drip moulding A roof trough to direct water from door openings.

Dryer A catalyst added to a paint to speed up the curing or drying time.

Drying The process in which a coat of paint changes from the liquid to the solid state due to evaporation of solvent, chemical reaction of the binding media, or a combination of these two.

Dry spray A granular or coarse textured finish with no gloss.

Electrode The usual term for the filler rod which is deposited when using the electric arc welding process.

Electrolyte A substance which dissolves in water to give a solution capable of conducting an electric current.

Enamel A type of paint that dries firstly by evaporation of the solvent and secondly by oxidation and/ or polymerization of the paint film.

Epoxy Based on an epoxy resin which is mixed with a hardener.

Evaporation A change of state from solid or liquid into vapour.

Expansion The increase in the dimensions of metals due to heat.

Extrude To draw into lengths.

Feather edge Tapering the edges of damaged paintwork.

Fender American term for wing.

Filler Inorganic types used to extend low-pressure resins, usually polyesters.

Filler metal Metal added to a weld in the form of a rod, electrode or coil.

Fillet weld A weld in which two surfaces at right angles to one another are welded together.

Film Formed by paint on the surface to which it is applied.

Firewall Panel dividing engine compartment from interior of body.

Fisheyes A paint defect usually circular and opalescent.

Flange A reinforcement on the edge of a panel formed at approximately right angles to the panel.

Flash off First stage of drying either lacquer or enamel paint during which some of the solvents evaporate.

Flashback Occurs when the flame disappears from the end of the welding tip and the gases burn in the torch.

Flat A panel is said to be flat when insufficient shaping has caused uneven contours and so flat areas are obvious.

Floor pan Main floor of the passenger compartment of an underbody assembly.

Flux A chemical material or gas used to dissolve and prevent the formation of surface oxides when soldering, brazing or welding.

Foams (flexible) A resin which is often used for cushioning in the automobile industries. These foams are usually urethanes.

Foams (rigid) A resin with a higher modulus than the flexible foams. These are also normally urethanes and are used in more structural applications such as cores in sandwich constructions.

Force-drying The application of heat to a painted surface to speed drying times.

Four-door Denotes the type of saloon body having four doors.

Frame gauges Self-centring alignment gauges which are hung from a car's underbody.

Friction The resistance to motion that a body meets when moving over another.

Fusion welding A process in which metals are welded together by bringing them to the molten state at the surface to be joined, with or without the addition of filler metal, and without the application of mechanical pressure or blows.

Galvanized Steel coated with zinc.

Gas welding A fusion welding process which uses a gas flame to provide the welding heat.

Gel Resin takes on a gel-like consistency (gels) usually within 10–15 minutes of being catalysed. At this point it is impossible to spray, paint or pour. Stored resin which has passed its shelf life may gel without being catalysed.

Gelcoat A thixotropic resin normally used without reinforcement and applied first to the mould. It forms the smooth shiny surface of the finished article.

Glass fibre Glass filaments drawn together into fibres and treated for use as reinforcement.

Gloss The surface shine on paintwork.

Ground coat The paint coat upon which the final coats will be applied.

Hardener A chemical curing or hardening agent.

Hardening Heating to a critical temperature followed by a relatively rapid rate of cooling.

Headlining The cloth or other material used to cover the inner surface of the car roof.

Heelboard The vertical board or panel under the rear seat which forms the support for the seat cushion.

Hinge pillar A pillar on which a door is swung.

Hood American term for bonnet.

Hydraulic pressure Pressure transmitted by a liquid.

Hydraulics The use of pressurized liquid to transfer force.

Impregnated The particles of one substance infused into that of another.

Independent front suspension Suspension system in which each wheel is independently supported by a spring.

Indirect damage Secondary damage found in the area surrounding the damage which caused it.

Inertia Property of an object by which it continues in its existing state of rest or motion in a straight line, unless that state is changed by an external force.

Insulation A material which is non-conductive of either heat or electricity.

Integral A necessary part to complete a whole unit.

Interchangeability The ability to substitute one part for another.

Isocyanate This is a principal ingredient in urethane hardeners and because it has a toxic effect, air-fed masks must be worn when using it.

Kerb weight The weight of an empty vehicle without passengers and luggage.

Kevlar A synthetic aramid fibre used as a reinforcement for resins. It is noted for its high impact resistance and is used in racing car bodywork.

Kinetic energy The energy of motion.

Lacquers A type of paint that dries by solvent evaporation.

Laminates A material composed of a number of layers.

Lap joint A form of joint obtained by overlapping the edges of two pieces of metal. The overlapping parts must be in the same plane.

Latex A natural rubber used for making flexible moulds. It is a liquid which solidifies in contact with air.

Lay-up Layers of glass fibres are laid on top of wet resin and then pressed down into the liquid resin.

Learning objectives A statement of desired or expected achievement on completion of a section or module of a programme of learning.

Leftward welding This is known as forehand welding.

Masking Using tape, paper, plastic sheeting or spray-on masking materials to protect an area that will not be painted.

Mass production Large-scale, high-speed manufacture.

Metal conditioner A chemical cleaner that removes rust and corrosion from bare metal and offers resistance to further rusting.

Metal fatigue A metal structural failure, resulting from excessive or repeated stress, finally resulting in a crack.

Metallic paint Finish paint that contains metallic flakes in addition to pigment.

Mist coat A lightly sprayed coat of high volume solvent for blending or gloss production.

Molecule A minute particle into which a substance can be divided and retain its properties.

Monomer A simple molecule capable of combining with itself, or a compatible similar chemical, to form a chain (polymer).

Moulding The resulting shape of a plastics material when it is removed from its mould.

Mould release A substance used to coat the mould to prevent sticking of the resin that will be used to make a part. It facilitates the removal of that part from the mould.

National vocational qualification A vocational qualification accredited by the National Council for Vocational Qualifications, and one that is based on employer-led standards of competence.

Near side The left-hand side of the vehicle as viewed from the driver's seat.

Neutral flame A balanced flame, indicating perfect combustion of both oxygen and acetylene gases.

Non-ferrous metals Metals which do not contain any ferrite or iron.

Normalizing Heating to a high temperature to produce a refinement of the grain structure of a metal or alloy.

Off side The right-hand side of a vehicle as viewed from the driver's seat.

Opaque Non-transparent.

Original finish The paint applied at the factory by the vehicle manufacturer. Also known as Original Engineering (O.E.) finish.

Overspray Paint which falls on an area adjacent to that being painted.

Oxidation Chemical reaction between oxygen and some other element resulting in oxides.

Oxidizing flame A gas welding flame which has an excess of oxygen when burning.

Paddle A wooden tool shaped for spreading body solder.

Paint remover A mixture of active solvents used to remove paint and varnish coatings.

Parent metal The material of a part to be welded.

Penetration Depth of fusion or weld penetration.

Performance criteria Statements by which an assessor judges the evidence that an individual can perform the activity specified in a competent element, to a level acceptable in employment.

Pickle To soak metal in an acid solution in order to free the surface of rust or scale.

Pigment Used to impart colour, opacity and other effects to paint.

Pillar A vertical support of a body frame.

Pillar face The front of a pillar visible when the door is opened.

Pinch weld Two metal flanges butted together and spot welded along the join of the flat surfaces.

Polyurethane A versatile material used for adhesives, paints, varnishes, resins and foam materials. These are often used in conjunction with polyester-based GRP.

Porosity The presence of gas pockets or inclusions within a weld.

Pot life The limited time a painter has to apply a paint finish to which a catalyst or hardener has been added.

Primer A first coat of paint applied to bare surfaces to promote the adhesion of later coats.

Prototype An original model.

Puddle The small body of molten metal created by the flame of a welding torch.

Quarter light The window directly above the quarter panel.

Quarter panel The side panel extending from the door to the rear end of the body (including rear wing).

Range statement Descriptors of the limits within which performance to the identified standard is expected if an individual is to be deemed competent.

Reinforcement Filler material added to plastics (resin) in order to strengthen the finished product.

Resin Resins occur in nature as organic compounds, insoluble in water, for example amber, shellac. Synthetic resins have similar properties and are normally converted to solids by polymerization.

Respirator A device worn over the mouth and nose to filter particles and fumes out of the air being breathed.

Retarder A slow evaporating thinner used to retard drying.

Return sweep A reverse curve.

Run-sags A well-defined local thickening of the paint film in the form of wavy lines.

Saloon An enclosed body not having a partition between the front and rear seats.

Scuttle panel The panel between the bonnet and windscreen.

Self-tapping screw A screw that cuts its own threads into a predrilled hole.

Silicone rubbers Used, among other applications, for sealants and flexible mould compounds. They are usually cold curing.

Solvent A chemical fluid which will dissolve, dilute or liquefy another material.

Specifications Information provided by the manufacturer on vehicle data in the form of dimensions.

Squab The rear seat back construction.

Streamlining The shaping of a vehicle body to minimize air resistance.

Subframe Members to which the engine and front-end assembly are attached.

Swage A raised form of moulding pressed into a piece of metal in order to stiffen it.

Swage line A design line on a vehicle body, caused by a crease or step in a panel.

Sweating Uniting two or more metal surfaces by the use of heat and soft solder.

Synthetic A substance produced artificially.

Tack cloth A specially impregnated cloth used to pick up dust and lint from the surface to be painted.

Temperature The measurement, in degrees, of the intensity of heat.

Template A form or pattern made so that other parts can be formed to exactly the same shape.

Tensile strength The resistance to breaking which metal offers when subject to a pulling stress.

Thermoplastic Plastic which can be softened by heating, and which still retains its properties after it has been cooled and hardened. Typical thermoplastics are polythene and PVC.

Thermosetting Plastic which hardens by non-reversible chemical reaction, initiated by heat and/or curing agents. Once hardened, it cannot be melted down without being destroyed.

Thinners A solvent combination used to thin lacquers and paints to spraying viscosity.

Thixotropic Generally used to describe substances which have a very high viscosity when stable, but low viscosity when stirred or brushed. 'Non-drip' paint is an obvious example; another is gelcoat resin.

Tint To add colour to another colour.

Tone A graduation of colour either in hue, tint or shade.

Top coat The last or final colour coat of paint.

Tunnel A raised floor panel section for driveshaft clearance.

Turret American term for roof.

Two-pack A paint supplied in two parts that must be mixed together in the correct proportions before use.

Underpinning skills and knowledge Identifies the knowledge and skill necessary to perform to the standards identified by the performance criteria in the contexts identified in the range statement.

Validation Tests, checks or assessment to ascertain whether a scheme, programme or package has achieved specified objectives.

Vaporization The conversion of solvents into gases during spray painting.

VIN Abreviation for Vehicle Identification Number.

Viscosity Consistency or body of a paint. The degree of resistance to flow.

Volatile Capable of evaporating easily.

Weathering Harm caused by exposure to weather.

Weld bead One single run of an electrode welding rod in manual metal arc welding.

Weld deposit Metal which has been added to a joint by one of the welding processes.

Wheel alignment The adjustment of a vehicle's wheels so that they are positioned to drive correctly.

Wheel arch Panel forming inner housing for rear wheels.

Wheelbase The distance between the centre lines of the front and rear wheels of a vehicle.

INDEX